International Association of Geodesy Symposia

Chris Rizos, Series Editor
Pascal Willis, Series Associate Editor

International Association of Geodesy Symposia

Chris Rizos, Series Editor
Pascal Willis, Series Associate Editor

Symposium 105: Earth Rotation and Coordinate Reference Frames
Symposium 106: Determination of the Geoid: Present and Future
Symposium 107: Kinematic Systems in Geodesy, Surveying, and Remote Sensing
Symposium 108: Application of Geodesy to Engineering
Symposium 109: Permanent Satellite Tracking Networks for Geodesy and Geodynamics
Symposium 110: From Mars to Greenland: Charting Gravity with Space and Airborne Instruments
Symposium 111: Recent Geodetic and Gravimetric Research in Latin America
Symposium 112: Geodesy and Physics of the Earth: Geodetic Contributions to Geodynamics
Symposium 113: Gravity and Geoid
Symposium 114: Geodetic Theory Today
Symposium 115: GPS Trends in Precise Terrestrial, Airborne, and Spaceborne Applications
Symposium 116: Global Gravity Field and Its Temporal Variations
Symposium 117: Gravity, Geoid and Marine Geodesy
Symposium 118: Advances in Positioning and Reference Frames
Symposium 119: Geodesy on the Move
Symposium 120: Towards an Integrated Global Geodetic Observation System (IGGOS)
Symposium 121: Geodesy Beyond 2000: The Challenges of the First Decade
Symposium 122: IV Hotine-Marussi Symposium on Mathematical Geodesy
Symposium 123: Gravity, Geoid and Geodynamics 2000
Symposium 124: Vertical Reference Systems
Symposium 125: Vistas for Geodesy in the New Millennium
Symposium 126: Satellite Altimetry for Geodesy, Geophysics and Oceanography
Symposium 127: V Hotine Marussi Symposium on Mathematical Geodesy
Symposium 128: A Window on the Future of Geodesy
Symposium 129: Gravity, Geoid and Space Missions
Symposium 130: Dynamic Planet - Monitoring and Understanding...
Symposium 131: Geodetic Deformation Monitoring: From Geophysical to Engineering Roles
Symposium 132: VI Hotine-Marussi Symposium on Theoretical and Computational Geodesy
Symposium 133: Observing our Changing Earth
Symposium 134: Geodetic Reference Frames
Symposium 135: Gravity, Geoid and Earth Observation
Symposium 136: Geodesy for Planet Earth
Symposium 137: VII Hotine-Marussi Symposium on Mathematical Geodesy
Symposium 138: Reference Frames for Applications in Geosciences
Symposium 139: Earth on the Edge: Science for a Sustainable Planet
Symposium 140: The 1st International Workshop on the Quality of Geodetic Observation and Monitoring Systems (QuGOMS'11)
Symposium 141: Gravity, Geoid and Height systems (GGHS2012)
Symposium 142: VIII Hotine-Marussi Symposium on Mathematical Geodesy
Symposium 143: Scientific Assembly of the International Association of Geodesy, 150 Years
Symposium 144: 3rd International Gravity Field Service (IGFS)

More information about this series at http://www.springer.com/series/1345

International Symposium on Geodesy for Earthquake and Natural Hazards (GENAH)

Proceedings of the International Symposium on Geodesy for Earthquake and Natural Hazards (GENAH), Matsushima, Japan, 22–26 July, 2014

Edited by

Manabu Hashimoto

Springer

Volume Editor
Manabu Hashimoto
Disaster Prevention Research Institute
Kyoto University
Uji, Kyoto
Japan

Series Editor
Chris Rizos
School of Civil & Environmental Engineering
University of New South Wales
Sydney
Australia

Associate Editor
Pascal Willis
Institut national de l'Information
géographique et forestière
Direction de la Recherche
et de l'Enseignement
Marne-la-Vallée
France

ISSN 0939-9585 ISSN 2197-9359 (electronic)
International Association of Geodesy Symposia
ISBN 978-3-319-39767-2 ISBN 978-3-319-39768-9 (eBook)
DOI 10.1007/978-3-319-39768-9

Library of Congress Control Number: 2016952234

Printed on acid-free paper

This Springer imprint is published by Springer Nature
The registered company is Springer International Publishing AG Switzerland

Preface

The world experienced an amazing development of science and technology related to natural disasters and their mitigation during the twentieth century, and people might expect that they were safer about natural disasters. However, the first two decades of the twenty-first century will be remarked as the decades of natural disasters. More than a million people were killed by natural hazards such as earthquakes, volcanic eruptions, storms, landslides, floods, etc. Among them, earthquakes caused the most serious impacts on the human society. There were several earthquakes of Mw 8 or larger in regions where such big ones were unexpected. Furthermore, recent researches on climate changes imply the increase of extreme weather. Hurricanes Katrina of 2005 and Sandy of 2012 and typhoons Talas of 2011 and Haiyan of 2014 may be its manifestation.

It is also true that geodesy is remarkably developing during a couple of decades. We can detect coseismic deformation in nearly real time with GNSS. We can also reveal changes in global mass distribution with gravimetric satellite missions. However, most knowledge was a posteriori obtained only after the disasters. We should step forward to contribute to the mitigation of natural disasters by utilizing products of geodetic researches. Fortunately, new observation systems are being deployed in space, on land, and on the seafloor. It is the time to share the knowledge and discuss how to apply geodesy to mitigate natural disasters.

For the above purpose, the International Symposium on Geodesy for Earthquake and Natural Hazards (GENAH2014) was organized by the Commission 3 of the International Association of Geodesy (IAG). The symposium was held in Matsushima, a small town on the Pacific coast of Japan, which suffered from a huge tsunami during the Tohoku-oki earthquake of March 11, 2011.

This volume contains 18 peer-reviewed papers presented at GENAH2014, which was held on July 22–26, 2014. 130 geodesists from 16 countries participated, to discuss the role of geodesy in earthquake and volcanic studies, natural hazard assessment, and disaster mitigation. In total, 83 oral and 50 poster presentations were made in seven sessions:

Session 1 Subduction Zone Earthquake
Conveners: Jeff Freymueller and Kelin Wang

Session 2 Geodetic Observations of the Earthquake Deformation Cycle:
Implication for Fault Mechanics, Crust/Mantle Rheology, and seismic hazards
Conveners: Simon McClusky and Cècile Lasserre

Session 3 Near Real-Time Warning
Conveners: Shuanggen Jin and Teruyuki Kato

Session 4 Interaction of Earthquake and Volcanoes
Conveners: Eric Fielding and Yusaku Ohta

Session 5 Impact of Great Earthquakes on Reference Frame
Conveners: Richard Gross and Kosuke Heki

Session 6 Geodetic Techniques in Volcanological Research
Conveners: Tim Wright and Takeshi Matsushima

Session 7 Natural Hazards
Conveners: Manabu Hashimoto and Urs Marti

I would like to express my gratitude to all those who have contributed to the success of GENAH2014, particularly the conveners who devoted a lot of time in organizing attractive sessions, the local organizing committee (LOC) members for their effort to prepare and organize the symposium, and all the participants. I must thank the Tokyo Marine Kagami Memorial Foundation, Association for Disaster Prevention Research, National Institute of Information and Communication Technology, Disaster Prevention Research Institute, Kyoto University, Science Council of Japan, Geodetic Society of Japan, Seismological Society of Japan, Volcanological Society of Japan, Japan Geoscience Union, and other related societies, agencies, and universities for their supports.

I also thank the associate editors of these proceedings who played a leading role in the peer review process. My sincere thanks should be given to the IAG proceedings editor Dr. Pascal Willis who kept on track the publication of these proceedings. I would like to thank all reviewers, who were devoted to the review of papers, as well. Finally, I would like to thank the Springer Publisher for their processes and cordial cooperation to publish this proceeding.

Kyoto, Japan
June 2015

Manabu Hashimoto

Contents

Part I

Subduction Zone Earthquake

Progress in the Project for Development of GPS/Acoustic Technique Over the Last 4 Years

Motoyuki Kido, Hiromi Fujimoto, Ryota Hino, Yusaku Ohta, Yukihito Osada, Takeshi Iinuma, Ryosuke Azuma, Ikuko Wada, Satoshi Miura, Syuichi Suzuki, Fumiaki Tomita, and Misae Imano

Abstract

GPS/Acoustic (GPS/A) survey is the most promising way to detect crustal deformation in the ocean far from the coast, where a dense onshore GPS network is not available. Monitoring seafloor deformation is crucial to understand the tectonic state in regions of geophysical significance such as subduction zones. We, Tohoku University, together with Nagoya University and Japan Coast Guard have been dedicated to GPS/A survey around the Japanese Islands and developing its instruments for more than a decade. Especially in 2010, a new project for the development of the GPS/A technique commenced, and since 2012 following the Tohoku earthquake, further acceleration of the project has been taken place. Tohoku and Nagoya Universities have been working on this project for 4 years. In the project, Tohoku University worked on several topics, such as realtime/continuous monitoring of crustal deformation using a moored buoy, automatic survey using an Autonomous Surface Vehicle (ASV), which makes the survey as efficient as possible, and constructing a new GPS/acoustic survey network along the Japan Trench and their intensive survey using a chartered ship. In this paper, we summarize the achievements in each of the topics above.

Keywords

Crustal deformation • GPS/acoustic • Seafloor geodesy • Tohoku-oki earthquake

1 Introduction

Since the idea of the GPS/A technique for seafloor geodesy was introduced by Spiess (1985) and its first successful result applied to the Juan de Fuca ridge was reported by Spiess et al. (1998), many efforts have been made on scientific surveys and improving its technology. Because geodetic data acquired at the seafloor is crucial to investigate great interplate earthquakes along the trenches surrounding Japan, the GPS/A technique has also been advancing in Japan. However, the basic style of GPS/A survey has been unchanged for a long time. It employs a campaign style (e.g., Spiess et al. 1998), in which a snap shot of positioning data is obtained only while a research vessel is at an observation site. In addition, the measurement must be continued for nearly 1 day at a given site to achieve centimeter-level determination of the seafloor position.

However, in the past several years, new approaches to GPS/A observation have been tested for next generation technologies. One is realtime/continuous measurement using a moored buoy, equipped with a local battery and a satellite communication modem for data and command transfer. The

M. Kido (✉) • R. Hino • T. Iinuma • I. Wada
International Research Institute for Disaster Science, Tohoku University, Sendai 980-0845, Japan
e-mail: kido@irides.tohoku.ac.jp

Y. Ohta • Y. Osada • R. Azuma • S. Miura • S. Suzuki • F. Tomita • M. Imano
Graduate School of Science, Tohoku University, Sendai, Japan

H. Fujimoto
National Research Institute for Earth Science and Disaster Prevention, Tsukuba, Japan

M. Hashimoto (ed.), *International Symposium on Geodesy for Earthquake and Natural Hazards (GENAH)*, International Association of Geodesy Symposia 145, DOI 10.1007/1345_2015_127

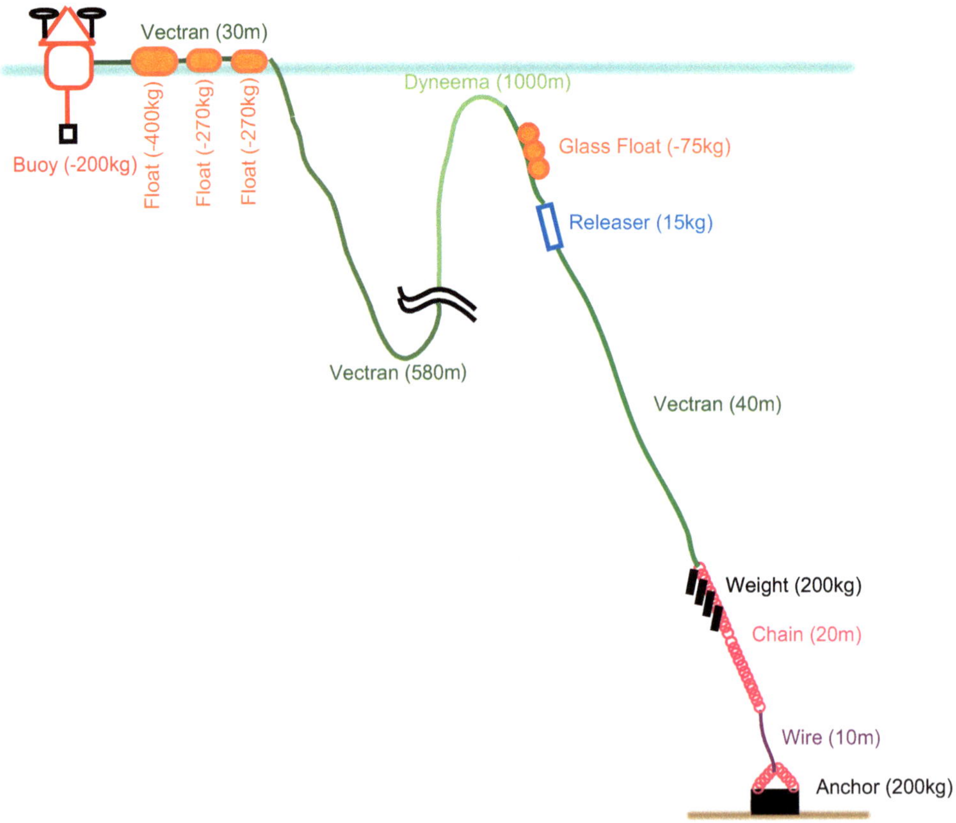

Fig. 1 Layout of the mooring system for the small buoy at an off-Miyagi site (not to scale). Total length of the tether is 1620 m for 1450 m depth

other is introducing an unmanned vehicle that automatically makes acoustic measurements along a pre-programmed survey line. Development of these technologies was conducted under the governmental project, named "Project for Development of GPS/Acoustic Technique", during 2010–2014. During this project, we encountered the devastating Tohoku-oki earthquake, which magnified the importance of seafloor geodetic survey in the earthquake source region. The project hence involved the activity for enhancement of GPS/A survey by newly constructing up to 20 observation sites along the Japan Trench. In this paper, we describe each of topics noted above in the following sections.

2 Moored Buoy System

Collaborating with Japan Agency for Marine-Earth Science and Technology (JAMSTEC) and Japan Aerospace eXploration Agency (JAXA), we have started long-term continuous and realtime seafloor geodetic surveys using a moored buoy at Kumano-nada, where a strong current, known as Kuroshio, exists. Preceding to this trial, in 2010, we carried out a short-term (2 days duration) sea-trial employing a small buoy (Fig. 1) as a mooring platform, which can be also used in towing survey, at off-Miyagi site (1,450 m in depth), where a modest current, ~1 kn, was dominant. The mooring system consists of a buoy, three floats, negatively buoyant vectran rope (Φ 9 mm), positively buoyant dyneema rope (Φ 9 mm), glass floats with an acoustic releaser for pup-up recovering, vectran rope, weights, chain rope, wire rope, and an anchor (Fig. 1). As shown in Fig. 2, the mooring system is designed to be slack-type mooring with the weights near the bottom, which shorten the system when current is weak so that the buoy is kept as close to the anchor as possible. With this system, the anchor needs to weigh only 200 kg. We equipped three floats (270–400 kg buoyancy for each) beside the buoy and found that even the furthest float was still at the surface when the current was at 1 kt (Fig. 3); i.e., tension of the system is less than 270 kgf.

Because the size of the battery is limited in the small buoy and power saving in the internal devices is not sufficient, the sea-trial lasted only for 2 days. However, using this platform, we developed an automatic ranging system and a simple on-demand operation technique via an UHF modem equipped on the buoy.

In 2012, we have started developing an automatic ranging system in realistic working condition using a time-proven platform, a k-TRITON buoy (Fig. 4), operated by JAMSTEC. Together with JAMSTEC and JAXA, the satellite communication component and the GPS positioning component have been shared for multi-purpose operations. The Tohoku University group is leading the development of the acoustic

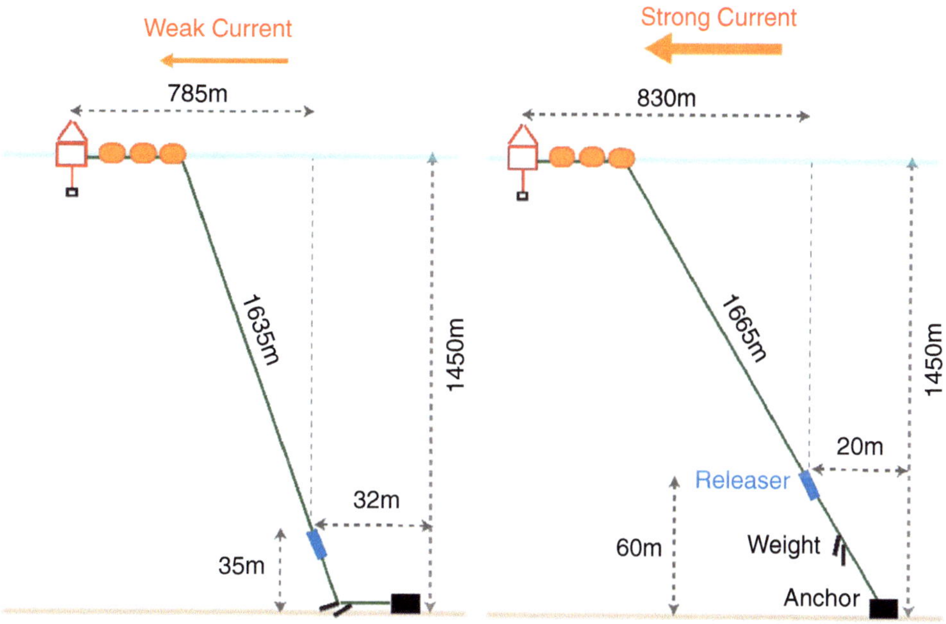

Fig. 2 Slack function of the system. The weight moves up and down depending on the speed of current, shortening the tether and bringing the buoy closer to the anchor

Fig. 3 Surface view of the small buoy. Three floats are still visible, indicating that only the third float of 270 kg buoyancy is sufficient to keep total tension of the system

ranging component and an onboard data processing approach to compute precise travel times within the buoy. Using limited onboard resources in the buoy, we have eliminated unnecessary and redundant procedures. Finally we achieved as small as 101 bytes of data for a single acoustic ranging measurement (including correlogram waveform) to be sent to the onshore station, which is small enough for Iridium Short Burst Data Transfer. At this moment, precise GPS data needs post-processing, however, it can be replaced with onboard real-time Precise Point Positioning (PPP) technique in the future. The first sea-trial was taken place in 2013 for 4 months and the second trial was in 2014 for 6 months. In the second trial, acoustic ranging data has been successfully transmitted to the onshore station every week and we can monitor it remotely from our laboratory. Further refinement of the system suited for practical use, e.g., pressure measurements, PPP analyses using Quasi-Zenith Satellite System (QZSS) or other services, and data transmission, are the subjects of the ongoing project. Details of the buoy operation and data processing are introduced in this issue by Takahashi et al. (2015) and Imano et al. (2015).

3 Autonomous Surface Vehicle

With the ASV system, we aimed to develop an unmanned survey system (autonomous or remotely controlled), which can also be used concurrently with other platforms, such as a research vessel, in order to improve the ranging accuracy with multi-acoustic paths. The ASV is like an unmanned

Fig. 4 Whole view of the k-TRITON buoy for the long-term Kumano-nada trials (*left*). An acoustic transducer is equipped at the bottom of the buoy (*right*)

Fig. 5 Deployment of the ASV. The acoustic transducer is at the lower end of the keel

boat (2.4-m long and 400 kg in weight) shown in Fig. 5, whose propulsion system is powered by large batteries that are charged by an onboard diesel generator. After tuning the system through several trial operations, we found that the ASV works well for GPS/A observation. Owing to the onboard diesel generator, electric power supplied to the survey system is nearly unlimited (up to 100 Wh) for our onboard devices as payload. Most of the power is consumed to drive the thrusters for its navigation. With standard sea condition, survey can be continued for 3–7 days with 80 L of diesel oil.

For navigation performance, the ASV can keep its position and move along given survey paths with 3 kt at maximum. Fig. 6 shows the track of the ASV during a sea trial. It moved around a 300-m radius circle three times and the difference in the circling path was less than 2 m as shown in the enlarged view. The holding ability at a fixed point was ± 3 m as shown in the enlarged view for 2 h fixed point survey at the center of the circle. The tracks in the southwest of the center were made during human operation by using a controller. We have also confirmed that the acoustic noise level is acceptable even when the generator working; most of the time, the generator is automatically stopped because the battery charged enough. Without the generator, the noise from the thruster was negligible as shown by clear wave signal in Fig. 7. As the ASV system demonstrates

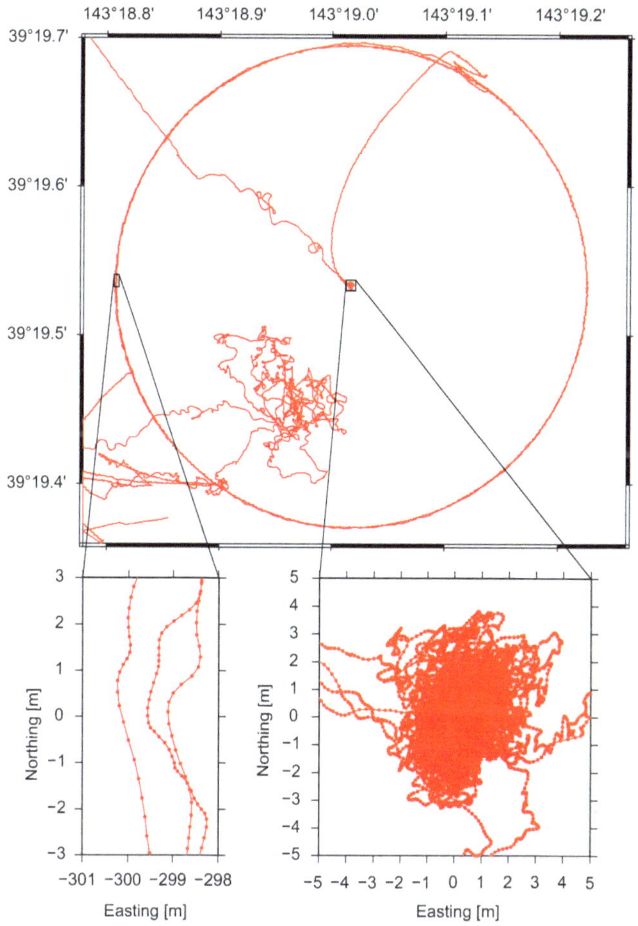

Fig. 6 Track of the ASV trial. Repeat surveys of a 300 m radius circle and point survey at the center of the circle. Enlarged views for circling and point survey are shown *lower panels* to highlight navigational and station keeping abilities of the ASV. The track data are obtained at 5 Hz sampling

sufficient performance for our use in GPS/A survey, it can be a candidate of multi surface platform for concurrent ranging to achieve high accuracy GPS/A measurement taking spatial sound speed variation into account with the utilization of multiple acoustic paths (e.g., Kido 2007). In the next project, we plan to gather enough data in various sea conditions and develop an optimization algorithm to deal with multi-acoustic-path survey for more practical applications of the ASV.

4 Tohoku-oki Earthquake

After the Tohoku-oki earthquake in 2011, the project expanded to include an extra mission that significantly enhanced the survey framework, especially over the deep seafloor (>5,000 m) near the trench axis. In this extra mission, we have developed a new type of seafloor transponders that work at over 5,000-m depth, and its acoustic communication range is greater than 15 km. We made 86 transponders in total and constructed 20 new GPS/A sites along the Japan Trench in 2012. Most of the sites are located on the seafloor near the trench axis at depths greater than 5,000 m because the Tohoku-oki earthquake revealed that the state of the plate interface near the trench axis plays important roll for earthquake generation.

A single site consists of three, four, or six transponders depending on seafloor morphology. It is quite difficult to find a flat terrace having enough area; in GPS/A positioning, the geometry of each transponder array has to remain fixed (Spiess 1985), and possible local faults should not cross the array. At most sites, the array lies on the edge of a terrace like in Fig. 8. We note that the required array dimensions proportionally increase with depth. The distribution of observation sites and their array geometries are visible in Fig. 9.

We chartered a research ship for about 50 days per year to construct and to observe the new observation sites. The initial survey was taken place just after the installation of the transponders in 2012. At present we have carried out five campaigns since the installation. We used to employ a towed buoy as a surface platform, but have transitioned to use a chartered ship equipped with a side-mount transducer system, which is much more efficient than towed buoy system, especially during a moving survey whilst underway in rough sea conditions, such as 2 m of wave height. At the new sites, we conducted both moving and point surveys; the former generally takes several hours and the latter 12 h at each site. We combine all the data obtained by both types of surveys to determine the array center position. We found a problem in the acoustic property of the new transponders, which can be corrected during post-processing shown in Azuma et al. (2014). Because the five campaigns span only about one and a half years, estimated errors are typically 10 cm, which are larger than obtained displacements at most sites. However, at a few sites, we detected significant post-seismic movements. These results were reported in Tomita et al. (2014). As the new transponders are hybrid type, Japan Coast Guard has also started to make measurements with their own surface platform at some of the new sites (Watanabe et al. 2014).

5 Summary

In the project for advanced GPS/Acoustic survey in the last 4 years, we have developed the key technologies for realtime monitoring of crustal deformation using a moored buoy and for a maneuverable autonomous surface platform using an ASV system. These technologies can be directly put to practical use in realistic conditions, such as long-lasting continuous operation even in hard sea environment, and in efficient GPS/A measurement for extensive surveys.

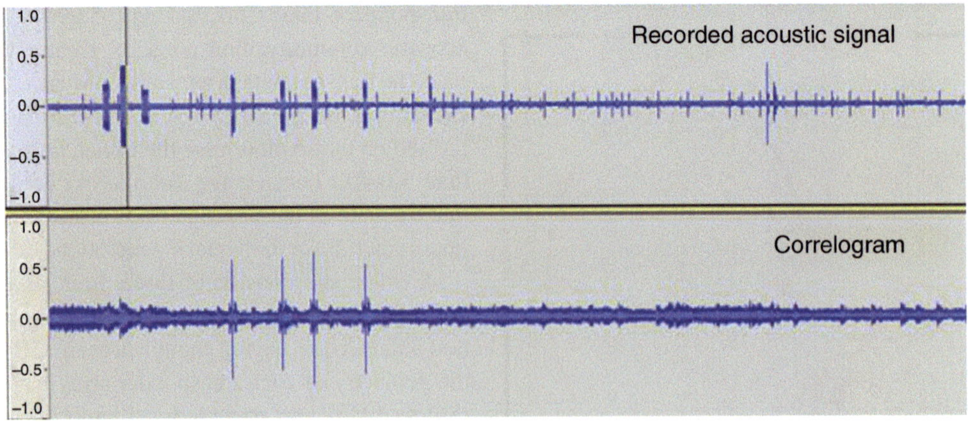

Fig. 7 Acoustic signal recorded on the ASV while running only by the thruster (*top*) and its correlogram (*bottom*) for 10 s. Four reply signals are clearly seen with a leading header for each

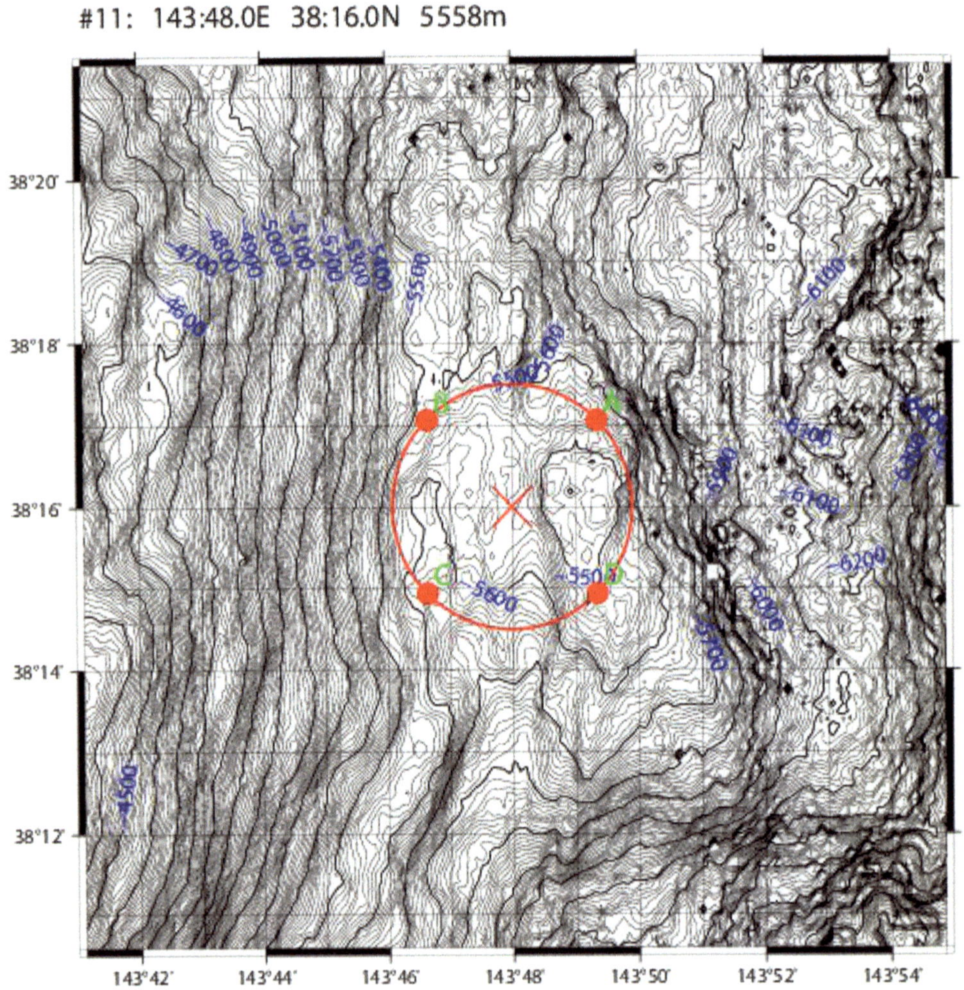

Fig. 8 Seafloor topography around G11 station. The transponder array lies on a narrow terrace and each of four transponders, indicated by *red dots*, are very close to adjacent steep cliff

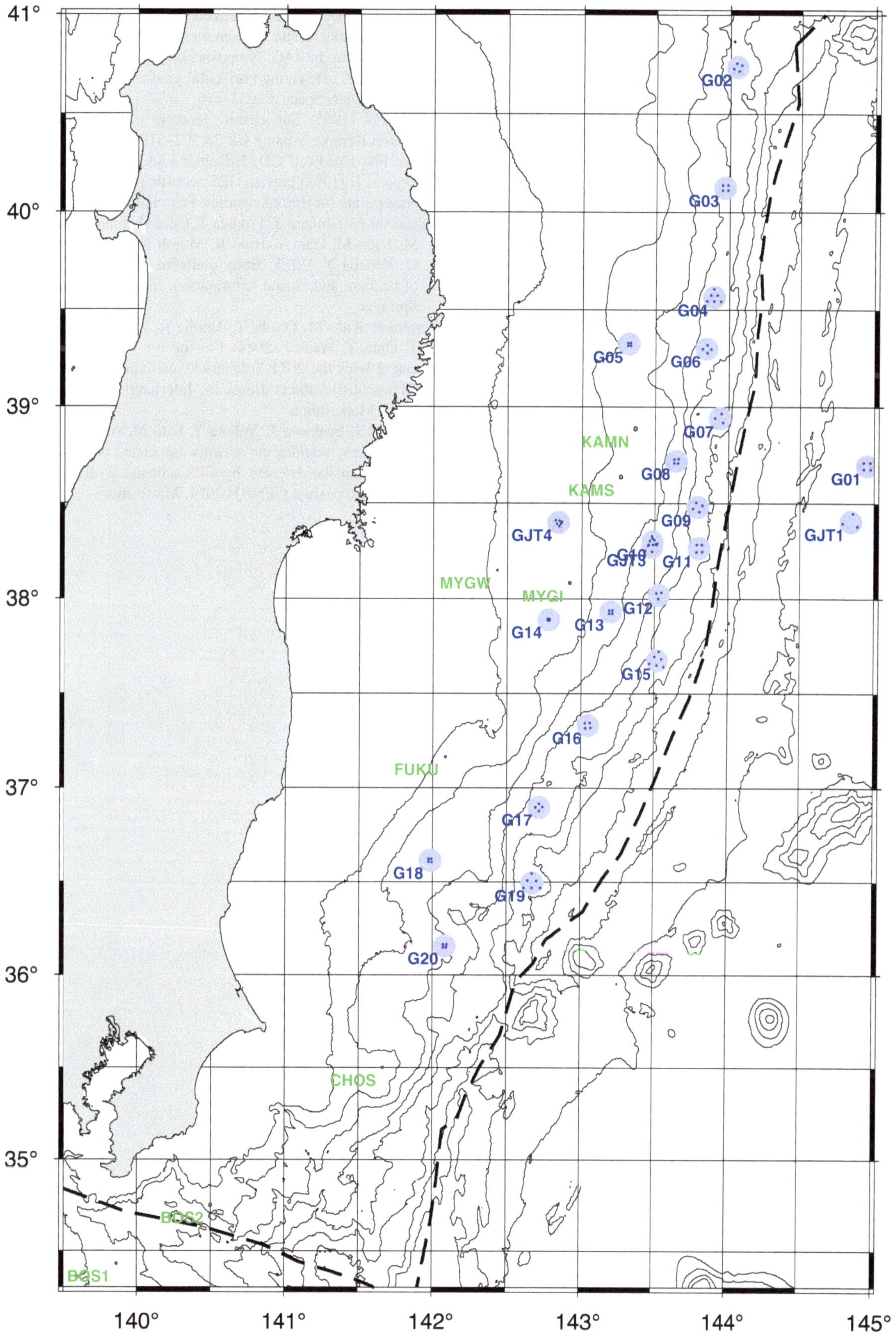

Fig. 9 Current GPS/A sites along Japan Trench. *Green labels* indicate sites of Japan Coast Guard. Labeled in *blues* are Tohoku University, newly installed G01–G20 and pre-existing GJT series. Individual transponders are indicated by *small dots*

In addition, the dense network of GPS/A observation sites constructed in this project and the subsequent extensive surveys lead us to systematic survey in order to maintain a sustainable framework that we lead to long-lasting geodetic surveys.

Acknowledgements This work has been supported by the MEXT through the Project for Development of GPS/Acoustic Technique. Stuffs in RCPEV and IRIDeS, Tohoku University, provided dedicated support and collaborative operation for this project, especially in the onboard surveys and management of onshore GPS reference data. The construction and surveys of the new observation sites were in collaboration with Nagoya University. We also thank the anonymous reviewers to provided intensive comments, which significantly improve the manuscript.

References

Azuma R, Tomita F, Iinuma, T, Hino R, Kido M, Fujimoto H, Osada Y, Ohta Y, Wada I (2014) Development and examination of new methods for traveltime detection in GPS/A geodetic data to high-precise and automated analysis. In: International Symposium GENAH 2014, Matsushima, p 38

Imano M, Kido M, Ohta Y, Fukuda T, Ochi H, Takahashi N, Hino R (2015) GPS/acoustic measurements using a multi-purpose moored buoy system. In: IAG Symposia Series, ibid. Springer

Kido M (2007) Detecting horizontal gradient of sound speed in ocean. Earth Planets Space 59:e33–e36

Spiess FN (1985) Suboceanic geodetic measurements. IEEE Trans Geosci Remote Sensing GE 23:502–510

Spiess FN, Chadwell CD, Hildebland JA, Young LE, Purcell Jr GH, Dragert H (1998) Precise GPS/acoustic positioning of seafloor reference points for tectonic studies. Phys Earth Planet Int 108:101–112

Takahashi N, Ishihara Y, Fukuda T, Ochi H, Tahara J, Mori T, Deguchi M, Kido M, Ohta Y, Hino R, Mutoh K, Mashimoto G, Motohashi O, Kaneda Y (2015) Buoy platform development for observation of tsunami and crustal deformation. In: IAG Symposia Series, ibid. Springer

Tomita F, Kido M, Osada Y, Azuma R, Fujimoto F, Hino R, Iinuma T, Ohta Y, Wada I (2014) Postseismic seafloor deformation associated with the 2011 Tohoku-oki earthquake revealed by repeated GPS/acoustic observations. In: International Symposium GENAH 2014, Matsushima

Watanabe S, Ishikawa T, Yokota Y, Sato M, Asada A (2014) Co- and postseismic seafloor movements associated with the 2011 Tohoku-oki earthquake detected by GPS/acoustic positioning. In: International Symposium GENAH 2014, Matsushima

Interplate Coupling in and Around the Rupture Area of the 2011 Tohoku Earthquake (M9.0) Before Its Occurrence Based on Terrestrial and Seafloor Geodetic Observations

Takeshi Iinuma, Ryota Hino, Motoyuki Kido, Yusaku Ohta, and Satoshi Miura

Abstract

In this study, we investigated the state of interplate coupling between the subducting Pacific plate and continental plate around the rupture area of the 2011 Tohoku Earthquake (M9.0) based on seafloor geodetic data as well as terrestrial GPS data obtained before the M9 earthquake in order to evaluate the potential for further seismic activity from the viewpoint of the slip deficit budget balance. Due to the difference between the site spatial densities on land and the seafloor observation networks, it was necessary to utilize a covariance matrix with non-zero off-diagonal components in the inversion to estimate the interplate coupling distribution on the plate interface and to account for the correlations among the observation errors for components and stations. We validated the implementation of the assumed covariance matrix by performing a computational test and we then applied the method to actual geodetic data obtained before the 2011 Tohoku Earthquake. The results demonstrate the persistence of the interplate coupling in the area of a very large coseismic slip near the Japan Trench.

Keywords

Acoustic ranging • Covariance • GPS • Interplate coupling • Seafloor geodesy • 2011 Tohoku Earthquake

1 Introduction

Monitoring the state of interplate coupling is important for understanding the earthquake generation cycle and to assess the risk of seismic and tsunami hazards caused by interplate earthquakes. Many previous studies have investigated interplate coupling based on terrestrial geodetic data around the Japanese Islands, especially beneath the Tohoku district in northeastern Japan, where large earthquakes have occurred repeatedly on the plate interface of the Pacific plate that subducts beneath the overriding continental plate (Okhotsk or North American plate) (e.g., El-Fiky and Kato 1999; Nishimura et al. 2004; Suwa et al. 2006; Hashimoto et al. 2009; Loveless and Meade 2010). However, the large coseismic slip along the trench axis during the 2011 off the Pacific coast of Tohoku Earthquake (M9.0, known as the Tohoku Earthquake) highlighted the need to re-examine

T. Iinuma (✉)
International Research Institute of Disaster Science, Tohoku University, 468-1, Aza-Aoba, Aramaki, Aoba-ku, Sendai 980-0845, Japan
e-mail: iinuma@jamstec.go.jp

Present Address: Research and Development Center for Earthquake and Tsunami, Japan Agency for Marine-Earth Science and Technology, 3173-25, Showa-machi, Kanazawa-ku, Yokohama 236-0001, Japan

R. Hino • M. Kido
International Research Institute of Disaster Science, Tohoku University, Sendai, Japan

Present Address: Graduate School of Science, Tohoku University, Sendai, Japan

Y. Ohta • S. Miura
Graduate School of Science, Tohoku University, Sendai, Japan

M. Hashimoto (ed.), *International Symposium on Geodesy for Earthquake and Natural Hazards (GENAH)*,
International Association of Geodesy Symposia 145, DOI 10.1007/1345_2015_48

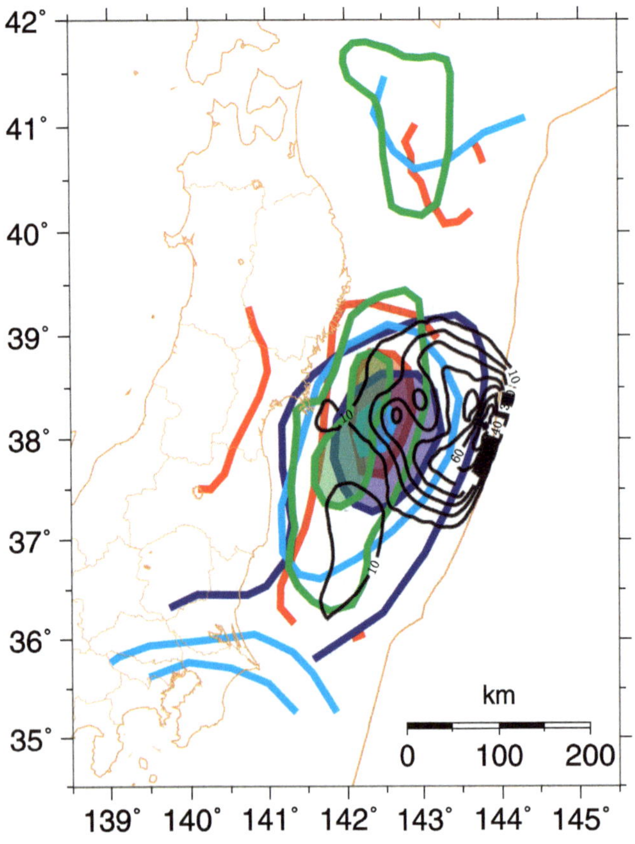

Fig. 1 Map of the northeastern Japanese Islands showing the slip deficit distributions estimated in previous studies. The *red*, *blue*, *cyan*, and *green* contours indicate the traces of the contours in the slip deficit distribution models of Ito et al. (1999), Nishimura et al. (2004), Suwa et al. (2006), and Hashimoto et al. (2009), respectively. The *black* contours represent the coseismic slip distribution of the 2011 Tohoku Earthquake estimated by Iinuma et al. (2012). The *brown solid line* in the Pacific Ocean indicates the axis of the Japan Trench

these estimates of interplate coupling. In general, the main rupture area estimated for the Tohoku Earthquake based on seismological, geodetic, or tsunami data corresponds to the area with strong interplate coupling off the Tohoku district in the direction parallel to the Japan Trench (e.g., Ide et al. 2011; Fujii et al. 2011; Iinuma et al. 2012), but the distribution differs in the downdip direction (Fig. 1). This discrepancy may be due to the lack of seafloor data in the estimates of interplate coupling distributions. Therefore, the interplate coupling along the shallowest portion of the megathrust before the Tohoku Earthquake must be the most important issue in this re-examination because the spatial resolution of the inversion analysis based only on terrestrial geodetic data is generally not sufficiently high to constrain the coupling state in the far offshore area.

Seafloor geodetic observation has been developed by connecting GPS positioning and acoustic ranging (GPSA) in seawater and applied off the Pacific coast of the Tohoku district (e.g., Fujita et al. 2006; Sato et al. 2011; Kido et al. 2011), where Sato et al. (2013) reported the secular displacement rate before the 2011 Tohoku Earthquake. These seafloor geodetic observation data are invaluable for estimating the interplate coupling beneath northeastern Japan, but they have not been used in any inversion analyses to estimate the distribution of the degree of interplate coupling. It has been difficult to include these seafloor data in an inversion analysis because the measurements contain large uncertainties and they are sampled more sparsely in both the temporal and spatial domains than terrestrial observations. If we simply give the seafloor data greater weights than the terrestrial data, very strong coupling is estimated in small patches nearest to the seafloor stations on the plate interface. By contrast, if we reduce the weights of the seafloor data, the spatial resolution of the inversion analysis is not improved sufficiently to detect the interplate coupling near the trench axis.

Table 1 Displacement rates at seafloor stations relative to the North American Plate reported by Sato et al. (2013)

Site	EW	NS	UD	Observation period
KAMN	−3.9 ± 0.2	1.8 ± 0.1	4.3 ± 0.2	2002/JUL–2010/NOV
MYGI	−5.6 ± 0.9	3.8 ± 0.7	6.8 ± 0.8	2002/MAY–2005/AUG
	−4.1 ± 0.4	2.1 ± 0.5	4.6 ± 0.4	2006/DEC–2011/FEB
MYGW	−4.2 ± 0.2	0.6 ± 0.4	4.2 ± 0.2	2006/DEC–2011/FEB
FUKU	−0.8 ± 0.4	−0.3 ± 0.3	0.8 ± 0.4	2002/JUL–2011/FEB
	−2.3 ± 0.7	0.0 ± 0.7	2.3 ± 0.7	2002/JUL–2008/MAY

The unit of the displacement rate is cm/year

In the present study, we address this problem by considering the correlations between the observed errors in the displacement rates in the inversion analysis using a covariance matrix of the observation errors but without assuming that the covariance matrix is a diagonal matrix. Instead, we assume that all of the observation errors are mutually independent in the diagonal covariance matrix. We also assume that the correlation between the displacement rates at two different GPS sites depends on the distance between them, and we configure the covariance between the different components, such as east-west (EW), north-south (NS) and up-down (UD), by applying the results obtained after raw GPS data processing.

2 Data

We used daily site coordinate time series data obtained from the GPS observation network of the Geospatial Information Authority of Japan (F3 solution) and secular velocities collected by seafloor GPSA stations of the Japan Coast Guard, which were reported by Sato et al. (2013), as shown in Table 1. The observation periods when the velocities were determined at the GPSA stations differed from each other; thus, we estimated the secular velocities at GPS sites for each period by assuming that a site coordinate time series comprised a linear trend (corresponding to the period-averaged velocity) with annual and semiannual variations, and steps due to earthquakes and antenna replacements. Thus, we had four velocity fields for four periods, i.e., Period 1: May 2002 to August 2005, Period 2: July 2002 to May 2008, Period 3: July 2002 to November 2010, and Period 4: December 2006 to February 2011.

3 Analysis

3.1 Method

We applied the inversion method summarized by Iinuma (2009) and Iinuma et al. (2012) to estimate the back slip distribution on the plate boundary fault based on the velocity field. Back slip was introduced by Savage (1983) to express the heterogeneous distribution of the degree of interplate coupling. In this inversion method, the weights of three constraint conditions, i.e., spatial smoothing, initial damping value, and the boundary condition, are optimized by minimizing Akaike's Bayesian Information Criterion (Akaike 1977, 1980). The initial value, which is configured for all of the subfaults to constrain the slip deficit to some positive value, and the weights of the constraint conditions are optimized simultaneously by a genetic algorithm. The boundary condition constrains the slip deficit at the north, south, and west (deep) edges of the modeled fault to be zero, whereas the slip deficit at the east (shallow) edge along the Japan Trench is not constrained by the boundary condition. The spatial distribution of back slip is expressed by the superposition of the dip-slip and strike-slip components on the triangulated tessellation of the plate interface geometry determined by Nakajima and Hasegawa (2006) in a homogeneous elastic half-space (Meade 2007). We virtually shallowed the depth of the plate interface to calculate Green's function for the seafloor sites based on their water depth in order to adjust the closeness to the plate boundary fault of such sites. For example, for the seafloor site MYGW, which is located on the seafloor at a water depth of 1.7 km, we subtracted 1.7 km from the original depth of each subfault at the plate interface in order to reposition the free surface of the homogeneous elastic half-space from the sea surface to the seafloor.

The covariance matrix of the data error vector must be employed in the inversion method. Frequently, the covariance matrix has been assumed to be diagonal in order to simplify the formulation and to reduce the calculation costs. However, because the site separations are very different for terrestrial and seafloor observation networks, i.e., less than 20 km for the terrestrial GPS network and larger than 50 km for the seafloor GPSA network, the use of a diagonal matrix as a covariance matrix is a poor solution. Because a diagonal covariance matrix can only tune the weights of the data, systematic residuals for GPS data and excessively large slip deficit patches beneath the seafloor stations will appear if we give greater weights to the seafloor data. By contrast, if we give greater weights to the terrestrial data, the excessively large residuals will remain for the GPSA stations and short wavelength perturbations in the slip deficit distribution will appear beneath the land. Therefore, the covariance of the observed displacement rate data needs to consider the non-diagonal components of the covariance matrix.

Ohta et al. (2010) performed a raw GPS data analysis and reported the noise characteristics, including the correlations between different components and stations. They found that the correlation coefficients between the EW and NS, EW and UD, and NS and UD components at an identical site were about 0.2, 0.0, and −0.2, respectively. They also reported the baseline length dependency of the correlation coefficient for a component observed at two different sites. The correlation coefficients decreases linearly from 1.0 at $l = 0$ km, where l is the baseline length, to 0.0 at $l = 200$ km. Thus, we configured the covariance matrices to reproduce these correlation coefficients and we examined the effect of including the non-diagonal covariance matrix by performing a numerical test, as described in the following subsection.

3.2 Numerical Test

We performed synthetic inversion tests to evaluate the effects of correlations between sites. In these tests, the displacement at each observation site was produced by adding uncorrelated or correlated observation errors to the calculated displacement according to the assumed slip distribution, as shown in Figs. 2a and 3a. Five sets of observation errors were generated using different pseudo-random numbers. The inversion analyses were performed using covariance matrices with zero or non-zero diagonal components in each case.

Figure 2 shows the results for the case where the observation errors were generated based on a diagonal covariance matrix, i.e., all of the errors are mutually independent. The displacement fields inverted to estimate the back slip distribution were identical for the results shown in Fig. 2b and g, c and h, d and i, e and j, and f and k, respectively. Figure 2b–f show the back slip distributions estimated using the same covariance matrix as that used to generate the errors. Figure 2g–k show the results obtained when we performed inversion analyses with a covariance matrix that contained non-zero non-diagonal components, as described in the previous section. In general, the extent of the estimated back slip distributions was similar in all cases and the magnitude of the back slip was relatively small for the results obtained using a covariance matrix that contained non-zero non-diagonal components.

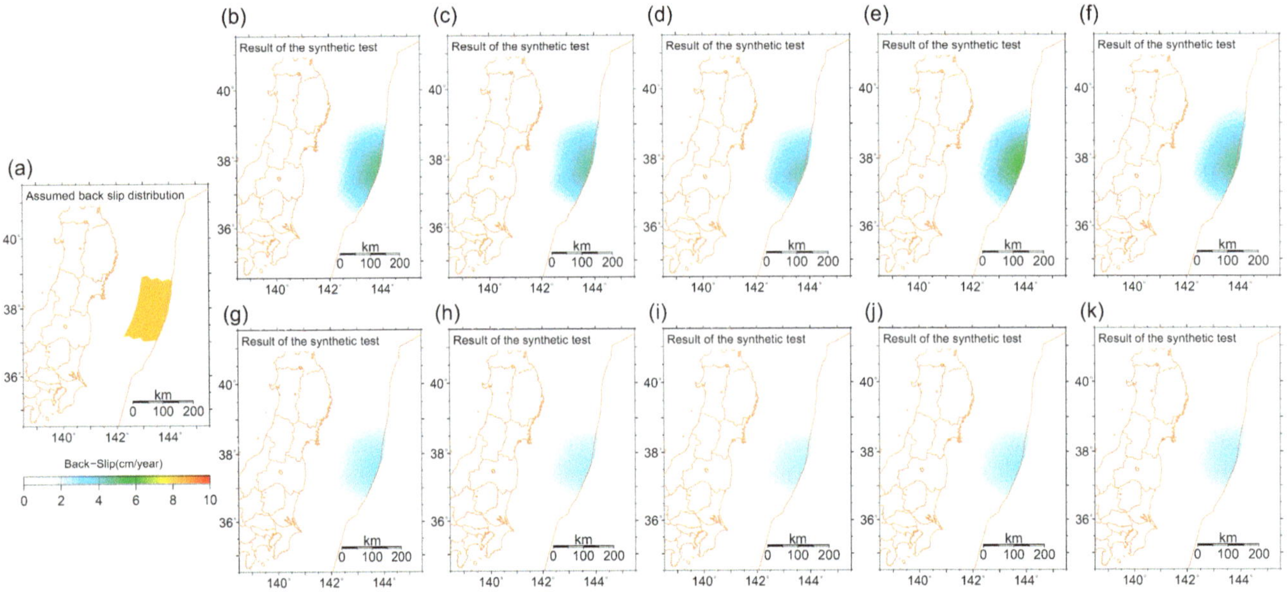

Fig. 2 Results of the numerical tests. Estimated back slip distributions based on the displacement fields calculated by adding uncorrelated observation errors to the displacements expected from the assumed back slip distribution (**a**), by performing inversion analyses involving covariance matrices with zero non-diagonal components (**b**–**f**), and with non-zero non-diagonal components (**g**–**k**). Each *column* corresponds to a different realization with random noise in the synthetic data

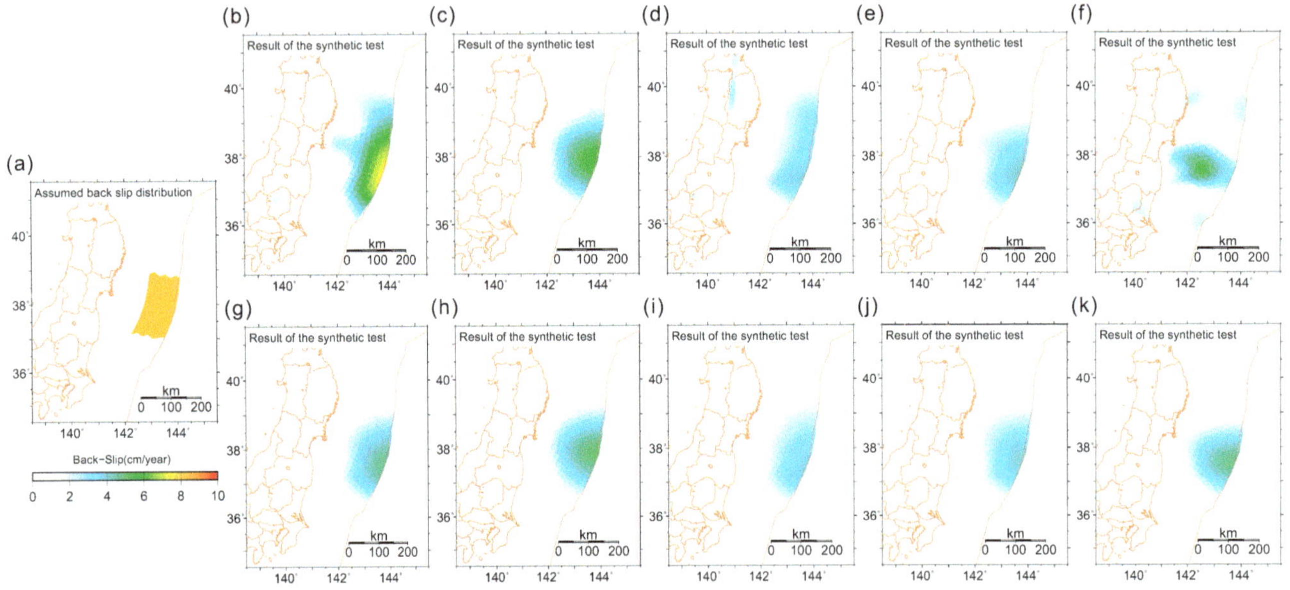

Fig. 3 Results of the numerical tests. Estimated back slip distributions based on displacement fields calculated by adding correlated observation errors to the displacements expected from the assumed back slip distribution (**a**), by performing inversion analyses involving covariance matrices with zero non-diagonal components (**b**–**f**), and with non-zero non-diagonal components (**g**–**k**). Each *column* corresponds to a different realization of random noise in the synthetic data

The results obtained when a covariance matrix that included inter-station and inter-component correlations was employed to generate observation errors are shown in Fig. 3. Figure 3b–f show the back slip distributions estimated using a diagonal covariance matrix. Figure 3g–k show the results obtained when we performed the inversion analyses with the same covariance matrix as that used to generate the errors. Clearly, the back slip distribution estimated using a diagonal matrix differed significantly from the back slip distribution given to synthesize the displacement field.

Table 2 shows the average moment accumulation due to the estimated slip deficit based on the moment magnitude scale. The moment accumulation due to the assumed slip deficit distribution was equivalent to Mw 7.08. The moments

Table 2 Average moment accumulation attributable to the estimated slip deficit distributions using the moment magnitude scale. The moment accumulation due to the given slip deficit distribution was equivalent to Mw 7.08

Data synthesizing	Covariance matrix in the inversion	
	Full correlation	No correlation
Full correlation	7.06	7.18
No correlation	6.98	7.06

of the estimated slip distributions were almost preserved when we applied covariance matrices identical to those used to synthesize the data as a matter of course. A slight underestimation of the moment accumulation could occur when we performed inversion analyses with a covariance matrix that represented the full correlations, although the errors were relatively small when none of the correlations reflected the magnitude of the slip. By contrast, we overestimated the moment accumulation when we used a diagonal covariance matrix for the correlated observation errors.

The results of these tests indicate that it is not possible to accurately estimate the spatial distribution of the back slip using a diagonal covariance matrix if the observation errors are correlated with the space and the components. However, it is possible to estimate the range of the area with back slip by applying a covariance matrix that assumes that the errors are correlated, even if the errors are mutually independent. Therefore, we conclude that it is necessary to use a covariance matrix with non-zero non-diagonal components to estimate the back slip distribution on the plate interface before the 2011 Tohoku Earthquake.

4 Results and Discussion

4.1 Estimated Back Slip Distribution

Figure 4 shows the distribution of the back slip on the plate interface during each period. The interplate coupling changed significantly in the two areas in different periods. The first area is near the coastline by Miyagi Prefecture, where M 7.5 class earthquakes have occurred repeatedly with a recurrence interval of about 40 years (Seno et al. 1980; Umino et al. 2006). On 16 August 2005, an M 7.2 earthquake ruptured part of the rupture area of the 1978 Miyagi-oki Earthquake (M 7.4) and a postseismic slip occurred around the source region (Miura et al. 2006; Miyazaki et al. 2006). Therefore, the interplate coupling near the coastline by Miyagi Prefecture vanished in Periods 2–4, whereas significant coupling existed in Period 1.

The second area with a significant change in back slip was estimated to be a region far from Fukushima and Ibaraki Prefectures on the shallow plate boundary. The change in the degree of coupling in this area corresponded to the change in the interplate slip rate before an M 7.0 earthquake off Ibaraki Prefecture, as reported by Uchida and Matsuzawa (2013).

The strong (almost full) interplate coupling far off Miyagi prefecture at the shallow portion of the plate interface was time independent. However, the temporal change in the back slip rate was not sufficiently large to discuss in detail considering the estimation error in the area near the trench. Figure 5 shows the back slip distribution throughout the different periods, which were calculated based on the results for all periods using weighted averages for the lengths of their periods. In general, the average back slip distribution corresponds to the area with strong interplate coupling that were previously estimated based only on the terrestrial GPS data. in the direction parallel to the Japan Trench, but the distribution differs in the downdip direction reflecting the improvement of the spatial resolution of the inversion analysis by using the seafloor data (Fig. 1). Loveless and Meade (2011) estimated the interplate coupling distribution with relaxing the assumption of zero coupling along the Japan Trench based only on the terrestrial data. Their result shows that the interplate coupling is strong at the area near the trench north to 39.5°N, while there is no strong coupling at the area in our estimates. This difference is caused by the use of the seafloor data, which improved evidently the spatial resolution of the inversion analysis.

After comparing the average back slip distribution with the coseismic slip distribution of the 2011 Tohoku Earthquake (Iinuma et al. 2012), we can conclude that the interplate coupling at the area with a very large (>50 m) coseismic slip was persistent before the M9 earthquake. This suggests that the 2011 Tohoku Earthquake released strain energy accumulated over about 1,000 years because the maximum coseismic slip was estimated at about 80 m and the plate convergence rate was about 8 cm/year (Altamimi et al. 2007). This estimate also means that the last previous earthquake to rupture this very shallow portion of the plate interface off Tohoku district was the 869 Jogan Earthquake (Minoura et al. 2001; Sawai et al. 2008).

Different estimates of the recurrence interval can be obtained from the budget of seismic moment. The moment accumulation due to the estimated back slip in the rupture area of the 2011 Tohoku Earthquake (dotted square in Fig. 5a) is equivalent to Mw 7.27/year, where the rigidity is assumed to be 30 GPa. If we divide the seismic moment of an M9.0 earthquake by this rate, the recurrence time is estimated

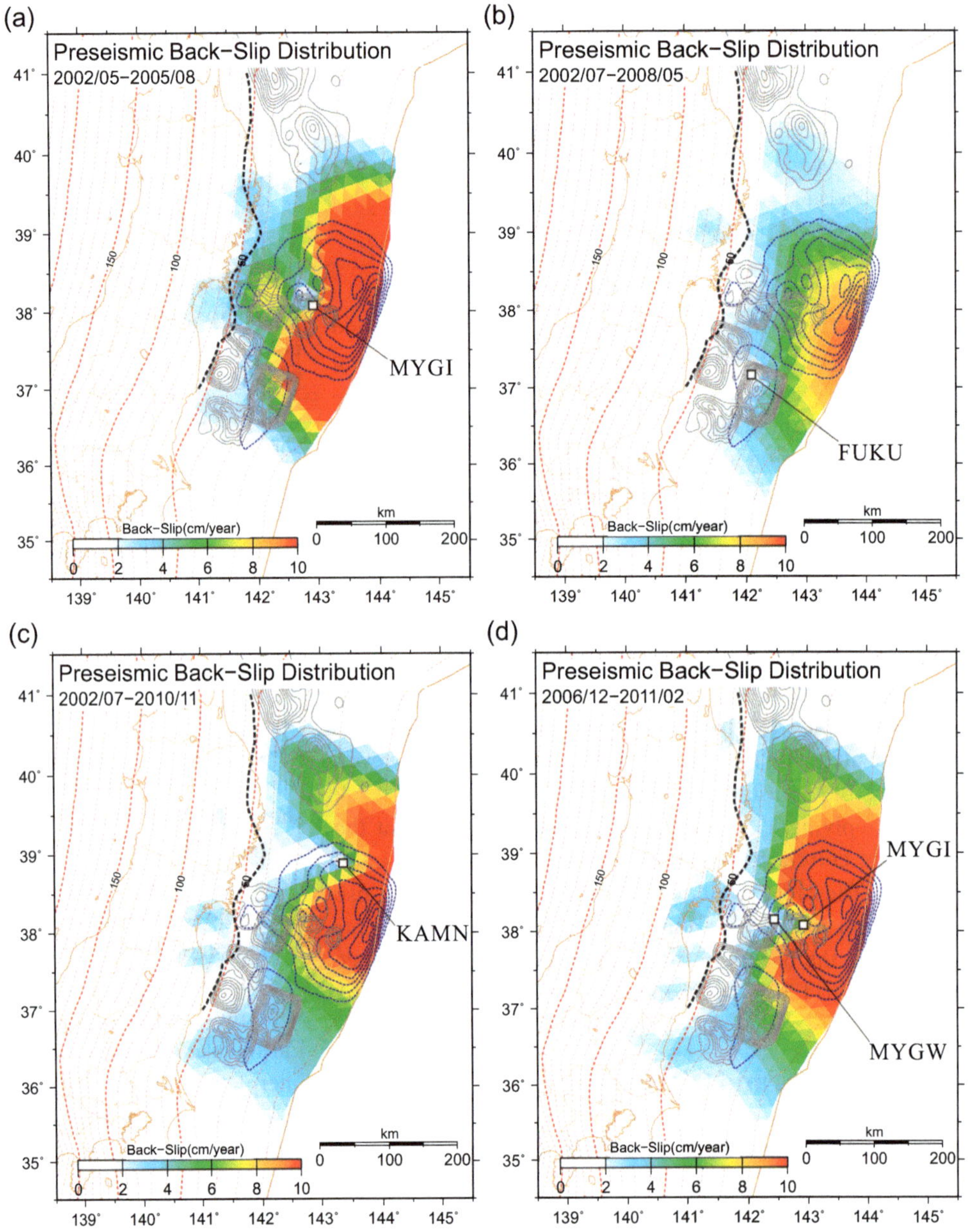

Fig. 4 Back slip distributions estimated based on terrestrial and seafloor geodetic data for Periods 1–4: (**a**) May 2002 to August 2005, (**b**) July 2002 to March 2008, (**c**) July 2002 to November 2010, and (**d**) December 2006 to February 2011

as about 350 years. This discrepancy cannot be explained by considering M < 8 earthquakes that occurred in this region because over 50 M8 earthquakes would be required every 1,000 years to release the moment accumulated over 650 years with Mw 7.27/year, but there is no evidence to support such frequent earthquake occurrences. However, the occurrence of aseismic slip on the plate interface before and after the main shock around the area of the very large coseismic slip, as clarified by Uchida and Matsuzawa (2013), can decrease the gap between the estimated recurrence interval for M9 earthquakes. Thus, it will also be necessary to consider the viscoelastic effect in future investigations to balance the accumulation and release of strain energy in the subduction zone.

Acknowledgements The authors would like to thank the Japan Nuclear Energy Safety Organization, the Geospatial Information Authority of Japan, and the National Astronomical Observatory of Japan for providing GPS data. The present study was supported through the Intramural Research Grant for Special Project Researches from International Research Institute of Disaster Science, Tohoku University. Three anonymous reviewers provided thoughtful comments, which improved the manuscript. The figures were created using GMT software (Wessel and Smith 1998).

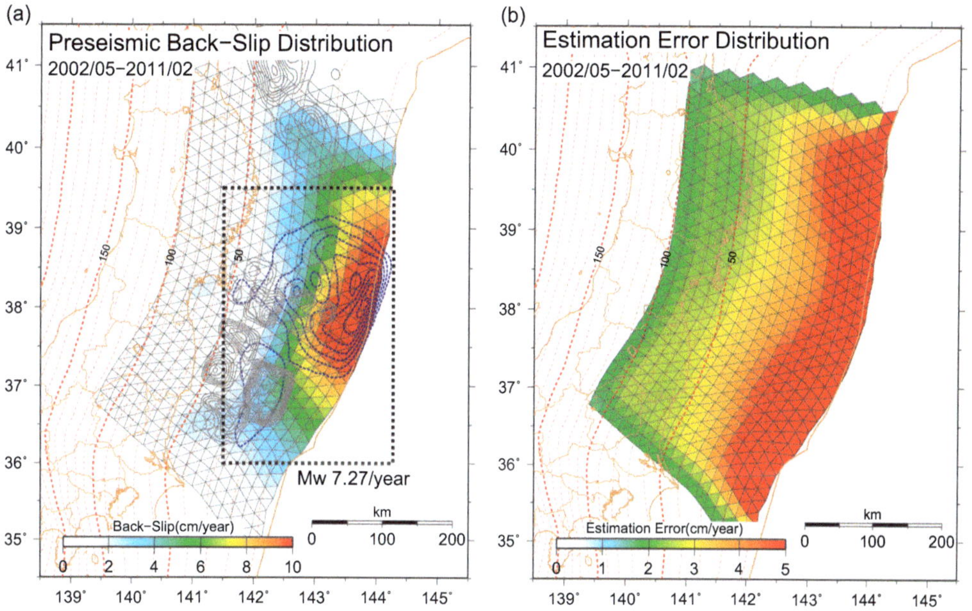

Fig. 5 Average back slip distributions (**a**) and their errors (**b**)

References

Akaike H (1977) On entropy maximization principle. In: Krishnaiah PR (ed) Application of statistics. North-Holland Publishing Company, Amsterdam, pp 27–41

Akaike H (1980) Likelihood and the Bayes procedure. In: Bernardo JM, DeGroot MH, Lindley DV, Smith AFM (eds) Bayesian statistics. University Press, Valencia, pp 143–166

Altamimi Z, Collilieux X, Legrand J, Garayt B, Boucher C (2007) ITRF2005: a new release of the International Terrestrial Reference Frame based on time series of station positions and Earth Orientation Parameters. J Geophys Res 112:B09401

El-Fiky GS, Kato T (1999) Interplate coupling in the Tohoku district, Japan, deduced from geodetic data inversion. J Geophys Res 104:20361–20377

Fujii Y, Satake K, Sakai S, Shinohara M, Kanazawa T (2011) Tsunami source of the 2011 off the Pacific coast of Tohoku Earthquake. Earth Planets Space 63:815–820

Fujita M, Ishikawa T, Mochizuki M, Sato M, Toyama S, Katayama M, Kawai K, Matsumoto Y, Yabuki T (2006) GPS/acoustic seafloor geodetic observation: method of data analysis and its application. Earth Planets Space 58:265–275

Hashimoto C, Noda A, Sagiya T, Matsu'ura M (2009) Interplate seismogenic zones along the Kuril-Japan trench inferred from GPS data inversion. Nat Geosci 2:141–144

Ide S, Baltay A, Beroza GC (2011) Shallow dynamic overshoot and energetic deep rupture in the 2011 Mw 9.0 Tohoku-Oki earthquake. Science 332(6036):1426–1429

Iinuma T (2009) Discussion on the rank deficiency of the representation matrix of the smoothing constraint in inversion methods using a Bayesian information criterion. J Geod Soc Jpn 55:345–353

Iinuma T, Hino R, Kido M, Inazu D, Osada Y, Ito Y, Ohzono M, Tsushima H, Suzuki S, Fujimoto H, Miura S (2012) Coseismic slip distribution of the 2011 off the Pacific coast of Tohoku earthquake (M9.0) refined by means of seafloor geodetic data. J Geophys Res 117:B07409

Ito T, Yoshioka S, Miyazaki S (1999) Interplate coupling in northeast Japan deduced from inversion analysis of GPS data. Earth Planet Sci Lett 176:117–130

Kido M, Osada Y, Fujimoto H, Hino R, Ito Y (2011) Trench-normal variation in observed seafloor displacements associated with the 2011 Tohoku-Oki earthquake. Geophys Res Lett 38:L24303

Loveless JP, Meade BJ (2010) Geodetic imaging of plate motions, slip rates, and partitioning of deformation in Japan. J Geophys Res 115:B02410

Loveless JP, Meade BJ (2011) Spatial correlation of interseismic coupling and coseismic rupture extent of the 2011 $M_w = 9.0$ Tohoku-oki earthquake. Geophys Res Lett 38:L17306

Meade BJ (2007) Algorithms for the calculation of exact displacements, strains, and stresses for triangular dislocation elements in a uniform elastic half space. Comput Geosci 33:1064–1075

Minoura K, Imamura F, Sugawara D, Kono Y, Iwashita T (2001) The 869 Jogan tsunami deposit and recurrence interval of large-scale tsunami on the Pacific coast of northeast Japan. J Nat Disaster Sci 23:83–88

Miura S, Iinuma T, Yui S, Uchida N, Sato T, Tachibana K, Hasegawa A (2006) Co- and post-seismic slip associated with the 2005 Miyagi-oki earthquake (M7.2) as inferred from GPS data. Earth Planets Space 58:1567–1572

Miyazaki S, Segall P, Fukuda J, Johnson KM, Kato T (2006) Poseseismic deformation following two thrust earthquakes at Kurile-Japan Trench: the 2003 Tokachi-oki and the 2005 Miyagi-oki earthquakes. EOS Trans AGU 87(52), Fall Meet. Suppl., Abstract G31A-05

Nakajima J, Hasegawa A (2006) Anomalous low-velocity zone and linear alignment of seismicity along it in the subducted Pacific slab beneath Kanto, Japan: reactivation of subducted fracture zone? Geophys Res Lett 33:L16309

Nishimura T, Hirasawa T, Miyazaki S, Sagiya T, Tada T, Miura S, Tanaka K (2004) Temporal change of interplate coupling in northeastern Japan during 1995–2002 estimated from continuous GPS observations. Geophys J Int 157:901–916

Ohta Y, Kobayashi T, Miura S, Takasu T (2010) Long term stabilities and its noise properties of the middle to long baseline RTK-GPS time series, SGD001-P02. In: Japan Geophysical Union Meeting 2010, Makuhari, Chiba, 23–28 May

Sato M, Saito H, Ishikawa T, Matsumoto Y, Fujita M, Mochizuki M, Asada A (2011) Restoration of interplate locking after the 2005 Off-Miyagi Prefecture earthquake, detected by GPS/acoustic seafloor geodetic observation. Geophys Res Lett 38:L01312

Sato M, Fujita M, Matsumoto Y, Ishikawa T, Saito H, Mochizuki M, Asada A (2013) Interplate coupling off northeastern Japan before the 2011 Tohoku-oki earthquake, inferred from seafloor geodetic data. J Geophys Res 118:3860–3869

Savage JC (1983) A dislocation model of strain accumulation and release at a subduction zone, J Geophys Res 88:4984–4996

Sawai Y, Fujii Y, Fujiwara O, Kamataki T, Komatsubara J, Okamura Y, Satake K, Shishikura M (2008) Marine incursions of the past 1500 years and evidence of tsunamis at Suijin-numa, a coastal lake facing the Japan Trench. Holocene 18:517–528

Seno T, Shimazaki K, Somerville P, Sudo K, Eguchi T (1980) Rupture process of the Miyagi-oki, Japan, earthquake of June 12, 1978. Phys Earth Planet Int 23:39–61

Suwa Y, Miura S, Hasegawa A, Sato T, Tachibana K (2006) Interplate coupling beneath NE Japan inferred from three-dimensional displacement field. J Geophys Res 111:B04402

Uchida N, Matsuzawa T (2013) Pre- and postseismic slow slip surrounding the 2011 Tohoku-oki earthquake rupture. Earth Planet Sci Lett 374:81–91. doi: 10.1016/j.epsl.2013.05.021

Umino N, Kono T, Okada T, Nakajima J, Matsuzawa T, Uchida N, Hasegawa A, Tamura Y, Aoki G (2006) Revisiting the three M $\sim$ 7 Miyagi-oki earthquakes in the 1930s: possible seismogenic slip on asperities that were re-ruptured during the 1978 M = 7.4 Miyagi-oki earthquake. Earth Planets Space 58:1587–1592

Wessel P, Smith W (1998) New, improved version of Generic Mapping Tools released. Trans Am Geophys Union (EOS) 79:579

Part II

Geodetic Observations of the Earthquake Deformation Cycle

Geodetic and Geomorphic Evaluations of Earthquake Generation Potential of the Northern Sumatran Fault, Indonesia

Takao Tabei, Fumiaki Kimata, Takeo Ito, Endra Gunawan, Hiroyuki Tsutsumi, Yusaku Ohta, Tadashi Yamashina, Yoshio Soeda, Nazli Ismail, Irwandi Nurdin, Didik Sugiyanto, and Irwan Meilano

Abstract

We have conducted geodetic and geomorphic observations in the northernmost part of Sumatra, Indonesia to monitor strain accumulation in the vicinity of the northern Sumatran fault. Evaluation of the earthquake generation potential in this region is highly urgent because of a large fault slip rate, absence of major earthquakes for more than 100 years and recent Coulomb stress increase on the fault due to the 2004 Sumatra-Andaman earthquake (M_w 9.2). We have deployed Aceh GPS Network for the Sumatran Fault System (AGNeSS) since 2005. The data collected have been used for estimating slip/locking distribution of the Sumatran fault and constructing a comprehensive model for postseismic deformation after the 2004 event. Tectonic geomorphic features are also important to reveal long-term slip history of the fault. We have used high-resolution stereo-paired ALOS (Advanced Land Observing Satellite) PRISM (Panchromatic Remote-sensing Instrument for Stereo Mapping) satellite images to map the surface trace of the Sumatran fault and conducted field observations to ensure the trace by geomorphic and geologic evidence. We introduce preliminary results of the fault mapping together with a brief description of the crustal deformation field detected by GPS.

Keywords

ALOS PRISM • Earthquake generation potential • GPS • Geomorphic observation • Sumatran fault

1 Introduction

The Sumatran fault is the longest trench-parallel strike-slip fault in Indonesia that accommodates a large amount of the right-lateral component of oblique convergence between the Indian-Australian and Sundaland plates at the Sunda trench (Fig. 1). The 1,900-km-long fault is divided into multiple

T. Tabei (✉) • T. Yamashina
Department of Applied Science, Kochi University, Kochi, Japan
e-mail: tabei@kochi-u.ac.jp

F. Kimata
Tono Research Institute of Earthquake Science, Mizunami, Japan

T. Ito
Graduate School of Environmental Studies, Nagoya University, Nagoya, Japan

H. Tsutsumi
Department of Geophysics, Kyoto University, Kyoto, Japan

Y. Ohta
Graduate School of Science, Tohoku University, Sendai, Japan

Y. Soeda
West Japan Engineering Consultants Inc., Fukuoka, Japan

N. Ismail • I. Nurdin • D. Sugiyanto
Department of Physics, Syiah Kuala University, Banda Aceh, Indonesia

E. Gunawan • I. Meilano
Faculty of Earth Science and Technology, Bandung Institute of Technology, Bandung, Indonesia

M. Hashimoto (ed.), *International Symposium on Geodesy for Earthquake and Natural Hazards (GENAH)*,
International Association of Geodesy Symposia 145, DOI 10.1007/1345_2015_200

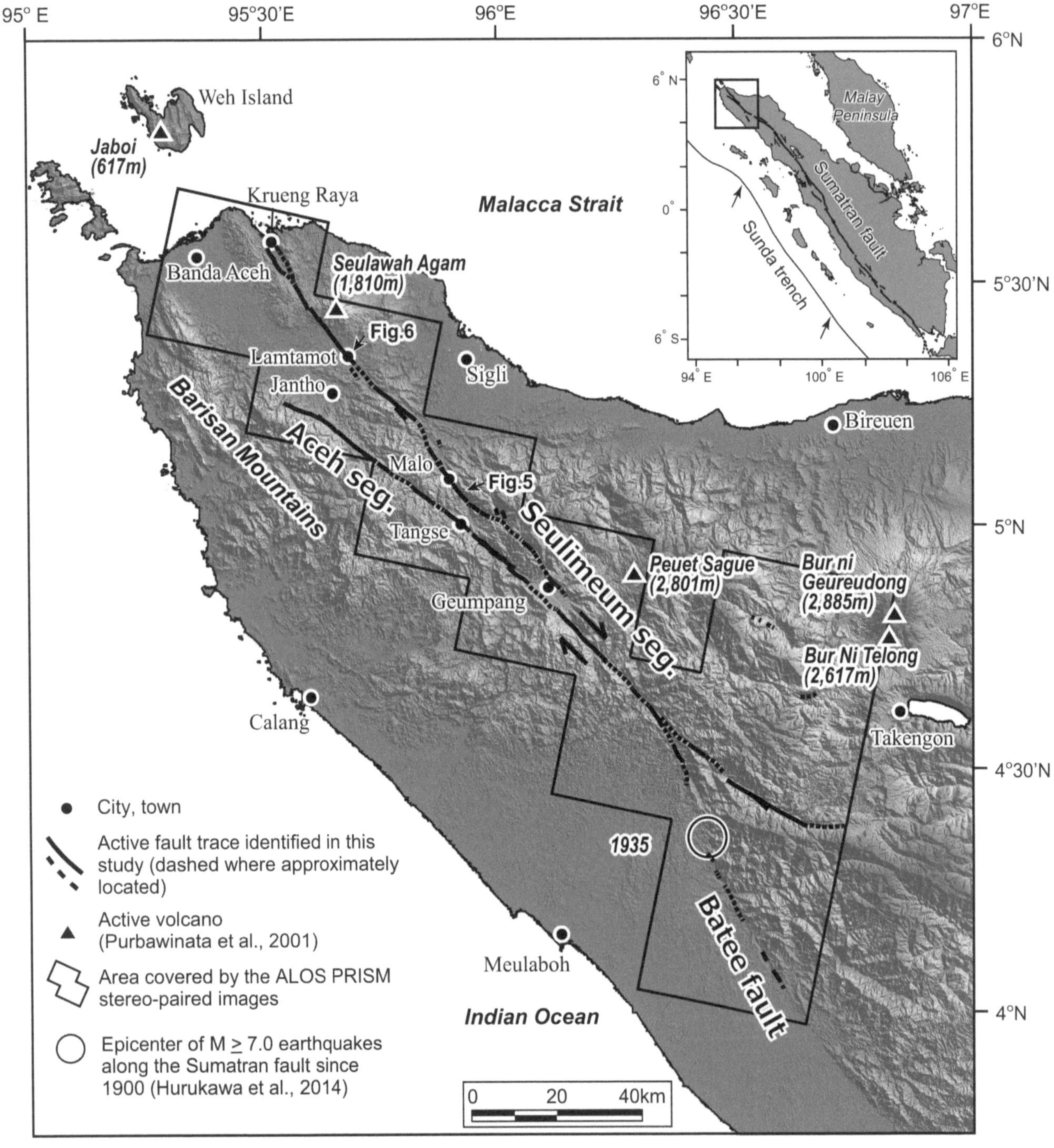

Fig. 1 Index map of the study area. *Solid lines* (*dashed line* where approximately located) are simplified surface traces of the Sumatran and Batee faults based on interpretation of stereo-paired ALOS PRISM images. Locations of Figs. 5 and 6 are shown. The *shaded* relief map was constructed using ASTER GDEM data (30 m mesh) from METI (Ministry of Economy, Trade and Industry of Japan) and NASA (National Aeronautics and Space Administration of the United States)

segments separated by geometric discontinuities (Sieh and Natawidjaja 2000). The long-term slip rate is estimated to increase from southeast to northwest as the obliquity of the plate convergence increases to northwest (compiled by Ito et al. 2012). In the northernmost part of Sumatra, two major segments of the Sumatran fault have not ruptured at least since 1892 (Hurukawa et al. 2014).

Recently the 26 December 2004 Sumatra-Andaman earthquake (M_w 9.2, referred to as SAE) and the subsequent 28 March 2005 Nias earthquake (M_w 8.6) ruptured the plate boundary at the Sunda subduction zone (Lay et al. 2005). They also altered the state of stress in the overriding plate. Coseismic Coulomb stress change is estimated to be a positive loading of up to 9 bars (0.9 MPa) (McCloskey et al. 2005; Nalbant et al. 2005) to 20 bar (2 MPa) (Cattin et al. 2009) for about 300-km-long portion of the northern Sumatran fault. Taking into consideration the large fault slip rate, absence of major earthquakes and recent Coulomb stress increase, evaluation of the earthquake generation potential of the northern Sumatran fault is highly urgent.

We have deployed Aceh GPS Network for the Sumatran Fault System (AGNeSS) since 2005 in Aceh Province in the northernmost Sumatra to infer slip/locking distribution of the Sumatran fault in addition to record a decaying phase of postseismic deformation due to the 2004 SAE (Ito et al. 2012; Gunawan et al. 2014). However, basic information on the Sumatran fault such as its exact surface trace location, slip rate, and recurrence interval is still missing due to insufficient geological studies in the area. This is partly due to the rugged mountainous topography in northern Sumatra but chiefly due to the unstable political condition before the 2004 SAE. We started precise mapping of the surface trace of the Sumatran fault using satellite images in 2011. Then we have conducted field observations to ensure the trace by geomorphic and geologic evidence.

In this paper we introduce preliminary results of the fault mapping based on satellite images and geologic field observations together with a brief description of the crustal deformation field detected by AGNeSS. Refer to Ito et al. (2012) for the evaluation of slip/locking distribution of the Sumatran fault and Gunawan et al. (2014) for a comprehensive model of the postseismic deformation due to the 2004 SAE.

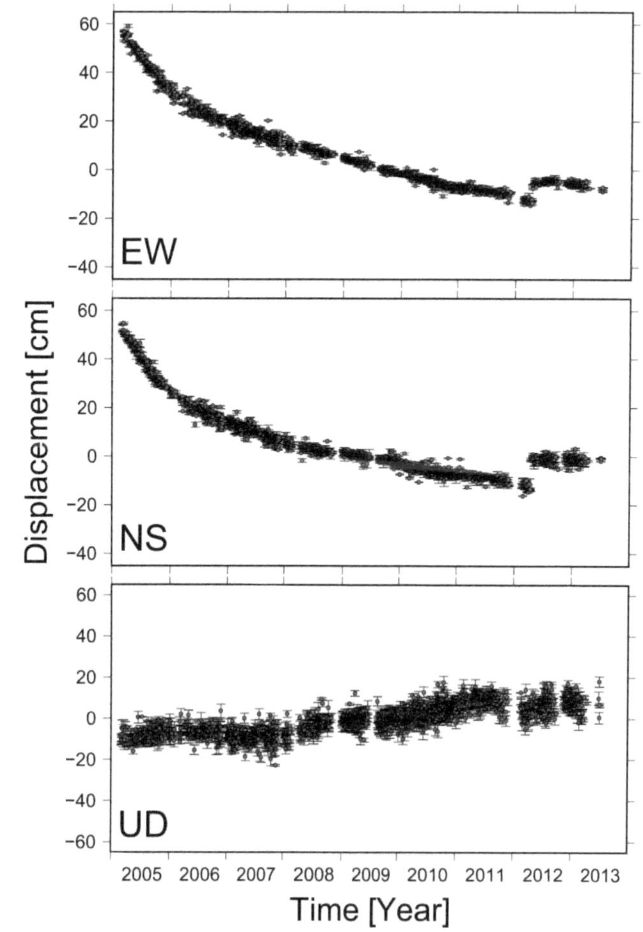

Fig. 2 Displacement time series of continuous GPS site ACEH (refer to Fig. 3 for the location). Decay times are different between horizontal and vertical components, implying multiple physical mechanisms for the postseismic deformation (Gunawan et al. 2014). Also clearly shown is the coseismic jump due to the 11 April 2012 Indian Ocean earthquakes (M_w 8.6 and 8.2)

2 GPS Observations and Results

In March 2005, immediately after the 2004 SAE, we started GPS observation in Banda Aceh city and its surroundings to detect postseismic deformation in the near field of the main rupture zone. Horizontal displacement detected at ACEH site in Banda Aceh in about 7 years until the 11 April 2012 Indian Ocean earthquakes (described later) reached approximately one meter trenchward (Fig. 2). Gunawan et al. (2014) have pointed out that the relaxation time in the vertical displacement is longer than the horizontal one and deduced that there are multiple physical mechanisms that control the postseismic deformation, referred to as afterslip and viscoelastic relaxation.

The GPS network has been expanded gradually since 2005. As the postseismic deformation has decayed with time, new GPS sites have been deployed so as to effectively monitor strain accumulation in the vicinity of the Sumatran fault. The network now consists of seven continuous and 17 campaign sites deployed along two local main roads across the Sumatran fault (as of the end of 2013, Fig. 3), covering roughly 200 × 200 km area in Aceh Province. At the continuous sites one-second-sampling data are recorded. Each campaign site has been reoccupied for a few days every year.

Ito et al. (2012) treated the postseismic viscoelastic effect on AGNeSS within the observation period of several years as a linear trend because viscoelastic relaxation has a long decay time and its wavelength is much longer than the spatial extent of AGNeSS. Removing the remaining afterslip effect using a parameterized afterslip model, Ito et al. (2012) inferred shallow aseismic creep and deeper locked zones for the northern Sumatran fault. Regardless of the shallow creep, the result shows that the locked zones have a width of 9.4–14.8 km and a slip deficit rate of 1.6–2.0 cm/year. Taking into account an empirical relation between length and

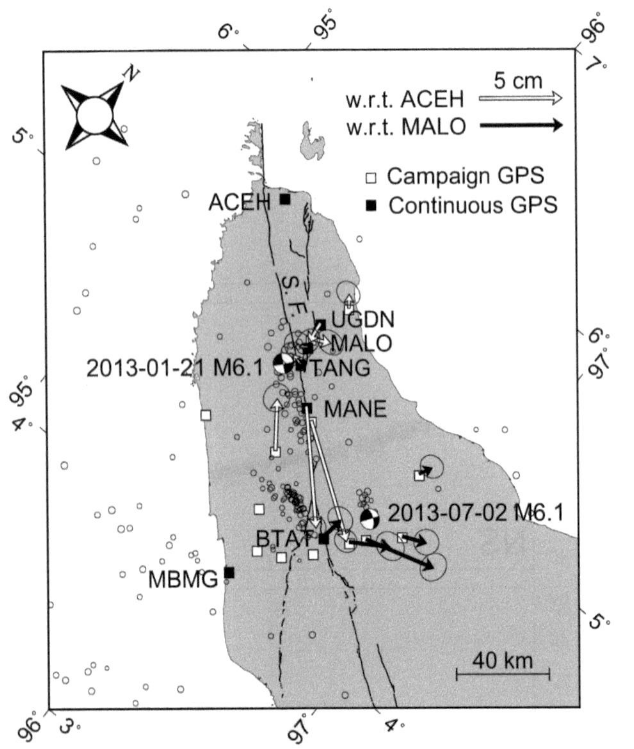

Fig. 3 Site distribution of AGNeSS as of the end of 2013 and coseismic displacements due to two M 6.1 earthquakes on January 21 and July 2, 2013. Displacement vectors for the first event (*open arrows*) are plotted with respect to ACEH and those for the second event (*closed arrows*) to MALO, respectively. Error ellipses denote uncertainties at 95% confidence level. Locations and focal mechanisms of the two events are based on the database of US Geological Survey while local seismicity data ($M > 2$, from January 2012 to August 2013) are taken from BMKG (Agency for Meteorology, Climatology and Geophysics) in Indonesia

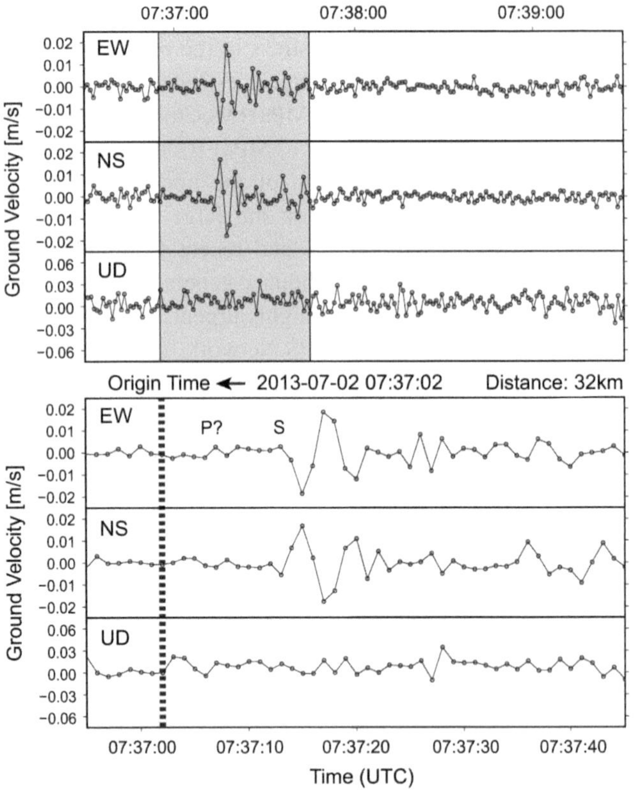

Fig. 4 Ground velocities detected at BTAT at the time of the July 2 earthquake (M 6.1). One-second-interval displacements with respect to MALO are converted to the velocities. Refer to Fig. 3 for the locations of epicenter and GPS sites

width of the strike-slip fault and a long quiescent period in Aceh Province, the above estimates suggest that the fault is capable of producing a M 7.0 event. To discuss more detail on the earthquake scenarios, knowledge of the fault segmentation and long-term slip history is essential in complement to geodetic estimate of current slip/locking distribution.

On April 11, 2012 the northern Sumatra was struck by the Indian Ocean earthquakes (M_w 8.6 and 8.2), which occurred 330 km west-southwest of the 2004 SAE epicenter and was the largest intraplate strike-slip event in the known history (Deleschluse et al. 2012). AGNeSS detected clear coseismic horizontal displacements of 12.1–14.1 cm in a direction of N36-40°E (Fig. 2), which were consistent with a rupture model from seismological data (Tabei et al. 2013).

On January 21 and July 2, 2013 two M 6.1 earthquakes occurred within AGNeSS (Fig. 3). Source location and mechanism of the January 21 event and resultant coseismic displacements strongly suggest a right-lateral strike-slip on the NW-SE trending Sumatran fault. The July 2 event occurred when GPS campaign observation was being conducted in the vicinity of the source region. At the nearest continuous GPS site BTAT where the epicentral distance was 32 km, horizontal ground shaking was clearly recorded in the one-second-sampled time series (Fig. 4). Northwestward initial motion due to the S wave arrival is very consistent with the mechanism solution by US Geological Survey (Fig. 3). The source mechanism of this event was very similar to that of the January event while the epicenter was about 30 km departed from the main trace of the fault. However coseismic displacements at the nearby GPS sites suggest a left-lateral strike-slip on a NE-SW trending fault plane. Tabei et al. (2013) estimate fault parameters of the two events using a Markov Chain Monte Carlo method and conclude that the July 2 event broke a conjugate fault plane. So far no evidence has been obtained that supports the existence of conjugate faults to the Sumatran fault system. Closer inspections are required from various kinds of geophysical and geological points of view.

3 Fault Trace Based on Satellite Image Interpretation

To map the surface trace of the Sumatran fault, we interpreted stereo-pairs of ALOS (Advanced Land Observing Satellite) PRISM (Panchromatic Remote-sensing Instrument for Stereo Mapping) images with a spatial resolution of 2.5 m. The ALOS had been operated from January 2006 to May 2011 by the Japan Aerospace Exploration Agency (JAXA). The PRISM had three independent optical systems to obtain along-track images for forward, nadir and backward views at the same time (JAXA 2007), which enables us to utilize stereo-pair images. We acquired nadir- and backward-looking PRISM pair images of northern Sumatra taken between 2007 and 2011, and processed them to make black-and-white ~1:30,000-scale stereo images. We mapped the faults in the area covered by AGNeSS; the area covered by the satellite images is shown in Fig. 1. The mapping was based on tectonic geomorphic features such as linear valleys (depressions), aligned saddles on mountain slopes, stream offsets, fault scarps and pressure ridges. We utilized 1:50,000-scale topographic maps published by Geological Agency of Indonesia and topographic contour maps prepared from ASTER GDEM (Global Digital Elevation Model) as base maps of our interpretation. By interpreting high-resolution satellite images, we were able to identify small-scale tectonic geomorphic features.

The study area includes the Aceh and Seulimeum segments of the Sumatran fault identified by Sieh and Natawidjaja (2000). The Aceh segment is the direct northwestern extension of the main fault trace southeast of the study area. The Aceh segment extends N50°W and bounds the southwestern margin of elongated intra-mountain basins such as the Geumpang and Tangse basins. The Aceh segment traverses mountainous areas and its trace is mainly characterized by linear valleys, systematic right-lateral deflections of stream channels, aligned saddles on mountain slopes, beheaded streams and shutter ridges. The Aceh segment is traced from southwest of Takengon at latitude of 4°23'N northwestward to west of Jantho at latitude of 5°15'N for a distance of ~160 km. There are no clear tectonic geomorphic features indicative of late Quaternary activity further to the north along the southwestern margin of the Banda Aceh basin.

The Seulimeum segment lies east of and branches northward from the Aceh segment (Fig. 1). The southeastern end of the Seulimeum segment is located ~4 km to the northeast of the Aceh segment. From there, the Seulimeum segment extends northwest (N40°W) for a distance of ~100 km. The fault traverses mountainous areas between Geumpang and Lamtamot and cuts the southwestern slope of the Seulawah Agam volcano before reaching to Krueng Raya. The Seulimeum segment appears to extend offshore and cuts across the Weh Island (Sieh and Natawidjaja 2000). The Seulimeum segment is accompanied by pronounced tectonic geomorphic features along its entire length and appears to accommodate much of the right-lateral displacement on the Sumatran fault in northernmost Sumatra.

Near the southeastern end of the Aceh segment, the Batee fault branches southward and is interpreted as a splay fault of the Sumatran fault (Sieh and Natawidjaja 2000). The Batee fault is located at the topographic boundary between mountains to the east and lowland to the west (Fig. 1). Aligned saddles and fault scarps on fluvial terrace surfaces are found along the fault.

Based on relocations of large historical earthquakes along the Sumatran fault, Hurukawa et al. (2014) concluded that the Aceh and Seulimeum segments have not ruptured since 1892 whereas the Tripa segment immediately to the south of the Aceh segment ruptured during the M 7.0 1935 and M 7.2 1936 earthquakes. The Batee fault, however, may be the source fault for the 1935 earthquake. Hurukawa et al. (2014) pointed out that the Aceh and Seulimeum segments form seismic gaps and may produce $M_s > 7.4$ earthquakes in the near future based on an empirical equation between the surface-wave magnitude and fault length of Abe (1981).

4 Geomorphic Field Observations

Since 2012, we conducted three campaigns of geological fieldwork along selected portions of the fault trace accompanied by clear tectonic geomorphic features. Here we describe two localities along the Seulimeum segment where clear geomorphic and geologic evidence of recent faulting is observed.

At Kuala Panteu and its vicinity between Tangse and Malo, a series of southwest-flowing stream channels are systematically offset right-laterally (Fig. 5). The amount of offset is 60–180 m. Sedimentary and metamorphic rocks along the fault are highly sheared and numerous landslides are observed in the area. At locality 1 in Fig. 5, cataclastic rocks are exposed on the floor of a stream channel. Riedel shears with right-lateral sense of slip are consistent with right-lateral deflections of stream channels.

Clear tectonic geomorphic features are observed at Lamtamot near the road connecting Banda Aceh and Sigli (Fig. 6). A ~1-km-long and southwest-facing fault scarp with a maximum height of 7–8 m cuts across a fluvial terrace (Fig. 7). Although there is no age control for the terrace, it mostly preserves original depositional surface and is interpreted to be late Quaternary in age. Immediately south of the scarp, there is a pressure ridge with its western flank bounded by the fault. Sedimentary rocks are folded into an

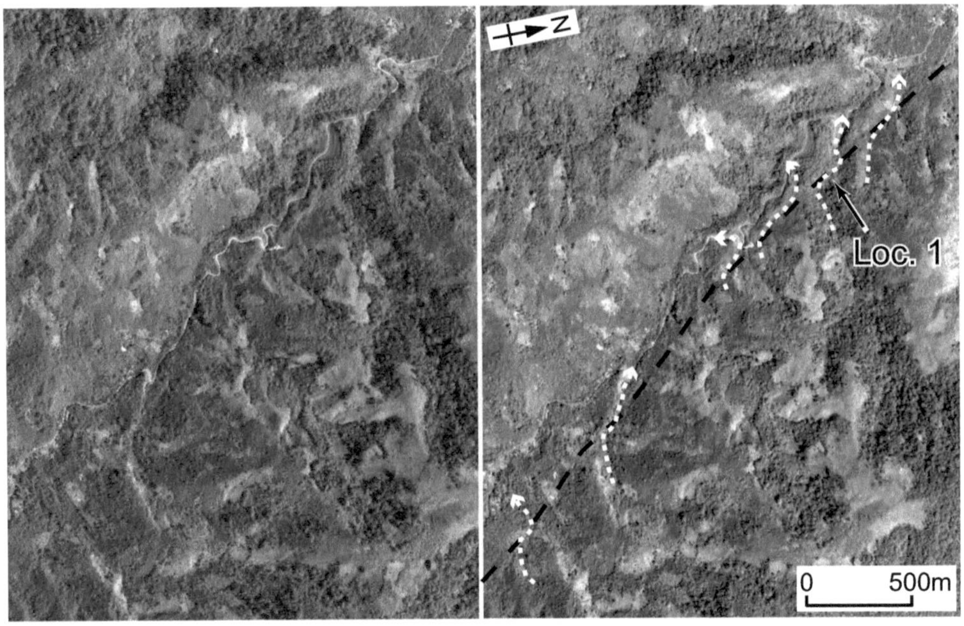

Fig. 5 Fault trace (*black dashed line*) and right-laterally offset streams (*white dashed line*) at Kuala Panteu shown on stereo-pairs of ALOS PRISM images (©JAXA, distributed by the Remote Sensing Technology Center of Japan (RESTEC))

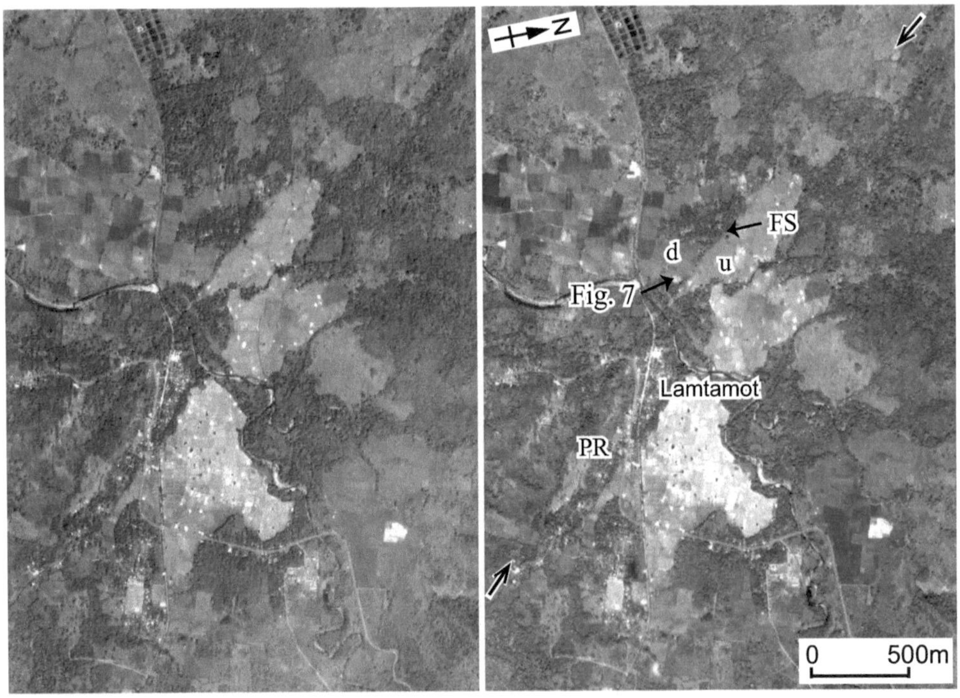

Fig. 6 Fault trace and tectonic geomorphic features at Lamtamot shown on stereo-pairs of ALOS PRISM images (©JAXA, distributed by RESTEC). The arrows indicate the location of the fault trace. PR denotes pressure ridge whereas "u" and "d" denote upthrown and downthrown sides of the fault scarp (FS), respectively

anticline and dip ~ 70°SW near the fault trace. Numerous normal faults with down-to-the-east sense of offset are observed on the western limb of the anticline. These normal faults offset modern soil, suggesting that they have moved during recent geologic time, probably in association with the main fault trace of the Seulimeum segment.

Fig. 7 Field photograph of the southwest-facing fault scarp (the base is denoted by the *white arrows*) cutting across a fluvial terrace at Lamtamot. The view is toward the north. The mountain in the background is the Seulawah Agam volcano

5 Conclusions and Future Works

We have deployed continuous and campaign GPS network since 2005 to monitor strain accumulation in the vicinity of the northernmost part of the Sumatran fault where high seismic potential is expected from large long-term fault slip rate, absence of recent major earthquakes, and local stress increase due to the 2004 Sumatra-Andaman event. Geodetic estimates such as fault locking depth and slip deficit rate suggest a capability of producing a large event in the near future. To discuss the earthquake scenarios more in detail, however, basic quantitative data on the fault slip history is essential. We map the surface trace of the fault using high-resolution ALOS PRISM images based on tectonic geomorphic features. We also conducted field observations along selected portions to ensure the fault trace by geomorphic evidence.

In order to geodetically detect surface creep on the fault, if any, we set up short-distance alignment arrays at four locations across the northern Sumatran fault and conducted the first measurements on November 2013. In the next few years, we will date geomorphic surfaces offset by the fault to calculate geological slip rate and conduct paleoseismic trenching to reveal the recurrence interval and the timing of the most recent surface-rupturing event on the different segments of the Sumatran fault. We will also continue GPS observation to constrain regional seismic potential.

Acknowledgements Our fieldwork was supported by many staff and students of Syiah Kuala University and Bandung Institute of Technology. We thank all collaborators and Indonesian authorities. Comments by N. T. Ramos, P. Willis (the editor) and two anonymous reviewers helped improve the manuscript. This work was supported by the Grant-in-Aid for Scientific Research (B) No.24403005 from Japan Society for the Promotion of Science.

References

Abe K (1981) Magnitudes of large shallow earthquakes from 1904 to 1980. Phys Earth Planet Int 27:72–92

Cattin R, Chamot-Rooke N, Pubellier M, Rabaute A, Delescluse M, Vigny C, Fleitout L, Dubernet P (2009) Stress change and effective friction coefficient along the Sumatra-Andaman-Sagaing fault system after the 26 December 2004 (M_w = 9.2) and the 28 March (M_w = 8.7) earthquakes. Geochem Geophys Geosyst 10. doi:10.1029/2008GC002167

Deleschluse M, Chamot-Rooke N, Cattin R, Fleitout L, Tuibienko O, Vigny C (2012) April 2012 intra-oceanic seismicity off Sumatra boosted by the Banda-Aceh megathrust. Nature 490:240–244. doi:10.1038/nature11520

Gunawan E, Sagiya T, Ito T, Kimata F, Tabei T, Ohta Y, Meilano I, Abidin HZ, Agustan, Nurdin I, Sugiyanto D (2014) A comprehensive model of postseismic deformation of the 2004 Sumatra-Andaman earthquake deduced from GPS observations in northern Sumatra. J Asian Earth Sci 88:218–229

Hurukawa N, Wulandari BR, Kasahara M (2014) Earthquake history of the Sumatran fault, Indonesia, since 1892, derived from relocation of large earthquakes. Bull Seismol Soc Am 104:1750–1762

Ito T, Gunawan E, Kimata F, Tabei T, Simons M, Meilano I, Agustam, Ohta Y, Nurdin I, Sugiyanto D (2012) Isolating along-strike variations in the depth extent of shallow creep and fault locking on the northern Great Sumatran Fault. J Geophys Res 117:B06409. doi:10.1029/2011JB008940

JAXA (Japan Aerospace Exploration Agency) (2007) NDX-070015 ALOS user handbook. http://www.eorc.jaxa.jp/ALOS/en/doc/alos_userhb_en.pdf

Lay T, Kanamori H, Ammon CJ, Nettles M, Ward SN, Aster RC, Beck SL, Bilek SL, Brudzinski MR, Butler R, DeShon HR, Ekström G, Satake K, Sipkin S (2005) The great Sumatra-Andaman earthquake of 26 December 2004. Science 308:1127–1133

McCloskey J, Nalbant SS, Steacy S (2005) Indonesian earthquake: earthquake risk from co-seismic stress. Nature 434:291–291

Nalbant SS, Steacy S, Sieh K, Natawidjaja D, McCloskey J (2005) Earthquake risk on the Sunda trench. Nature 435:756–757

Purbawinata MA, Mulyadl E, Wahyudin D, Bacharudin R, Hadisantono RD (2001) Distribution map of active volcanoes in Indonesia, Directorate of Volcanology and Geologic Hazard Mitigation, Ministry of Energy and Mineral Resources

Sieh K, Natawidjaja D (2000) Neotectonics of the Sumatran fault, Indonesia. J Geophys Res 105:28,295–28,326

Tabei T, Ito T, Kimata F, Ohta Y, Gunawan E, Ismail N, Sugiyanto D, Nurdin I (2013) Crustal deformation detected by GPS observation network across the Sumatran fault system in northwestern Sumatra, Indonesia. American Geophysical Union 2013 Fall Meeting, 9–13 December, San Francisco, USA

Virtual Quake: Statistics, Co-seismic Deformations and Gravity Changes for Driven Earthquake Fault Systems

Kasey W. Schultz, Michael K. Sachs, Eric M. Heien, Mark R. Yoder, John B. Rundle, Don L. Turcotte, and Andrea Donnellan

Abstract

With the ever increasing number of geodetic monitoring satellites, it is vital to have a variety of geophysical simulations produce synthetic datasets. Furthermore, just as hurricane forecasts are derived from the consensus among multiple atmospheric models, earthquake forecasts cannot be derived from a single comprehensive model. Here we present the functionality of Virtual Quake (formerly known as Virtual California), a numerical simulator that can generate sample co-seismic deformations, gravity changes, and InSAR interferograms in addition to producing probabilities for earthquake scenarios.

Virtual Quake is now hosted by the Computational Infrastructure for Geodynamics. It is available for download and comes with a user manual. The manual includes a description of the simulator physics, instructions for generating fault models from scratch, and a guide to deploying the simulator in a parallel computing environment. http://geodynamics.org/cig/software/vq/.

Keywords

Earthquake forecasting • GENAH • Virtual California • Virtual quake

1 Introduction

The purpose of this paper is to introduce a subset of the wide range of applications of the Virtual Quake (VQ) earthquake simulator, originally developed by Rundle (1988a,b). Virtual Quake has been developed to characterize the space-time patterns and correlations that emerge from the underlying stress and strain dynamics. Virtual Quake simulates hundreds of thousands of years of interactions on any fault system given information about the fault geometry and the observed long-term slip rates.

Over the past decade, Virtual Quake has been part of a multi-disciplinary effort to better understand, predict, and respond to earthquake hazards known as QuakeSim (Donnellan et al. 2006). As part of the QuakeSim team, Virtual Quake was selected as a co-winner of NASA's 2012 Software of the Year award. Tullis et al. (2012a) confirmed that the Virtual Quake (then called Virtual

K.W. Schultz (✉) • M.K. Sachs • M.R. Yoder
Department of Physics, University of California-Davis, One Shields Avenue, Davis, CA 95616, USA
e-mail: kwschultz@ucdavis.edu

J.B. Rundle
Department of Physics, University of California-Davis, One Shields Avenue, Davis, CA 95616, USA

Department of Earth and Planetary Sciences, University of California-Davis, One Shields Avenue, Davis, CA 95616, USA

D.L. Turcotte • E.M. Heien
Department of Earth and Planetary Sciences, University of California-Davis, One Shields Avenue, Davis, CA 95616, USA

A. Donnellan
Jet Propulsion Laboratory, California Institute of Technology, Pasadena, CA, USA

M. Hashimoto (ed.), *International Symposium on Geodesy for Earthquake and Natural Hazards (GENAH)*, International Association of Geodesy Symposia 145, DOI 10.1007/1345_2015_134

California) simulations are consistent with both the observed seismicity in California and with independent California fault system simulations. The analysis compared the simulated earthquake catalogs to the observed earthquake rates in California and to observed scaling relations for magnitude, rupture area, and mean slip. Yikilmaz et al. (2010) also showed that Virtual Quake simulations of the Nankai Trough in Japan are consistent with observed earthquake sequences.

The output of a Virtual Quake simulation is a physics-based catalog of earthquakes. In Sect. 4 we discuss how this simulated earthquake catalog can be used to compute conditional probabilities of earthquake scenarios. In Sect. 3 we show how the simulated co-seismic slip distributions can be used to generate spacetime patterns using Green's functions. These surface patterns can be compared to geodetic data like InSAR interferograms, co-seismic displacements, and co-seismic gravity changes.

2 Virtual Quake Simulator

Virtual Quake (VQ) is a boundary element code that simulates realistically driven fault systems based on stress interactions. VQ is designed to explore the long term statistical behavior of topologically complex fault networks (Rundle 1988a,b; Rundle et al. 2006a,b,c). The most recent version of VQ is a modern scientific code that simulates earthquakes in a high performance computing environment (Sachs et al. 2012; Heien and Sachs 2012). VQ simulates many thousands of years of seismic history on fault networks with any physically realizable geometry.

VQ consists of three major components: a fault model, simulation physics, and an event model. The fault model is the simulation input and it can be changed to any arbitrary fault geometry and function properly the simulation physics and event model. The simulation physics is based on fault stress via the accumulation of a slip deficit – the amount of slip each fault should move in a given time period given the long term slip rate. Actual values of stress are computed by a set of quasi-static elastic interactions given in Rundle et al. (2006a).

VQ initiates simulated earthquakes using both static and dynamic friction laws. Slip during a simulated earthquake is triggered by the stress on a fault element reaching the failure threshold as specified in the fault model, this is static failure. Elements on the same fault are also allowed to slip – even if they have not reached failure stress – if the slip from the initial failed element(s) increases their stress by at least 50%. This dynamic failure condition can be tuned and is used to control rupture propagation for simulated earthquakes.

Hereafter Virtual Quake simulations of the California fault system are referred to as Virtual California (VC). For example, if we simulated the Japan fault system then we would refer to that application of Virtual Quake as Virtual Japan. For a detailed explanation of the properties of modern earthquake simulators and a comparison of Virtual California to other simulators see Tullis et al. (2012b). For a comparison of Virtual California simulations to those of other earthquake simulators using the same fault model see Tullis et al. (2012a).

3 Surface Patterns

Virtual Quake calculates co-seimic deformations, InSAR (Synthetic Aperture Radar Interferometry) patterns, co-seismic gravity changes and more for arbitrarily complex fault geometries (Rundle et al. 2006c; Sachs et al. 2012; Schultz et al. 2014). VQ calculates these surface patterns by using fault geometry and co-seismic slips as input to a specific set of Green's functions. Since Virtual Quake partitions the fault system into finite boundary elements embedded in an elastic half-space, the Green's functions are a logical extension of the simulator's capability. For surface deformation and InSAR calculations we use the Green's functions presented in Okada (1992), and for gravity and potential changes we use the Green's functions presented in Okubo (1992).

These surface patterns reveal changes in the dynamic variables associated with the earthquake cycle that are inherently unobservable like the stress field. Rundle et al. (2000) developed a technique for describing the evolution of space-time patterns as a "pattern dynamics". Tiampo et al. (2002) applied this analysis on seismicity data in southern California and argued for the development of more sophisticated computer simulations to carry out a more systematic analysis. Our goal is to provide a tool for generating physics-based catalogs of geophysical fields arising only from fault-fault interactions across entire fault networks. Utilizing these catalogs, future studies carrying out a detailed pattern dynamics analysis may help reveal characteristic patterns associated with high seismic risk and help guide fault monitoring methods.

3.1 Gravity Green's Functions

Following Hayes et al. (2006) we have added a more generalized method for modeling co-seismic gravity changes for simulated earthquakes on arbitrary faults with Virtual Quake (Schultz et al. 2014). We utilize a custom version of the gravity Green's functions presented in Okubo (1992) for faulting in an elastic half-space. Though VQ does not yet support a three dimensional layered earth, these gravity Green's functions have been extended for a dislocations in

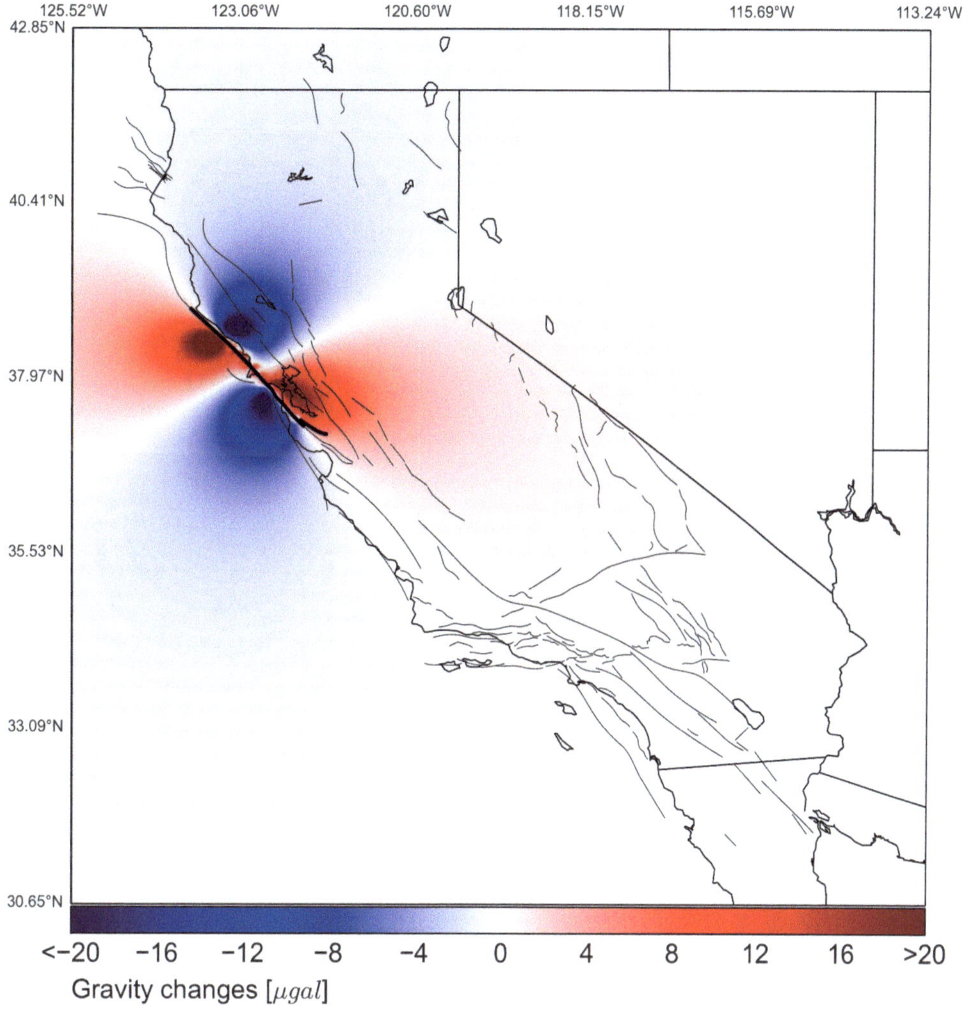

Fig. 1 Gravity Green's function solutions for co-seismic slips from a magnitude 7.69 VC simulated earthquake on the San Andreas Fault. The color unit is microGals (10^{-8}m/s^2), fault rupture is shown in *dark black*

a three dimensional earth and are presented in Sun et al. (2009). These equations have shown good agreement with observed co-seismic gravity signals observed by the GRACE satellite for the 2011 Tohoku-Oki earthquake (Matsuo and Heki 2011) and for the 2010 Central Chile earthquake (Heki and Matsuo 2010).

We have implemented a custom version of Okubo's gravity equations in Virtual Quake that allow modeling of gravity changes for arbitrary fault geometry and slip. For a detailed explanation of the Virtual Quake implementation of the Okubo's equations see Schultz et al. (2014).

Figure 1 shows the co-seismic gravity pattern produced by Virtual California for a M=7.69 simulated earthquake rupturing the northern San Andreas Fault. This is a model of the total gravity signal at the surface including contribution from surface displacement and from compression and dilatation. The dark black line indicates the fault sections that slipped during the simulated event. Our model is a single layer elastic half-space model and it does not account for viscoelasticity, elastic discontinuities, or the three dimensional layered earth.

3.2 Displacement Green's Functions and InSAR

Virtual Quake employs a custom version of the displacement Green's functions given in Okada (1992) to produce simulated co-seismic deformation maps and InSAR interferograms. Using simulated co-seismic slip distributions and the Okada Green's functions, Virtual Quake can generate entire catalogues of InSAR interferograms and surface deformation maps for arbitrary observing angles. Figure 2 shows the modeled InSAR interferogram for the same $M = 7.69$ simulated earthquake shown in Fig. 1, as seen by observing with an azimuthal angle of 30° and elevation angle of 40°.

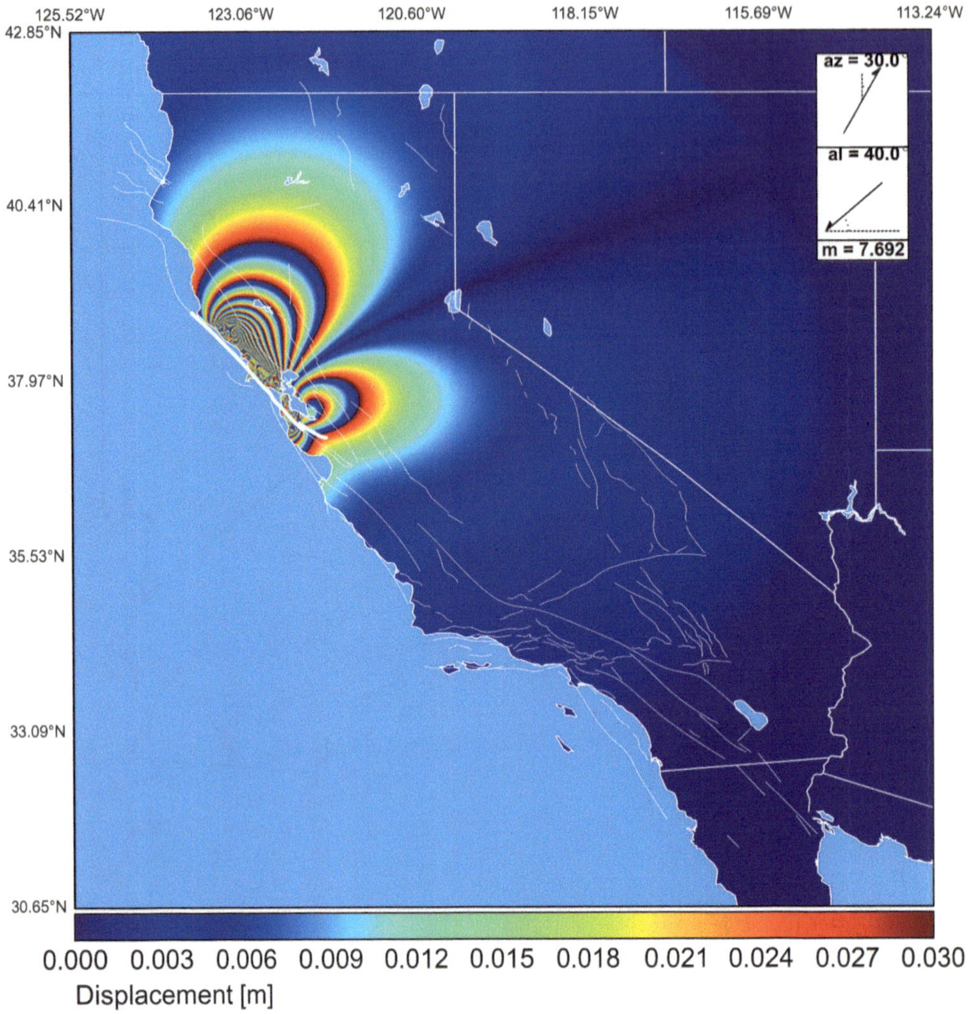

Fig. 2 Simulated InSAR interferogram given co-seismic slips at the surface from a magnitude 7.69 VC simulated earthquake on the San Andreas Fault. View angles: azimuth = 30°, elevation = 40°, fault rupture is shown with *bold white lines*

4 Virtual California Forecasts

Van Aalsburg et al. (2007, 2010) first studied the feasibility of computing earthquake probabilities with an early version of Virtual California and compared the forecasting method with forecasts produced by the Working Group on California Earthquake Probabilities. Yikimaz et al. (2011) made a detailed comparison of observed northern California seismicity and simulated seismicity by adding background seismicity to an early Virtual California simulation.

Here we derive the probabilities for a northern California earthquake scenario using a recent 50,000 year Virtual California simulation of the allcal2 California fault system model. The fault model, which is partitioned into more than 14,000 3 km × 3 km square boundary elements, is described in detail in Tullis et al. (2012b). We follow Rundle et al. (2005) and illustrate our approach using the Virtual California simulation to obtain recurrence statistics and construct the probability distributions. We then compare earthquake probabilities generated by Virtual California to an independent forecasting method.

4.1 Weibull Distribution

A probability distribution that is used frequently for earthquake recurrence statistics is the Weibull distribution (Rundle et al. 2005). Yakovlev et al. (2006) and Abaimov et al. (2008) showed that the Weibull distribution fit early Virtual California simulations of the San Andreas Fault. The Weibull distribution specifies the fraction of recurrence times $P(t)$ that are less than t, and is expressed as

$$P(t) = 1 - \exp\left[-\left(\frac{t}{\tau}\right)^{\beta}\right], \qquad (1)$$

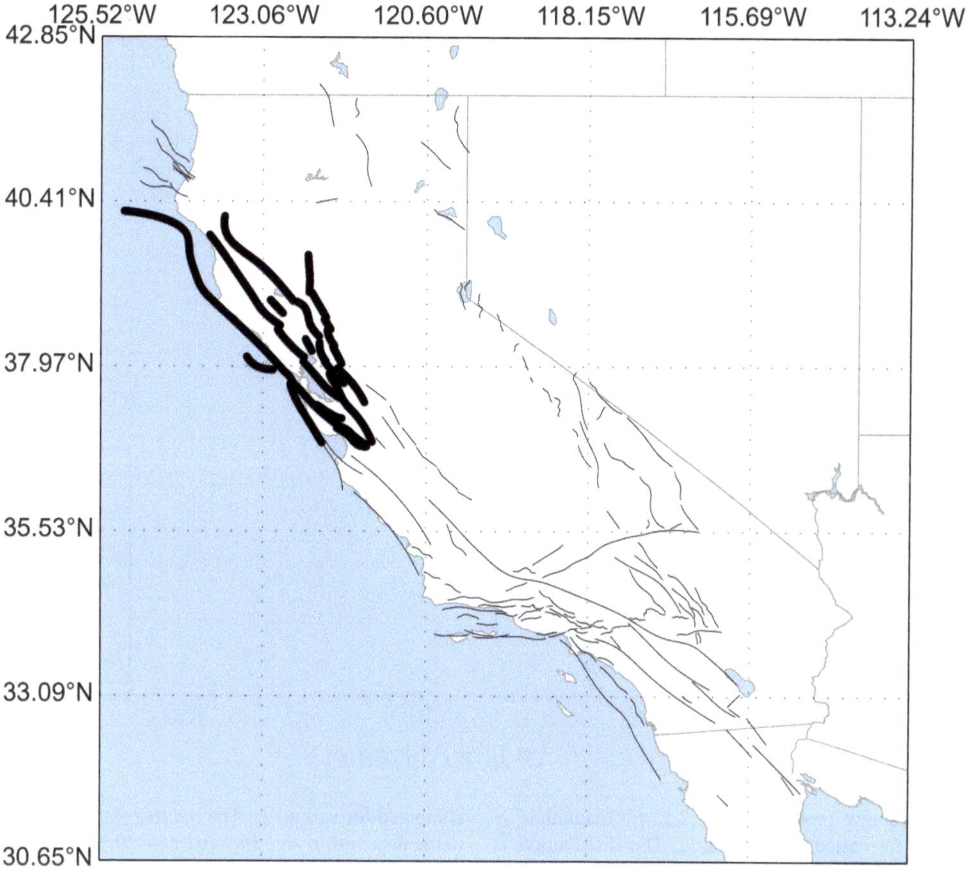

Fig. 3 Virtual California fault sections targeted for the northern California forecast, including the northern San Andreas Fault

where β and τ are fitting parameters (Sieh et al. 1989; Sornette and Knopoff 1997). The Weibull distribution is also extended to a cumulative conditional distribution

$$P(t, t_0) = 1 - \exp\left[\left(\frac{t_0}{\tau}\right)^{\beta} - \left(\frac{t}{\tau}\right)^{\beta}\right]. \tag{2}$$

Equation (2) defines the cumulative conditional probability that an earthquake will have occurred at time t after the last earthquake given that the last earthquake occurred a time t_0 years ago.

4.2 Northern California Forecast M ≥ 7.0

The Virtual California simulation data that are fundamental in generating earthquake probabilities are the recurrence times, defined as the time between successive large earthquakes. Following the methods outlined in Rundle et al. (2005), we compute the cumulative probability distributions and waiting times until the next earthquake on a selected subset of faults directly from a set of simulated recurrence times.

We consider earthquakes on the major faults in northern California, including the northern San Andreas and surrounding faults, shown in Fig. 3. Over the 50,000 year simulation there are 2,463 earthquakes with moment magnitudes $M \geq 7.0$ that cause slip on at least one of these faults. These earthquakes have an average recurrence time of 20.2 years, with a maximum of 190.4 years. From these data we construct the distribution of recurrence times t, defined as the time between successive earthquakes on the selected faults with $M \geq 7.0$. The cumulative conditional probability of an $M \geq 7.0$ earthquake occurring on the forecasted faults $P(t, t_0)$ at a time t given that the last $M \geq 7.0$ earthquake on these faults occurred t_0 years ago is plotted in Fig. 4 for multiple values of t_0. The conditional Weibull distribution (2) is fit to the simulation-derived distribution and is shown as the smooth black curves in Fig. 4, with $\beta = 1.089$ and $\tau = 20.92$ years.

In addition to the distribution of recurrence times, we compute the distribution of waiting times Δt until the next large earthquake. The waiting time Δt is measured forward from the present, such that $t = t_0 + \Delta t$. We express our results in terms of the conditional cumulative probability

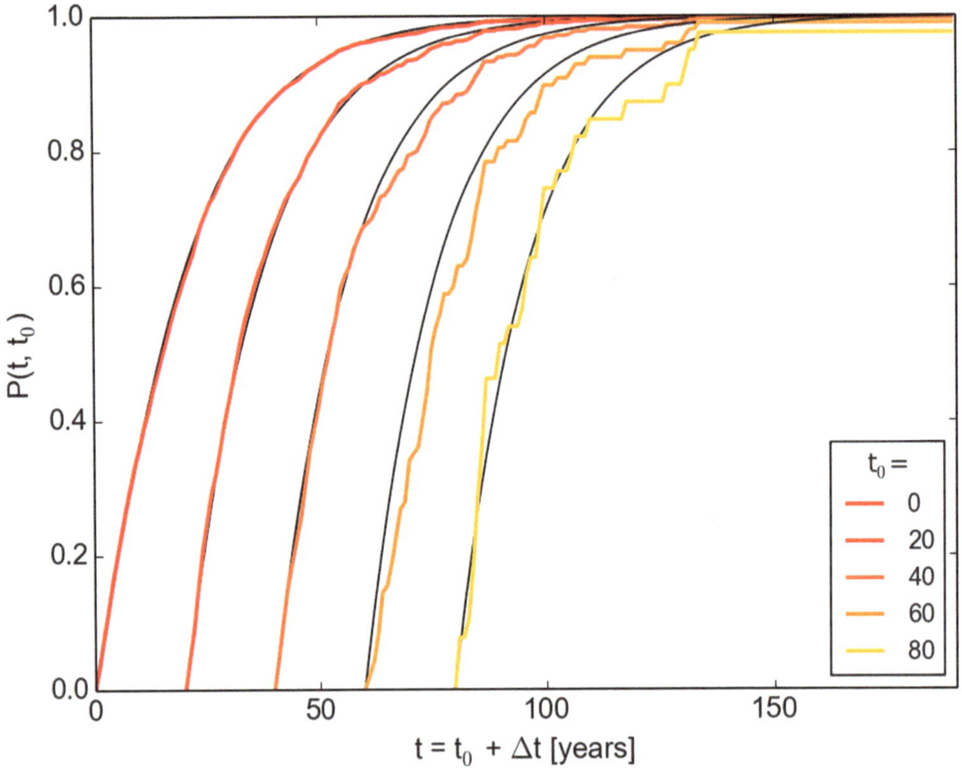

Fig. 4 Conditional cumulative probability $P(t, t_0)$ of a magnitude $\geq$ 7.0 earthquake on the forecasted faults (Fig. 3). The distribution is evaluated at $t = t_0 + \Delta t$, with the last earthquake occurring t_0 years ago, computed for various t_0. The *red line* shows the Weibull distribution fit to the data with $\beta = 1.089$ and $\tau = 20.92$ years

$P(t, t_0)$ that an earthquake will occur in the waiting time $\Delta t = t - t_0$, given that the last major earthquake occurred t_0 years ago, these distributions are shown in Fig. 5 for waiting times with 25%, 50%, and 75% probability.

If we define the last observed earthquake with $M \geq 7.0$ in northern California to be the magnitude 7.0 Loma Prieta earthquake in 1989, then that determines $t_0 = 2015 - 1989 = 26$ years, shown as the vertical dashed line in Fig. 5. The relatively constant waiting times as a function of t_0 combined with the fitted Weibull parameter $\beta = 1.089$ being nearly 1 implies that the $M \geq 7.0$ earthquakes on faults shown in Fig. 3 are occurring randomly. A Weibull distribution with $\beta \sim 1$ behaves like a Poisson process with the time to the next event independent of previous events (Sornette and Knopoff 1997). This behavior is expected as the subset of faults is weakly correlated and contains multiple major faults capable of $M \geq 7.0$ earthquakes, each with its own recurrence interval.

4.3 Comparing to Independent Method

We have selected to use the earthquake forecasting website OpenHazards.com to compute earthquake probabilities and compare with our Virtual California forecast. The website computes the probabilities of earthquakes of various magnitudes within a selected region by employing the natural time Weibull method (NTW). This model uses small earthquakes to forecast the occurrence of large earthquakes. The basic idea of their method is to compute large earthquake probabilities using the number of small earthquakes (updated daily from the ANSS catalog) that have occurred in a region since the last large earthquake. This forecasting method is described in detail in Rundle et al. (2012) and Holliday et al. (2014).

Figure 6 shows the Northern California region selected for comparison to the Virtual California forecast, with the same fault model used in the VC simulation shown in

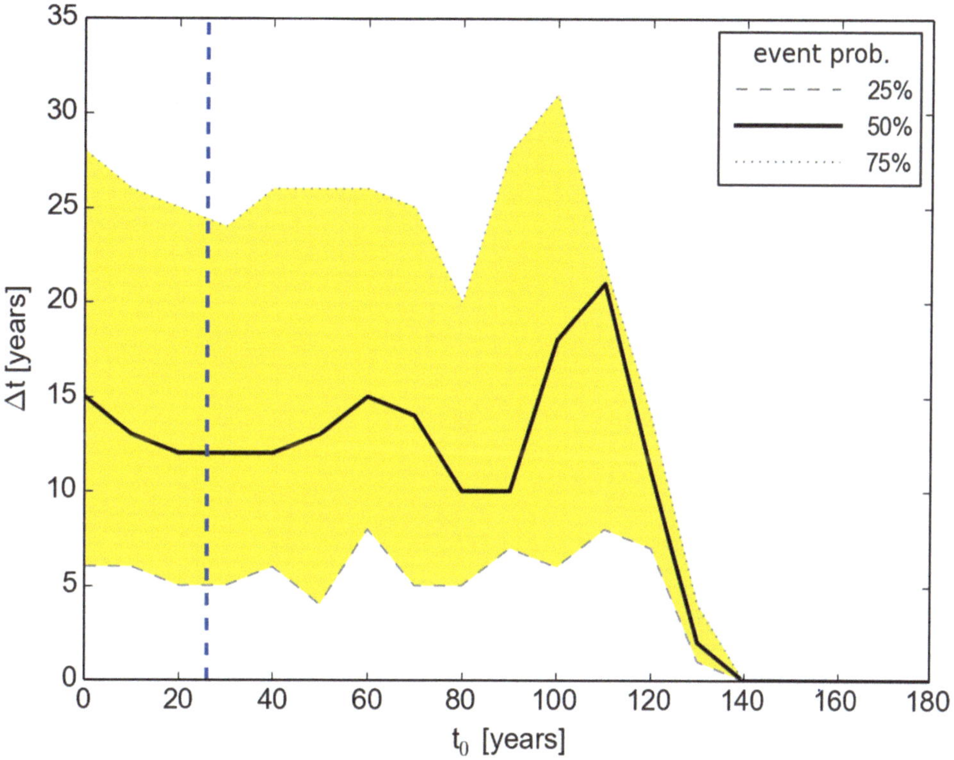

Fig. 5 The waiting times for the next magnitude $\geq$ 7.0 earthquake on the northern California faults, as a function of time since the last earthquake, t_0. The *dark line* indicates the median waiting time (50% probability), and the *upper* and *lower edges* of the *yellow band* represent the waiting times with 75% and 25% probability respectively. The *vertical dashed line* denotes $t_0 = 26$ years, the time elapsed since the M=7.0 Loma Prieta earthquake in 1989

red. The forecast generated from their Hazards Viewer – www.openhazards.com/viewer – is completely independent from the physics governing the Virtual California simulator and completely independent from how VC computes probabilities. Taking the VC conditional probabilities from Fig. 4, we find there is a 14.2% chance of a $M \geq 7.0$ earthquake occurring on the faults shown in Fig. 3 in the next 3 years. OpenHazards computes a 20.3% probability for a $M \geq 7.0$ earthquake occurring anywhere within the blue region shown in Fig. 6 within the next 3 years.

5 Conclusions and Discussion

Here we have introduced and highlighted a few applications of Virtual Quake. We have shown that Virtual Quake can produce earthquake probabilities that are fairly consistent with a more sophisticated time-dependent forecasting method. We have also illustrated how Virtual Quake can be used to generate catalogs of thousands of maps of geophysical observables like co-seismic gravity changes and InSAR interferograms.

Tullis et al. (2012b) provided a concise overview of today's earthquake simulators, including Virtual Quake. Tullis et al. (2012a) compared the output of Virtual Quake to observed seismicity and to other earthquake simulators. These papers provide very general validation of the Virtual Quake method and statistics, but there are still many questions that Virtual Quake can answer. How does the inclusion of an Epidemic Type Aftershock (ETAS) model affect fault interactions and earthquake statistics? How would analytically-derived values of the stress drop for faults affect earthquake statistics compared to those used in the Tullis et al. (2012a) study (derived by tuning the stress drops to match observed recurrence intervals using one particular simulator)? How sensitive are simulation-derived conditional probabilities to the model parameterization? What percentage of simulated earthquakes have slip that jumps between adjacent faults, and what are the statistics of these jumps? These are but a few examples of the research pathways afforded by the Virtual Quake simulator.

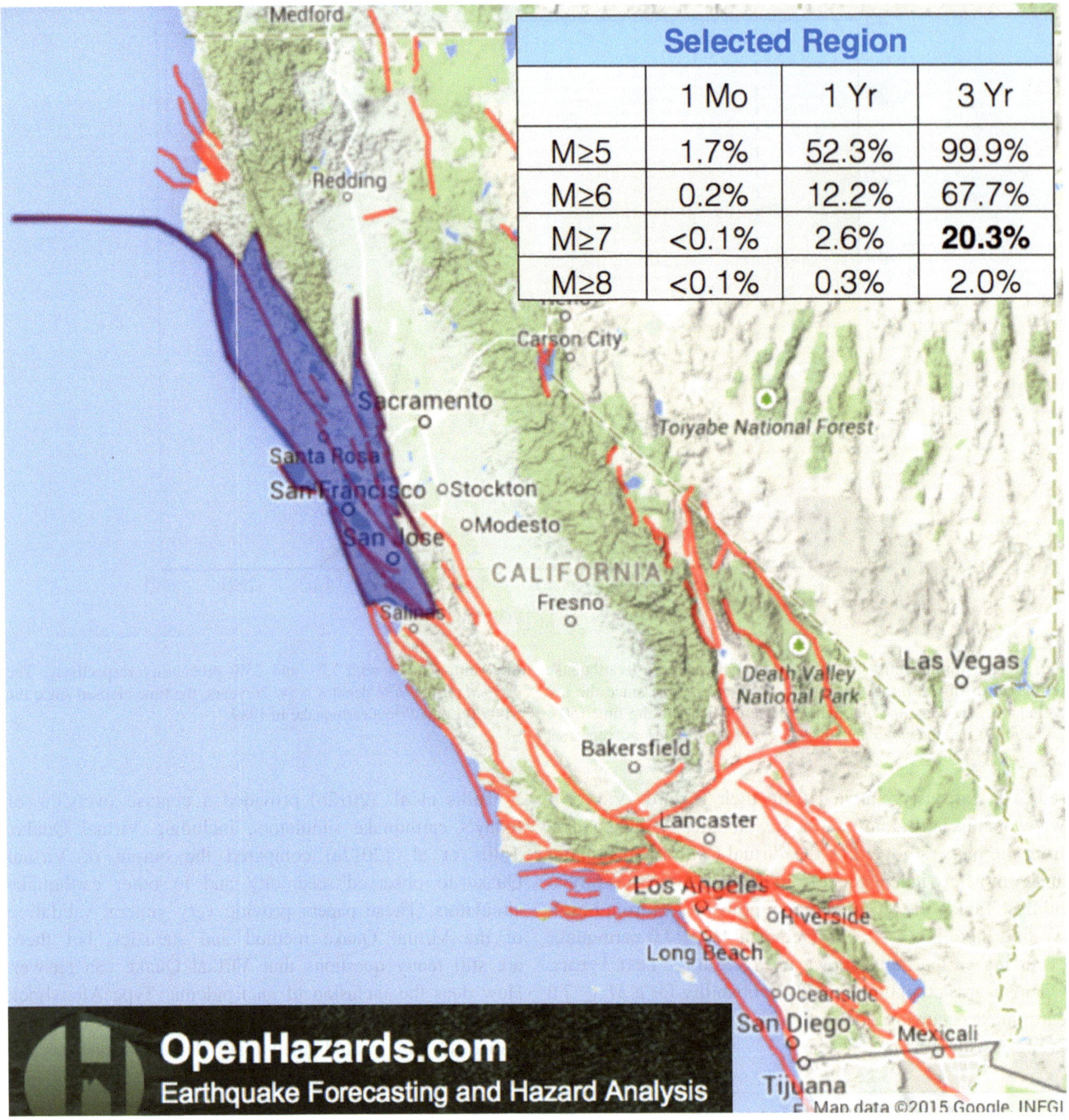

Fig. 6 The northern California forecast as computed by OpenHazards (openhazards.com/viewer) for the region selected in *blue*; VC faults are in *red*. This forecast states there is a 20.3% chance of a $M \geq 7.0$ earthquake within 3 years, compared to the VC computed probability of 14.2%. OpenHazards data last accessed Jan 7, 2015

Acknowledgements This research was supported by National Aeronautics and Space Administration (NASA) Earth and Space Science fellowship number NNX11AL92H. The release version of Virtual California and the Users' Manual are hosted by the Computational Infrastructure for Geodynamics which is supported by NSF Grant EAR-0949446.

References

Abaimov S, Turcotte D, Shcherbakov R, Rundle J, Yakovlev G, Goltz C, Newman W (2008) Earthquakes: recurrence and interoccurrence times. Pure Appl Geophys 165(3–4):777–795

Donnellan A, Rundle J, Fox G, McLeod D, Grant L, Tullis T, Pierce M, Parker J, Lyzenga G, Granat R, Glasscoe M (2006) Quakesim and the solid Earth research virtual observatory. Pure Appl Geophys 163 (11–12):2263–2279

Hayes TJ, Tiampo KF, Rundle JB, Fernández J (2006) Gravity changes from a stress evolution earthquake simulation of California. J Geophys Res Solid Earth 111(B9)

Heien E, Sachs M (2012) Understanding long-term earthquake behavior through simulation. Comput Sci Eng 14(5):10–20

Heki K, Matsuo K (2010) Coseismic gravity changes of the 2010 earthquake in central chile from satellite gravimetry. Geophys Res Lett 37(24)

Holliday J, Graves W, Rundle J, Turcotte D (2014) Computing earthquake probabilities on global scales. Pure Appl Geophys. doi: 10.1007/s00024-014-0951-3

Matsuo K, Heki K (2011) Coseismic gravity changes of the 2011 tohoku-oki earthquake from satellite gravimetry. Geophys Res Lett 38(7)

Okada Y (1992) Internal deformation due to shear and tensile faults in a half-space. Bull Seismol Soc Am 82(2):1018–1040

Okubo S (1992) Gravity and potential changes due to shear and tensile faults in a half-space. J Geophys Res Solid Earth 97(B5):7137–7144

Rundle JB (1988a) A physical model for earthquakes: 1. Fluctuations and interactions. J Geophys Res Solid Earth 93(B6):6237–6254

Rundle JB (1988b) A physical model for earthquakes: 2. Application to southern California. J Geophys Res Solid Earth 93(B6):6255–6274

Rundle JB, Klein W, Tiampo K, Gross S (2000) Linear pattern dynamics in nonlinear threshold systems. Phys Rev E 61:2418–2431

Rundle JB, Rundle PB, Donnellan A, Turcotte DL, Shcherbakov R, Li P, Malamud BD, Grant LB, Fox GC, McLeod D, Yakovlev G, Parker J, Klein W, Tiampo KF (2005) A simulation-based approach to forecasting the next great San Francisco earthquake. Proc Natl Acad Sci USA 102(43):15363–15367

Rundle JB, Rundle PB, Donnellan A, Li P, Klein W, Morein G, Turcotte D, Grant L (2006a) Stress transfer in earthquakes, hazard estimation and ensemble forecasting: inferences from numerical simulations. Tectonophysics 413(1–2):109–125

Rundle P, Rundle J, Tiampo K, Donnellan A, Turcotte D (2006b) Virtual California: fault model, frictional parameters, applications. In: Computational earthquake physics: simulations, analysis and infrastructure, part I. Pageoph topical volumes. Birkhäuser-Verlag, Basel, pp 1819–1846

Rundle P, Rundle J, Tiampo K, Donnellan A, Turcotte D (2006c) Virtual california: fault model, frictional parameters, applications. Pure Appl Geophys 163(9):1819–1846

Rundle JB, Holliday JR, Graves WR, Turcotte DL, Tiampo KF, Klein W (2012) Probabilities for large events in driven threshold systems. Phys Rev E 86:021106

Sachs MK, Heien EM, Turcotte DL, Yikilmaz MB, Rundle JB, Kellogg L (2012) Virtual California earthquake simulator. Seismol Res Lett 83(6):973–978

Schultz K, Sachs M, Heien E, Rundle J, Turcotte D, Donnellan A (2014) Simulating gravity changes in topologically realistic driven earthquake fault systems: first results. Pure Appl Geophys. ISSN 0033-4553. doi: 10.1007/s00024-014-0926-4

Sieh K, Stuiver M, Brillinger D (1989) A more precise chronology of earthquakes produced by the san andreas fault in southern California. J Geophys Res Solid Earth 94(B1):603–623

Sornette D, Knopoff L (1997) The paradox of the expected time until the next earthquake. Bull Seismol Soc Am 87(4):789–798

Sun W, Okubo S, Fu G, Araya A (2009) General formulations of global co-seismic deformations caused by an arbitrary dislocation in a spherically symmetric Earth model-applicable to deformed Earth surface and space-fixed point. Geophys J Int 177(3):817–833

Tiampo KF, Rundle JB, McGinnis S, Gross SJ, Klein W (2002) Eigenpatterns in southern California seismicity. J Geophys Res Solid Earth 107(B12):2429–2467

Tullis TE, Richards-Dinger K, Barall M, Dieterich JH, Field EH, Heien EM, Kellogg LH, Pollitz FF, Rundle JB, Sachs MK, Turcotte DL, Ward SN, Burak Yikilmaz M (2012a) A comparison among observations and earthquake simulator results for the allcal2 California fault model. Seismol Res Lett 83(6):994–1006

Tullis TE, Richards-Dinger K, Barall M, Dieterich JH, Field EH, Heien EM, Kellogg LH, Pollitz FF, Rundle JB, Sachs MK, Turcotte DL, Ward SN, Yikilmaz MB (2012b) Generic earthquake simulator. Seismol Res Lett 83(6):959–963

Van Aalsburg J, Grant LB, Yakovlev G, Rundle PB, Rundle JB, Turcotte DL, Donnellan A (2007) A feasibility study of data assimilation in numerical simulations of earthquake fault systems. Phys Earth Planet Inter 163(14):149–162 [Computational Challenges in the Earth Sciences]

Van Aalsburg J, Rundle JB, Grant LB, Rundle PB, Yakovlev G, Turcotte DL, Donnellan A, Tiampo KF, Fernandez J (2010) Space- and time-dependent probabilities for earthquake fault systems from numerical simulations: feasibility study and first results. In: Seismogenesis and earthquake forecasting: the frank evison, volume II. Pageoph topical volumes. Springer, Basel, pp 113–123. ISBN 978-3-0346-0499-4

Yakovlev G, Turcotte DL, Rundle JB, Rundle PB (2006) Simulation-based distributions of earthquake recurrence times on the San Andreas Fault system. Bull Seismol Soc Am 96(6):1995–2007

Yikilmaz MB, Turcotte DL, Yakovlev G, Rundle JB, Kellogg LH (2010) Virtual California earthquake simulations: simple models and their application to an observed sequence of earthquakes. Geophys J Int 180(2):734–742

Yikimaz MB, Heien EM, Turcotte DL, Rundle JB, Kellogg LH (2011) A fault and seismicity based composite simulation in northern California. Nonlinear Processes Geophys 18(6):955–966

Dynamic Strain in a South African Gold Mine Produced by the 2011 Tohoku Earthquake

Makoto Okubo, Hiroshi Ogasawara, Shigeru Nakao, Osamu Murakami, Hiroshi Ishii, and Anthony K. Ward

Abstract

The 2011 Tohoku earthquake (2011/03/11 05:46:18 UT, M_W 9.0) produced huge permanent displacements of up to 50 m and dynamic strain of up to 10^{-5} in the near-field. We observed dynamic strain (10^{-7}) produced by this earthquake at a depth of 1 km in a gold mine in the Republic of South Africa at a distance of more than 14,000 km (125.25°) from epicenter. The dynamic strain was observed by two Ishii-type borehole strainmeters about 30 m apart on opposite sides of a fault filled with gouge of several decimeters thickness, allowing the response of the fault to be investigated. The Tohoku earthquake seismic waves passed through the gold mine outside blasting hours, allowing us to analyze the tele-seismic body waves and the surface waves that circled the globe several times ($R_3 - R_5$ and $G_3 - G_5$, $\sim 10^{-8}$). We discuss the fault deformation associated with the dynamic strains by the wave packets. The seismograms of some sub-parallel components of the two strainmeters installed on the opposite sides of the fault appeared similar, although with differences on the order of $\sim 10^{-8}$. These differences may imply a complex response of the fault (fault rheology; i.e. not only elastic, but perhaps plastic or even indicating the effects of fluid).

Keywords

Fault rheology • Ishii-type Borehole strainmeter • Rock mass deformation • Surface waves • Teleseismic waves

1 Introduction

In the Republic of South Africa, Ritsumeikan University was leading the projects "Multidisciplinary monitoring of preparation and generation of earthquakes at M2 sources in South African gold mines," funded by a Grant-in-aid from the Japan Science Promotion Society (*JSPS*) and "Observational Studies in South African Mines to Mitigate Seismic Risks" sponsored by the Japan Science and Technology Agency (*JST*) and Japan International Cooperation Agency

M. Okubo (✉)
Tono Research Institute of Earthquake Science, Association for the Development of Earthquake Prediction, 1-63 Yamanouchi, Mizunami 509-6132, Japan

Research and Education Faculty, Natural Sciences Cluster, Kochi University, 2-17-47 Asakura Honmachi, Kochi 780-8073, Japan
e-mail: okubo@kochi-u.ac.jp

H. Ogasawara • O. Murakami
College of Science and Engineering, Ritsumeikan University, 1-1-1 Noji Higashi, Kusatsu 525-8577, Japan; JICA-SATREPS

S. Nakao
Graduate School of Science and Engineering, Kagoshima University, 1-21-35 Korimoto, Kagoshima 890-0065, Japan

H. Ishii
Tono Research Institute of Earthquake Science, Association for the Development of Earthquake Prediction, 1-63 Yamanouchi, Mizunami 509-6132, Japan

A.K. Ward
Ezulwini Mine (Cooke #4), Sybanyegold, Westonaria, South Africa

Seismogen CC, Westonia, South Africa

29 Carlton Jones Street, Carletonville, South Africa

M. Hashimoto (ed.), *International Symposium on Geodesy for Earthquake and Natural Hazards (GENAH)*,
International Association of Geodesy Symposia 145, DOI 10.1007/1345_2015_207

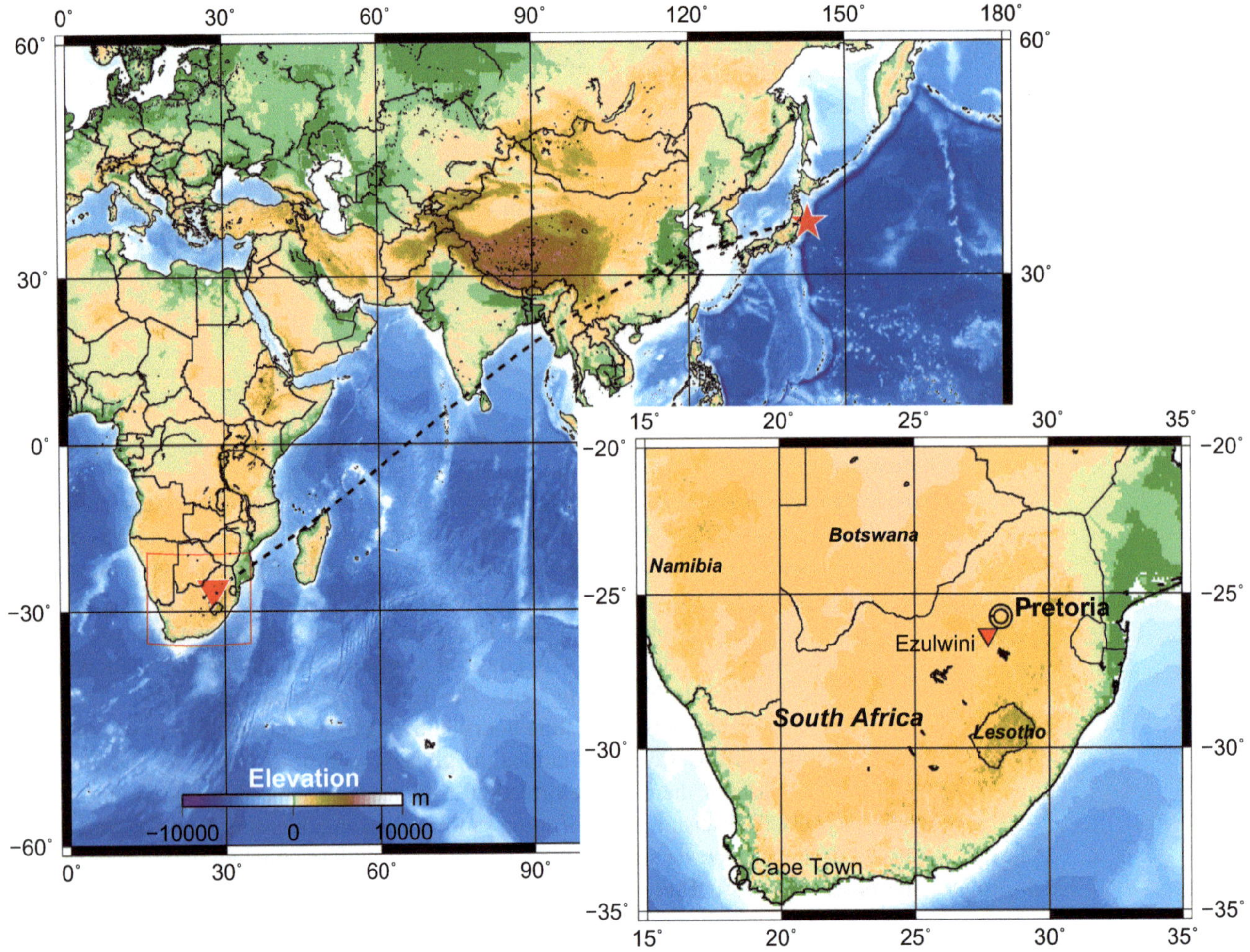

Fig. 1 Locations of the 2011 Tohoku earthquake epicenter (*star*) and the Ezulwini gold mine in South Africa (*triangle*). The great circle path from the epicenter to the mine is shown as a *broken line*

(*JICA*) as a Science and Technology Research Partnership for Sustainable Development (*SATREPS*). Ten Ishii-type borehole strainmeters were installed in 2010–2013 at depths of 1.0–3.0 km from the earth's surface in four deep gold mines to monitor rock the mass response to mining and to allow mitigation of seismic hazard (Ogasawara et al. 2014a). In previous projects in deep South African gold mines, these instruments showed that mining induced strain changes ($<\sim10^{-3}$; Ishii et al. 1998) and coseismic strain changes ($<\sim10^{-4}$; Ogasawara et al. 2005) caused by mine earthquakes ($M < 2$), and episodic strain changes ($<\sim10^{-7}$; Naoi et al. 2006).

The Ezulwini gold mine, which is located southwest of Pretoria in South Africa (Fig. 1), is one of the *SATREPS* observational sites. At the Ezulwini mine, the entire ore body, except for the shaft pillar (about 500 m × 400 m), was mined out before 2006, and increased the stress at the remnant, namely, at the shaft pillar. Ogasawara et al. (2014b) measured a vertical maximum principal stress of ∼130 MPa in the eastern half of the shaft pillar, after the completion of mining at the western half of the shaft pillar. In order to investigate and monitor the time-dependent deformation mainly caused by the inelastic response of the highly stressed or already-fractured rock mass, two Ishii-type borehole strainmeters were installed about 30 m apart on either side of a fault at 1 km depth. The 2011 Tohoku earthquake took place when the mining was still going on in the remaining northwestern quarter, with a fault located at the eastern edge of mining.

The 2011 Tohoku earthquake ($M_W = 9.0$; JMA 2011) took place at 05:46:18 UT (14:46:18 JST on 11 March 2011). A maximum fault slip of 50 m was estimated (e.g. Fujiwara et al. 2011; Simons et al. 2011; Ohta et al. 2012; Suzuki et al. 2011). At the GSI (Geospatial Information Authority of Japan) GPS stations on the Japanese main island, the maximum vertical displacement of −1.14 m was observed at Ishinomaki, and the maximum horizontal movement at Onagawa (5.85 m eastward; GSI 2011). Furthermore, large amplitude dynamic strain variations were observed. For example, at the Byobusan crustal activity observatory, which is located in Gifu 500 km to the west of the epicenter, a body wave with a maximum amplitude of 10^{-6} was observed (Okubo et al. 2011). We also observed tele-seismic dynamic strain variations ($\sim 10^{-7}$) in South Africa, despite a great circle distance of 14,000 km (125.25°).

Our interest is how the fault responds to longer wavelength variations in strain or stress, e.g. tele-seismic body waves, surface waves and tidal motions. In this paper, we introduce a few interesting observations in South Africa deep gold mine and discuss some unusual phenomena around the fault that can not be explained with elastic theory.

2 Observations

The Ezulwini gold mine, which is located southwest of Pretoria in South Africa (Fig. 1), is one of the *SATREPS* observational sites. Table 1 shows the latitude, longitude and elevation (above mean sea level) of the shaft collar of the mine. In this mine, two Ishii-type borehole strainmeters (Ishii et al. 2002) were installed in 60 mm-diameter boreholes (BH) drilled from a horizontal tunnel at a depth of 1 km from the earth's surface. The trends and plunges of the BHs are given in Table 1.

These strainmeters, named ST1212 and STxc04, are 26.2 m apart and located on opposite sides of a normal fault call the 'Zebra fault'. Each strainmeter has three components on a plane perpendicular to BH axis, and CH1 was set downward. CH2 and CH3 were set on the same plane with clockwise angle intervals of 120° from CH1 and CH2, respectively. The Analog strainmeter outputs were continuously converted to 24 bit resolution digital value at a sampling frequency of 40 Hz and stored in digital WIN format (Urabe and Tsukada 1992) with a logger, Datamark LS-7000XT (Hakusan Corporation 2013). Although the data logger can automatically synchronizes its clock with the GPS time, this function was not available at underground.

Table 1 Location of the Ezulwini gold mine and the installation details of the strainmeters

Location	Latitude	Longitude	Elevation
Ezulwini gold mine shaft collar	26.3621°S	27.7085°E	1693 m
Installation	*Sensor code*	*BH trend*	*BH plunge*
	ST1212	N347°E	29° down
	STxc04	N151°E	22° down

3 In Situ Sensor Calibration

An Ishii-type borehole strainmeters are designed for the monitoring of long-term crustal deformation associated with tectonic and volcanic activity (c.f. Ishii et al. 2002). The original version of the Ishii-type strainmeter is sensitive enough to detect strain changes as small as 10^{-9} or less. However, the strainmeter was too sensitive, and the required hole diameter ($\sim$100 mm) was too large to drill easily, quickly, and safely in South African gold mines. Therefore, we used the downsized strainmeters into 42 mm in diameter and 1000 mm long (e.g. Ogasawara et al. 2014b), which decreased the minimum sensitivity but increased the upper measurement limit upto 10^{-3}.

The sensitivity of a BH strainmeter is strongly affected by installation conditions, such as differences in rigidity between the host rock and grout, BH diameter, and damage, even when sensor is exactly centered. Ishii-type BH strainmeters were generally grouted with expansive cement in order to adapt to the environmental pre-stress. Consequently, in-situ sensitivity sometimes differs from the manufacturer's factory specification by ten percent or more. Furthermore, the effect of a thin, tabular and extensively excavated stope (typically of 1 m height) must also be evaluated, strictly speaking.

In order to calibrate the strainmeter sensitivity, we compared the observed earth tide with the theoretically estimated earth tide by using '*GOTIC2*' (Matsumoto et al. 2001). We also compared the observed strain seismogram with a simulate tele-seismic strain seismogram by using '*synthe_me*' (e.g. Kamigaichi 1998). These tools, *GOTIC2* and *synthe_me*, can calculate the time series of horizontal strain variations at an arbitrary position on earth, except the vertical component. However, the CH1 component of each strainmeter is sub-vertical, because BH plunge angles of ST1212 and STxc04 are 29° and 22°, respectively (Table 1; plunge angles of CH1 components are 61° and 68°). Rigorously speaking, even for an isotropic homogeneous medium, all of the six independent components of the strain tensor are required to calculate the strain for an arbitrary 3-D direction. However, for simplicity, we assumed that vertical strain is one of the principal strains. Then the vertical component (S_V) can be expressed as follows:

$$S_V = -\nu \times (S_{AB} + S_{BT}), \quad (1)$$

where ν is Poisson's ratio and S_{AB} and S_{BT} are horizontal extensions and contractions in the direction parallel and

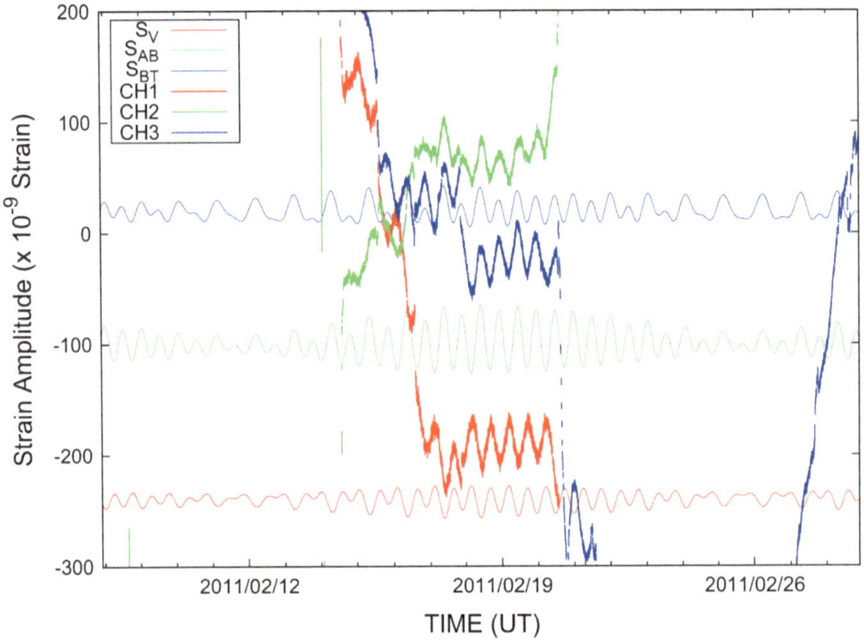

Fig. 2 The earth tide observed by the strainmeter ST1212 (CH1, CH2, and CH3) and the earth tide theoretically estimated by *GOTIC2* (S_V: vertical, S_{AB}: horizontal in line with the BH trend, S_{BT}: horizontal normal to the BH trend)

perpendicular to the BH trend, respectively. All of the three strainmeter components are on the plane perpendicular to the BH axis. CH1 is sub-vertical and the strain change should be mainly similar to S_V. CH2 and CH3 are sub-horizontal, with a strain change that should be similar to S_{BT}. Additionally all the components of strainmeters include S_{AB} and S_V to some extent, because the BH of ST1212 and STxc04 is inclined with 29° and 22°, respectively.

Figure 2 shows the ST1212 observations (CH1, CH2, and CH3) and the theoretical tidal variations (S_V, S_{AB}, and S_{BT}). We used $\nu = 0.25$ and the sensitivity (23.5×10^3 V/strain) recommended by the manufacturer. Blasting, which is usually performed on weekdays, causes significant sudden change in strain and induces seismic events, followed by significant time-dependent plastic deformation. Thus, we cannot observe a continuous and stable earth tide in weekday records (e.g., here, 21–25 February 2011, which corresponds to Monday–Friday). However, on a quiet weekend (e.g. 18–20 February), the sub-vertical component of ST1212 (CH1) had an amplitude approximately double that of S_V, but did not exceed the peak-to-peak amplitude of S_{AB}. Furthermore, the amplitudes of CH2 and CH3 were within the range of the peak-to-peak amplitudes of S_{AB} and S_{BT}. The observed variations were almost in phase with the theoretically calculated earth tide. This consistency indicates that the sensors were acceptably coupled with the host rock and that the assumed sensitivity was also acceptable for tidal frequency ranges.

Next, we compared observations with the theoretical areal strain variations of the 2011 Tohoku earthquake calculated by using *synthe_me* with periods of 5–5000 s. We used the earth velocity structure of the PREM (Dziewonski and Anderson 1981) and the earthquake parameters given by Nettles et al. (2011), and the results are compared in Fig. 3. By assuming that the depths of the strainmeters are shallower than the surface wave skin depth and that the periphery rock is homogeneous and isotropic, we compared the observed areal strain variations with the synthetic areal strain variations on the horizontal plane. The synthetic areal strain on the horizontal plane (H. Dilatation) was calculated as the sum of 'Radial' and 'Transverse' component. The observed areal strains were slightly larger than the synthetic strains for the first large packet (probably SS waves), but they were smaller for the surface waves. However, the strain variations were in reasonable agreement; the dips of the plane for areal strain were not identical, and the Poisson's ratio may have been different from the assumed value.

From these considerations, we concluded that the output amplitude (23.5×10^3 V/strain) of strainmeters ST1212 and STxc04 were stable at the same value for the analyzed period range from the seismic (5–5000 s) up to the tidal (about a half a day: ~40,000 s). We also concluded that we could not find significant amplification of strain variation in the discussed period, suggesting the strainmeters were installed so as to be isolated from the cavity effects of tunnels or thin tabular stopes. An attenuation or amplification in seismogram amplitude would be observed only at a much higher frequency range (several Hz–kHz) when the configuration of cavities or scatterers as the source and receiver was critical.

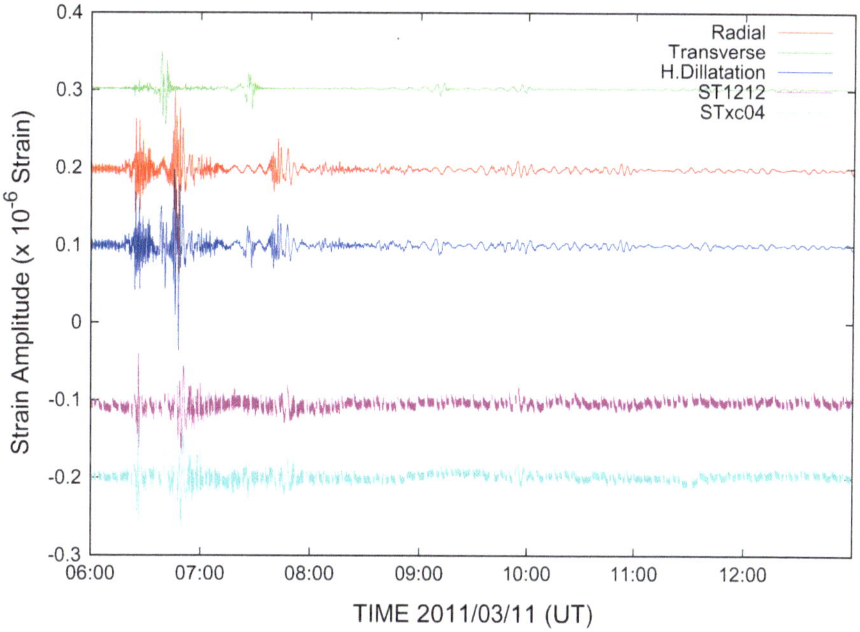

Fig. 3 Synthetic and observed dynamic strain seismograms of the 2011 Tohoku earthquake at the Ezulwini gold mine. The observed areal strains on the inclined observational planes are shown in *magenta* and *cyan* for ST1212 and STxc04, respectively. Synthetic 'Radial', 'Transverse', and their sum 'H. Dilatation' are also shown. The *synthe_me* tool was used to calculate the synthetic seismograms

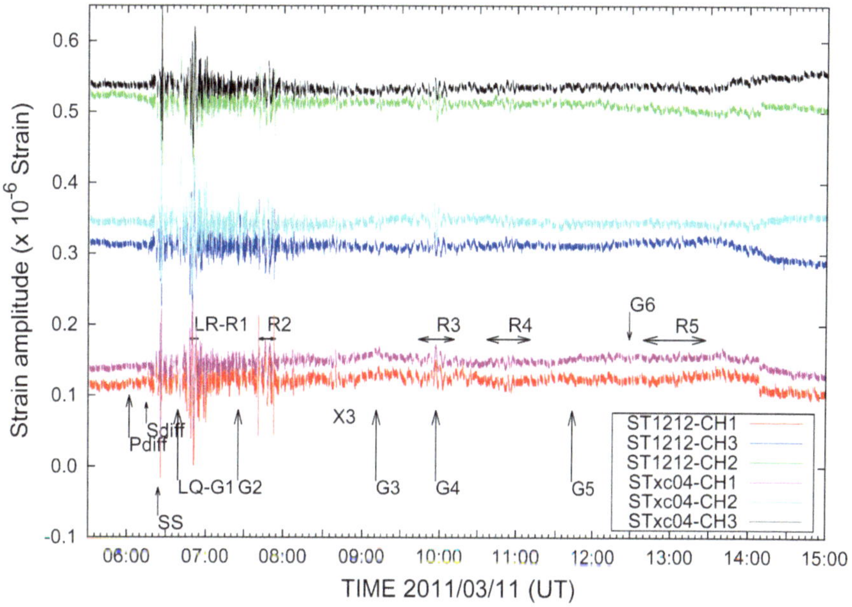

Fig. 4 Strain seismograms of the 2011 Tohoku earthquake. *Red* and *magenta*: sub-vertical components (CH1). *Black* and *green*: components oriented in a similar direction (ST1212$_{CH2}$ and STxc04$_{CH3}$). *Arrows* with phase notation indicate theoretical arrival time as calculated by Tau-P (Crotwell et al. 1999)

4 Observation Summaries

We tested the sensitivities of the two strainmeters at the Ezulwini, as described in the previous section. Because mining over months causes strain changes as large as 10^{-3}, we removed the DC offsets and linear trends from the raw strain records. Figure 4 shows the strain seismograms of the 2011 Tohoku earthquake. Daily blasting at 13:00–16:00 (UT) every weekday caused sudden and significant change. However, the seismic waves of the Tohoku earthquake passed through the Ezulwini mine earlier than the blasting on the Friday.

As shown in Fig. 4, we calculated the theoretical arrival times of phases with the Tau-P utility (Crotwell et al. 1999) by assuming the PREM earth velocity structure (Dziewonski and Anderson 1981). The first arrival phases (P_{diff}) were not clear because of high-frequency electric noise (corresponding to $\sim 10^{-8}$). However, the S_{diff} and SS packets were clear, with amplitudes of $\sim 10^{-8}$ and several 10^{-7}, respectively.

Table 2 Dihedral angles in degree of each strainmeter component and the normal vector of the Zebra fault plane

		Strainmeter STxc04			Zebra
		CH1	CH2	CH3	normal
Strainmeter ST1212	CH1	50°	77°	66°	73°
	CH2	65°	57°	10°	73°
	CH3	79°	38°	55°	55°
Zebra normal		42°	20°	78°	

P_{diff} and S_{diff} are wave packets that have diffracted once at the core-mantle boundary (CMB), whereas *SS* packets have reflected once on the earth's surface. Several tens of minutes after the arrival of the first packet, many wave packets with larger amplitudes, which were surface waves, arrived. Love (LQ) and Rayleigh (LR) waves had the largest amplitudes (maximum of several times of 10^{-7}), ten times larger than the earth tide. LQ and LR were followed by some more Love waves ($G_3 \sim G_5$) and Rayleigh waves ($R_3 \sim R_5$), which circled the globe several times.

$ST1212_{CH2}$ and $STxc04_{CH3}$ have a dihedral angle of 10° to each other, is similar observational directions (Table 2). The variations in their strain (green and black in Fig. 4) were also similar. The slight difference might have been influenced by the existence of the Zebra fault between the strainmeters ST1212 and STxc04. We discuss the composite strainmeter recordings of ST1212 and STxc04 in the following section.

5 Preliminary Strain Analyses Around the Fault

Table 2 shows the dihedral angle between each component of each strainmeter, and the dihedral angle of every component with respect to the normal vector of the Zebra fault. $STxc04_{CH3}$ (78°) and $STxc04_{CH2}$ (20°) were almost sub-parallel and sub-normal to the Zebra fault, respectively. The dihedral angle between $ST1212_{CH2}$ and $STxc04_{CH3}$ was small (10°), although the two strainmeters were located on opposite sides of the Zebra fault. Thus, the difference in these components can be treated as a shear strain on the fault plane in strike slip mode (hereinafter the fault shear).

The strain variations of these components, $STxc04_{CH2}$, $STxc04_{CH3}$, and ($ST1212_{CH2}$ − $STxc04_{CH3}$), are shown in Fig. 5. We can clearly identify both Rayleigh and Love surface wave phases in the fault perpendicular component ($STxc04_{CH2}$), and small Love wave phases in the fault parallel components ($STxc04_{CH3}$); the *SS* phases can be identified clearly in both. The difference in responses for shear (*SS* packets and Love waves) might be related to their incident angles to the Zebra fault, and their wave lengths.

Long-term variations with an amplitude of 10^{-8} from 06:00 to 08:30 in the fault shear component (black in Fig. 5) were observed, after the largest wave packets passed. These variations were not evident in the fault sub-normal ($STxc04_{CH2}$; red in Fig. 5) or sub-parallel component ($STxc04_{CH3}$; blue in Fig. 5). These variations are unlikely to be a result of noise because subtraction cancels common-mode noise, as a waveform cleaner; we propose that they are related to the fault. Similar variations might also occur during other periods, for example from 09:00 to 14:00.

To test this hypothesis, we investigated at different composite variations by using the strainmeter components on the opposite sides of the fault. The dihedral angle was 77° between $STxc04_{CH2}$ and $ST1212_{CH1}$, and 79° between $STxc04_{CH1}$ and $ST1212_{CH3}$ (Table 2), it shows that these components are almost perpendicular to each other. Thus, the areal strain can be calculated approximately for the planes that include the fault. Table 3 shows the dihedral angles of the planes with respect to the Zebra fault. The configuration of the planes and the Zebra fault is also schematically illustrated in perspective view in Fig. 6. Figure 7 shows the time variations of these approximate areal strains in cyan and magenta, which are the same colors as in Fig. 6. Again, we can easily see that the largest changes ($\sim 10^{-8}$) are associated with the most significant seismic phases after 06:30 UT. We cannot find any variation beyond the magnitude of electrical noise, in the magenta component (approximate areal strain inclined to the Zebra fault, $STxc04_{CH1}$ + $ST1212_{CH3}$). In contrast, there was clear long-term variation, which was similar to the fault shear component with the opposite sign, in the cyan component ($STxc04_{CH2}$ + $ST1212_{CH1}$; approximately fault normal areal strain). The long-term variation, which was related to the fault plane, was small and could be observed only by comparing the components of different strainmeters. Thus, these long-term strain variations seem likely to be a behavior of the fault plane because the strain component outside the fault plane did not vary and the areal and differential components, which include the fault plane, did vary. However, it was not possible to show strong evidence indicating behavior specific to the fault plane in this study. The hour after 13:00 UT (15:00 SAST) is scheduled for blasting. Therefore, the significant change ($\sim 10^{-7}$) seen in all strain variations was probably associated with blasting some distance from the strainmeters.

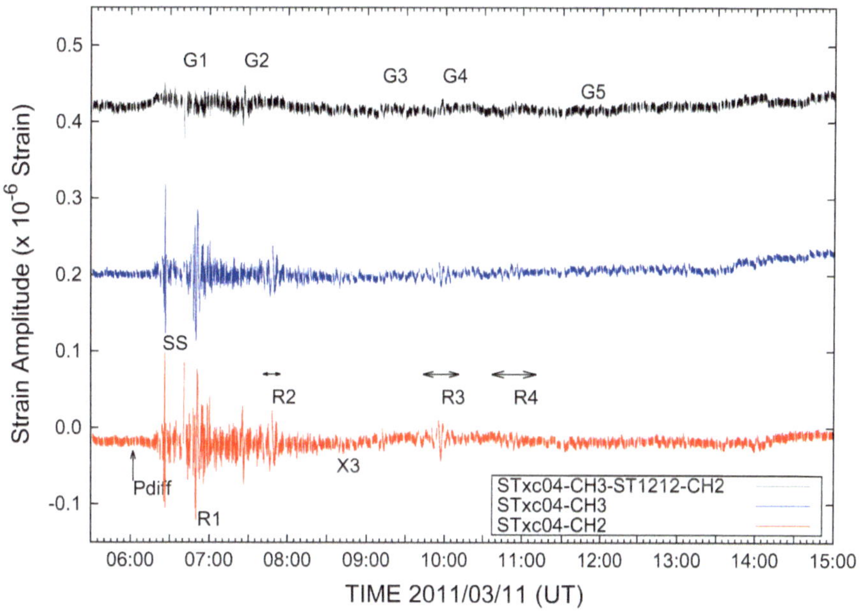

Fig. 5 Fault sub-parallel ($STxc04_{CH3}$), sub-normal ($STxc04_{CH2}$), and shear ($STxc04_{CH3} - ST1212_{CH2}$) components are shown in *red*, *blue*, and *black*, respectively. Phase notation is the same as in Fig. 4

Table 3 Dihedral angles of the planes on which the approximate areal strain was calculated with respect to the Zebra fault plane

	Approximate areal strain	
	$STxc04_{CH2} + ST1212_{CH1}$	$STxc04_{CH1} + ST1212_{CH3}$
Zebra fault plane	71°	59°

6 Conclusions

In the Ezulwini gold mine, South Africa, tidal strain changes ($\sim 10^{-8}$) are usually masked by mining-induced rock deformations and local earthquakes that frequently accompany blasting. However, we observed dynamic strains on the order of 10^{-7} produced by the 2011 Tohoku earthquake at the mine at a great circle distance of 14,000 km (125.25°) from the epicenter. The amplitudes of the tele-seismic *SS* and R_1 wave packets were the largest ($\sim 10^{-7}$), and strain variations lasted for about 7 h until G_5 ($\sim 10^{-8}$). Considering the dynamic strain amplitude observed at the Japanese crustal deformation observatories ($\sim 10^{-5}$) and global attenuation, the observed dynamic strain amplitude ($\sim 10^{-7}$) in South Africa was reasonable. In other words, if a M8 class earthquake occurs somewhere in the world, a dynamic strain with an amplitude larger than the electrical noise level will always be observed in the Ezulwini mine. In this study, we did not observe the onset of rock mass deformation in the mine associated with the passage of the tele-seismic waves. However, we observed unusual behavior around the fault that could not be explained with elastic theory. We observed some variations that may be related to the fault behavior, but did not obtain qualitative or quantitative evidence indicating

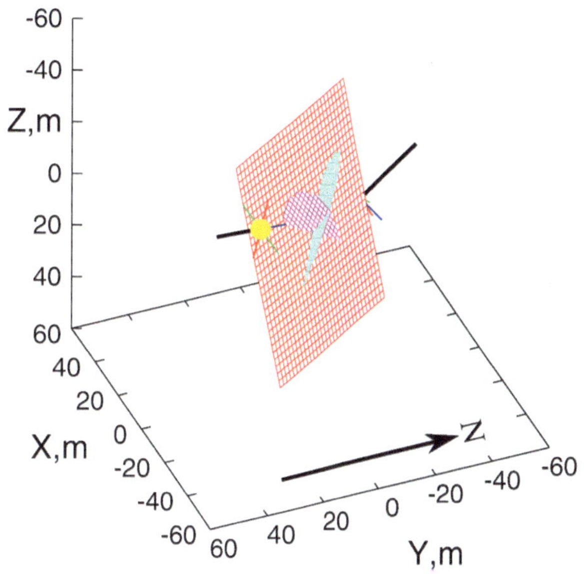

Fig. 6 Perspective view from ESE to WNW showing the configuration of the Zebra fault plane (*red mesh parallelogram*), two strainmeter boreholes (*black lines*), three strainmeter components (*red*, *green*, and *blue lines*), and two planes where approximate areal strain was calculated (*cyan* and *magenta* mesh ellipsoids). The strainmeters in front of and behind the Zebra fault are STxc04 and ST1212, respectively. *Cyan* indicates the pair $STxc04_{CH2}$ and $ST1212_{CH1}$, and *magenta* indicates the pair $STxc04_{CH1}$ and $ST1212_{CH3}$. Strain variations in *magenta* and *cyan* in Fig. 7 are for approximate areal strain calculated using the pairs. The *cyan* plane ($STxc04_{CH2} + ST1212_{CH1}$) is nearly perpendicular to the Zebra fault (71°; Table 3)

a fault-plane specific behavior. However, further observations and analysis should allow qualitative and quantitative discussion of the fault behavior. Our analysis of the strain variation data recorded by these densely distributed borehole

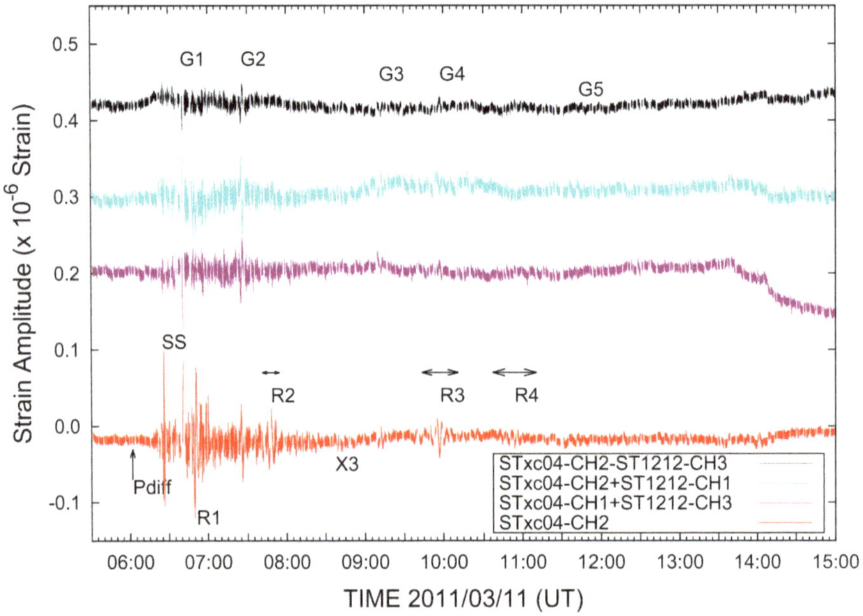

Fig. 7 Strain variations including the Zebra fault plane region. Two approximate areal strains, ($STxc04_{CH2}$ + $ST1212_{CH1}$ and $STxc04_{CH1}$ + $ST1212_{CH3}$), are shown in *cyan* and *magenta*, respectively. For reference, we plotted variations in the fault shear and the fault perpendicular components in *black* and *red*, respectively, as shown in Fig. 5

strainmeters will help us understand the in-mine deformation and fault behavior better, and help to mitigate mine seismic hazards.

Acknowledgements We want to thank Gilbert Morema, Thabang Masakale, Sylvester Morema, and Luiz Ribeiro for their great efforts in installing and maintaining the monitoring equipment. This study was supported by the JSPS Grant-in-aid (No. 21224012) and JST-JICA SATREPS program. This manuscript was greatly improved by the careful reviews and appropriate comments of reviewers and editors.

References

Crotwell HP, Owens TJ, Ritsema J (1999) The TauP toolkit: flexible seismic travel-time and ray-path utilities. Seismol Res Lett 70:154–160

Dziewonski AM, Anderson DL (1981) Preliminary reference earth model. Phys Earth Planet Inter 25:297–356

Fujiwara T, Kodaira S, No T, Kaiho Y, Takahashi N, Kaneda Y (2011) The 2011 Tohoku-Oki earthquake: displacement reaching the Trench Axis. Science 334(6060):1240. doi:10.1126/science.1211554

GSI (2011) http://www.gsi.go.jp/sokuchikijun/sokuchikijun60011.html in Japanese. Accessed 23 Sept 2015

Hakusan Corporation (2013) http://www.hakusan.co.jp/products/keisoku/main/ls7000xt.shtml in Japanese. Accessed 23 Sept 2015

Ishii H, Ohkura T, the Research Group (1998) Semi-controlled earthquake generation experiment in South African gold mines, strain monitoring and earthquake generation in South African gold mine. Earth Monthly 229:419–420 (in Japanese)

Ishii H, Yamauchi T, Matsumoto S, Hirata Y, Nakao S (2002) Development of multi-component borehole instrument for earthquake prediction study: some observed examples of precursory and co-seismic phenomena relating to earthquake swarms and application of the instrument for rock mechanics. In: Ogasawara H, Yanagidani T, Ando M (eds) Seismogenic process monitoring. Balkema, Leiden, pp 365–377

JMA (2011) The 2011 Great East Japan Earthquake-Portal. http://www.jma.go.jp/jma/en/2011_Earthquake/2011_Earthquake.html. Accessed 23 Sept 2015

Kamigaichi O (1998) Green functions of the earth at borehole sensor installation depths for surface point load. Pap Meteorol Geophys 48(4):89–100

Matsumoto K, Sato T, Takanezawa T, Ooe M (2001) GOTIC2: A program for computation of oceanic tidal loading effect. J Geod Soc Jpn 47:243–248

Naoi M, Ogasawara H, Takeuchi J, Yamamoto A, Shimoda N, Morishita K, Ishii H, Nakao S, Aswegen GV, Mendecki AJ, Lenegan P, Ebrahim-Trollope R, Iio Y (2006) Small slow-strain steps and their forerunners observed in gold mine in South Africa. Geophys Res Lett 33:L12304. doi:10.1029/2006GL026507

Nettles M, Ekström G, Koss HC (2011) Centroid-moment-tensor analysis of the 2011 off the Pacific coast of Tohoku Earthquake and its larger foreshocks and aftershocks. Earth Planets Space 63:519–523

Ogasawara H, Takeuchi J, Shimoda N, Ishii H, Nakao S, van Aswegen G, Mendecki AJ, Cichowicz A, Ebrahim-Trollope R, Kawakata H, Iio Y, Ohkura T, Ando M, the Research Group for Semi-controlled Earthquake-Generation Experiments in South African Deep Gold Mines (2005) High-resolution strain monitoring during M∼2 events in a South African deep gold mine in close proximity to hypocentres. In: Proceedings of the 6th international symposium on rockburst and seismicity in mines, pp 385–391

Ogasawara H, Katsura T, Hofmann G, Yabe Y, Nakatani M, Naoi M, Ishii H, Roberts D, Nakao S, Okubo M, Wienand J, Lenegan P, Ward AK (2014a) In-situ monitoring of rock mass response to mining in South African gold mines using the Ishii strainmeters. In: Proceedings of the sixth South African rock engineering symposium 2014. Southern African Institute of Mining and Metallurgy, Johannesburg, pp 21–34

Ogasawara H, Nakatani M, Durrheim RJ, Naoi M, Yabe Y, Moriya H, Hofmann GF, Stander C, Roberts DP, de Bruin P, Oelofse J, Kato H, Cichowicz A, Birch D, Ngobeni D, Milev A, Kgarume T, Satoh T, Horiuchi S, Kawakata H, Murakami O, Yoshimitsu N, Ward AK, Wienand J, Lenegan P, Yilmaz H, Mngadi S, Piper PS, Clements TN, Nakao S, Okubo S, Ishii H, Visser AV (2014b) Observational studies of the rock mass response to mining in highly stressed gold mines in South Africa. In: Hudyma M, Potvin Y (eds) Proceedings of 7th international congress on deep and high stress mining, Sudbury, Canada, 16–18 September 2014. Australian Centre for Geomechanics, Crawley, pp 123–137

Ohta Y, Kobayashi T, Tsushima H, Miura S, Hino R, Takasu T, Fujimoto H, Iinuma T, Tachibana K, Demachi T, Sato T, Ohzono M, Umino N (2012) Quasi real-time fault model estimation for near-field tsunami forecasting based on RTK-GPS analysis: application to the 2011 Tohoku-Oki earthquake (Mw 9.0). J Geophys Res 117:B02311. doi:10.1029/2011JB008750

Okubo M, Saiga A, Nakashima T (2011) The 2011 off the Pacific coast of Tohoku Earthquake inferred from TRIES observation network. Japan Geoscience Union Meeting 2011, MIS036-P27

Simons M, Minson SE, Sladen A, Ortega F, Jiang J, Owen SE, Meng L, Ampuero J-P, Wei S, Chu R, Helmberger DV, Kanamori H, Hetland E, Moore AW, Webb FH (2011) The 2011 magnitude 9.0 Tohoku-oki earthquake: mosaicking the megathrust from seconds to centuries. Science 322:1421. doi:10.1126/science.1206731

Suzuki W, Aoi S, Sekiguchi H, Kunugi T (2011) Rupture process of the 2011 Tohoku-Oki mega-thrust earthquake (M9.0) inverted from strong-motion data. Geophys Res Lett 38:L00G16. doi:10.1029/2011GL049136

Urabe T, Tsukada S (1992) win - a workstation program for processing waveform data from microearthquake networks. Prog Abstr Seismol Soc Jpn 2:331

Twenty-Two Years of Combined GPS Products for Geophysical Applications and a Decade of Seismogeodesy

Yehuda Bock, Sharon Kedar, Angelyn W. Moore, Peng Fang, Jianghui Geng, Zhen Liu, Diego Melgar, Susan E. Owen, Melinda B. Squibb, and Frank Webb

Abstract

Continuous GPS monitoring on global and regional scales has become an essential component of geophysical and meteorological infrastructure for studying fundamental Earth processes that drive natural hazards, weather, and climate. The NASA-funded "Solid Earth Science ESDR System (SESES)" project provides long-term Earth Science Data Records (ESDRs), the result of a combined solution of independent GPS analyses by the Jet Propulsion Laboratory and Scripps Institution of Oceanography using a common source of metadata archived at the Scripps Orbit and Permanent Array Center. The project has now produced up to twenty-two years of consistent, calibrated and validated ESDR products for over 2,500 GPS stations in western North America, other plate boundaries, and global networks. We describe the methodology to estimate a single set of time series with 24-h resolution of station displacements in north, east and vertical components. This is followed by a time series analysis for velocities, coseismic offsets, postseismic deformation, seasonal signals and nuisance offsets, primarily due to GPS antenna changes. Realistic one-sigma velocity are on the order of 0.03–0.05 mm/year in horizontal components and 0.1–0.3 mm/year in the vertical based on time series of 10–20 year duration. We present examples of time series that are well modeled by this parameterization and of time series that exhibit residual transient motions exhibiting episodic tremor and slip (ETS) processes. The project also catalogs seismic displacement and velocity waveforms estimated for a set of historical earthquakes in Japan and the U.S. through a seismogeodetic combination of GPS and collocated strong-motion accelerometers.

Keywords

Continuous GPS • GPS data products • GPS time series • Seismogeodesy

1 Project Overview

The NASA-funded "Solid Earth Science ESDR System (SESES)" project provides long-term Earth Science Data Records (ESDRs), the result of a combined solution of independent GPS analyses by the Jet Propulsion Laboratory and Scripps Institution of Oceanography using a common source of metadata archived at the Scripps Orbit and Permanent Array Center. The project produces and archives a hierarchy of data products including station displacements, troposphere zenith delay parameters for calibrating atmospheric delay errors in Interferometric Synthetic Aperture Radar analysis, and precipitable water vapor time series for use in Probable Maximum Precipitation studies, historical weather event analysis, and studies of long-term water vapor trends for climate studies. Here we focus on the station displacement time series that span up

Y. Bock (✉) • P. Fang • J. Geng • D. Melgar • M.B. Squibb
Scripps Institution of Oceanography, La Jolla, CA, USA
e-mail: ybock@ucsd.edu

S. Kedar • A.W. Moore • Z. Liu • S.E. Owen • F. Webb
Jet Propulsion Laboratory, Pasadena, CA, USA

M. Hashimoto (ed.), *International Symposium on Geodesy for Earthquake and Natural Hazards (GENAH)*,
International Association of Geodesy Symposia 145, DOI 10.1007/1345_2016_220

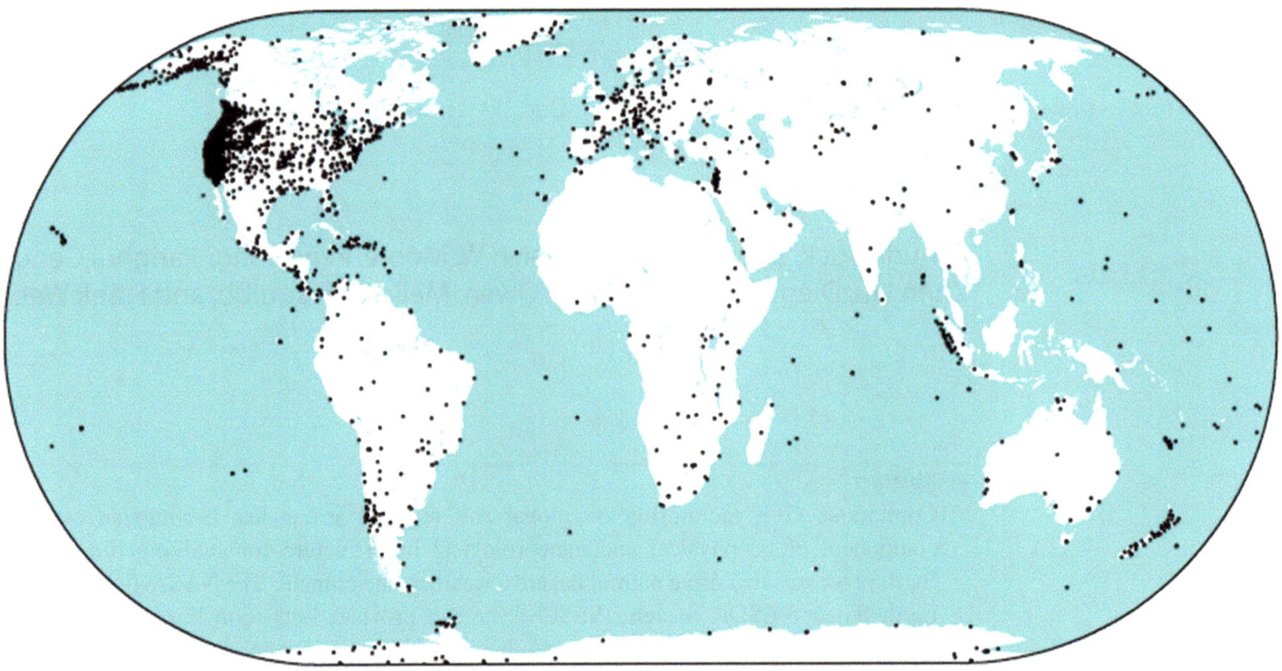

Fig. 1 Global distribution of stations independently processed on a daily basis by Caltech's Jet Propulsion Laboratory and Scripps Institution of Oceanography's Orbit and Permanent Array Center. The displacement time series are combined to support a hierarchy of combined data products (Fig. 2). The high-level data products and the IT data exploration and distribution mechanisms were developed through the NASA-funded REASoN program (2001–2005), and were tested and implemented for the 250-station Southern California Integrated GPS Network (SCIGN), and subsequently for the Plate Boundary Observatory and other regional networks

to twenty-two years for up to about 2,500 GPS stations in western North America, other plate boundaries, and globally distributed (Fig. 1). The displacements are operationally provided at a resolution of 24 h; in addition, once per second displacements and seismic velocity waveforms (not to be confused with station velocities primarily due to tectonic motion and subsidence) for select historical earthquakes are also provided. The daily time series of raw displacements is modeled for 3-D station velocities, coseismic displacements, postseismic parameters, seasonal terms (annual and semiannual), and non-tectonic offsets primarily due to changes in antenna models from different manufacturers. The residual time series may contain remaining transient signals; a few examples are provided in this paper.

2 Data Products (ESDRs) Generation

The process flow of the SESES data product generation system is shown in Fig. 2. At its core, the process relies on a routinely maintained unified database at SOPAC, including relevant station metadata. The left side of the figure shows the basics of our technical approach and methodology for products that are derived from standard 15–30 s-sampled continuous GPS data; the right side shows the methodology for data products derived from high-rate GPS (typically, 1 Hz) and seismic (typically, 100 Hz accelerometer) data. The center of the figure indicates that all ESDRs flow to our extensible, and widely-used GPS Explorer data portal, a personalized on-the-fly interactive research environment (http://geoapp.ucsd.edu/).

The products are continuously improved and updated and extended every week to maintain a consistent long-term data record. As shown in Fig. 2, this is an iterative process that includes validation of relevant metadata, automatic and manual quality control for the individual time series, identification of instrumental offsets, optimal combination of the individual time series using JPL's st_filter software (http://qoca.jpl.nasa.gov/), choice of proper spatio-temporal filtering to remove common mode error (Dong et al. 2006), appropriate fitting/modeling of the displacement time series (Nikolaidis 2002) using JPL's PCA and analyze_tseri software and SOPAC's administrator interface to perform detailed quality control and to improve time series models. The process begins with a new week of daily SOPAC "h-files" produced through GAMIT software (http://www-gpsg.mit.edu/~simon/gtgk/) and JPL "STACOV" files independently produced through GIPSY software (https://gipsy-oasis.jpl.nasa.gov/) containing station adjustments and covariance matrices. Each week, the *entire* time series of h-files and STACOV files are combined by the st_filter software to improve the model. The updated time

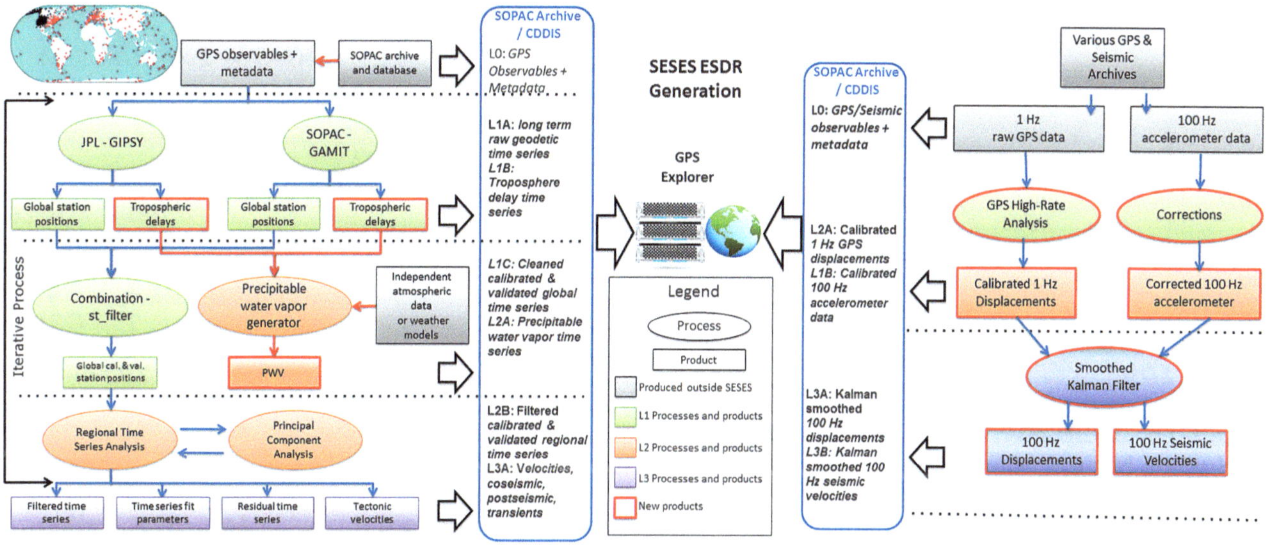

Fig. 2 Schematic of SESES system

series model is then used to generate updated true-of-date coordinates through SOPAC's SECTOR utility (http://sopac.ucsd.edu/processing/coordinates/sector.shtml) that can be used as input to the next individual GAMIT and GIPSY analyses, as well as the next st_filter and analyze_tseri (ATS) iteration. This optimal approach is necessary in the presence of significant seismic events with large coseismic offsets and significant postseismic deformation. Examples are given in the next section. Through this iterative process we are essentially maintaining our own global terrestrial reference frame, using SOPAC orbits and earth orientation parameters (EOP). Both JPL and SIO submit products to the International GNSS Service (IGS) by minimizing the deviations of station coordinates with respect to a preferred set of global coordinates and velocities, currently in the International Terrestrial Reference Frame 2008 (http://itrf.ensg.ign.fr/ITRF_solutions/2008/ITRF2008.php), for a core of about 300 globally distributed stations. Note that the SESES process is independent of our individual IGS products processes. By obtaining an optimal combination of independent JPL and SIO daily solutions and fitting the resulting time series for typical earthquake cycle signatures, the residual time series then can be mined for transient deformation including tectonic sources such as episodic tremor and slip observed at subduction zones (e.g., Rogers and Dragert, 2003), and non-tectonic effects resulting from, for example, hydrologic and anthropogenic sources (e.g., Dong et al. 2002; Argus et al. 2014).

We are selective in the choice of GPS stations to include in our analysis. We prefer to choose stations with reliable metadata and stable monumentation (this is the case for most stations in Western North America) in order to better resolve physically meaningful parameters and to root out systematic effects. Occasionally we will exclude a station altogether or a portion of the station's data when it is clear that there is a documented instrumental or site problem, or large gaps. We also choose stations that are processed by other efforts such as the UNAVCO-operated Plate Boundary Observatory (PBO) in the U.S. and those analyzed by our IGS analysis center counterparts. Over the last two years we have added nearly 1,000 stations to our joint analysis. Other efforts use our ESDRs, for example JPL's ARIA project uses our true-of-date stations coordinates to correct orbit-error induced long wavelength features in InSAR studies of surface deformation over a region of interest. We also would like, as resources permit, to include data products from other analysis efforts (e.g., PBO, University of Nevada/Reno) either in the existing combination or just to display the data products in our GPS Explorer interface.

3 Examples of SESES Data Product Applications

In general, we achieve, based on time series of 10–20 year duration, realistic one-sigma velocity uncertainties on the order of 0.03–0.05 mm/year in horizontal components and 0.1–0.3 mm/year in vertical components, with consideration of temporal correlations (Zhang et al. 1997; Williams et al. 2004) and with insignificant spatial cross correlations.

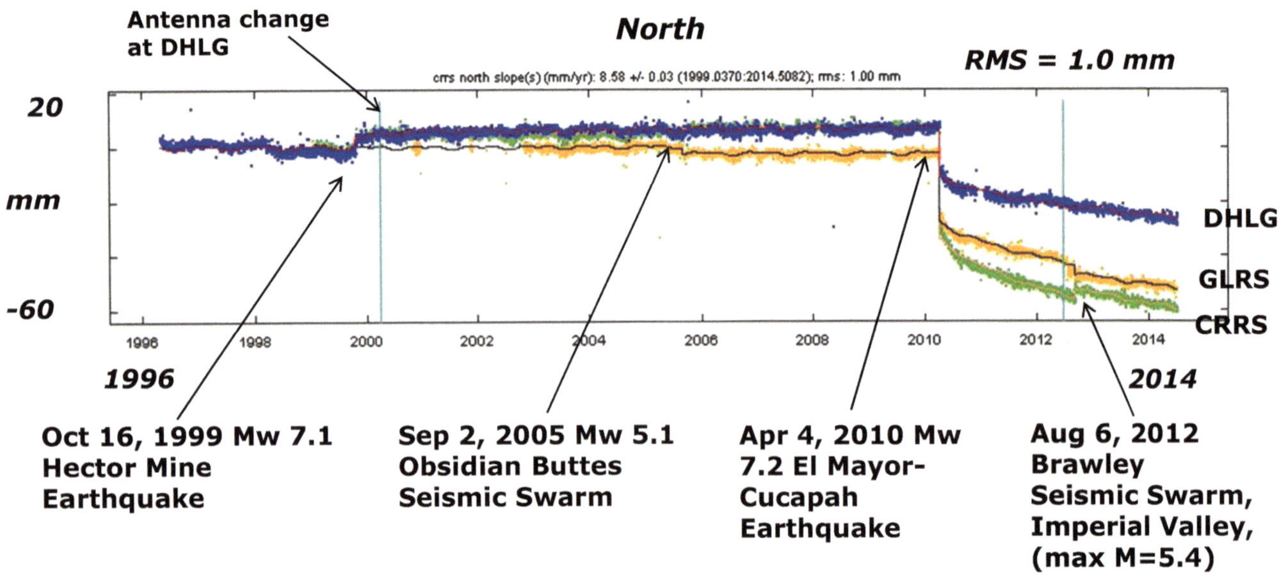

Fig. 3 Daily combined north component time series from three stations in southern California reveal earthquake cycle (interseismic, coseismic, postseismic) deformation for two significant strike slip earthquakes, and two seismic swarms

3.1 Earthquake Cycle Deformation

One of the primary goals of the SESES project is to support studies of plate boundary deformation. We focus on the Pacific Rim (Western North America, New Zealand, Japan, South America), Sumatra, and the Caribbean (Fig. 1). In Figs. 3 and 4 we show typical time series.

3.2 Transient Deformation: Detection of Episodic Tremor and Slip

Often the term "transient deformation" (e.g., Lohman and Murray 2013) is applied to a more restricted class of aseismic motions (including "slow earthquakes" and "slow slip") that deviate from the basic earthquake cycle model shown in Figs. 3 and 4. One of the earliest discoveries of transient deformation and an active area of research is episodic tremor and slip (ETS) identified first in the Cascadia subduction zone (Rogers and Dragert 2003) (Fig. 5). SESES products provide basic data for ETS research.

3.3 Seismogeodesy: Rapid modeling of the 2011 Mw 9.0 Tohoku-oki Earthquake

The SESES project produces seismic and velocity waveforms, typically at 100 Hz sampling rate, for significant historical earthquakes, from an optimal combination of high-rate (1–5 Hz) GPS and very-high-rate (100 Hz) accelerometer data (Bock et al. 2011; Geng et al. 2013). Rapid characterization of finite fault geometry and slip for large earthquakes is important for mitigation of seismic and tsunamigenic hazards (Melgar and Bock 2013). Saturation of near-source weak motion (Crowell et al. 2013) and problematic integration of strong-motion data into displacements make this difficult, particularly in real time (Melgar et al. 2013a). Combining GPS and accelerometer data to estimate seismogeodetic displacement and velocity waveforms overcomes these limitations. Currently our database of earthquake seismogeodetic waveforms includes the 2003 Mw 8.3 Tokachi-oki, Japan earthquake (Crowell et al. 2012); 2010 El Mayor-Cucapah, northern Baja California, Mexico, earthquake (Crowell et al. 2012); 2011 Mw 9.0 Tohoku-oki, Japan earthquake (Melgar et al. 2013b); 2012 Brawley seismic swarm, southern California, U.S (Geng et al. 2013); 2014 Napa earthquake, northern California.

4 Data Accessibility

Users can access and visualize SESES-combination displacement time series and associated metadata, levels 1–3 ESDR products through an interactive data portal GPS Explorer (http://geoapp.ucsd.edu/). The time series is permanently archived at NASA's Crustal Dynamics Data Information System (http://cddis.nasa.gov).

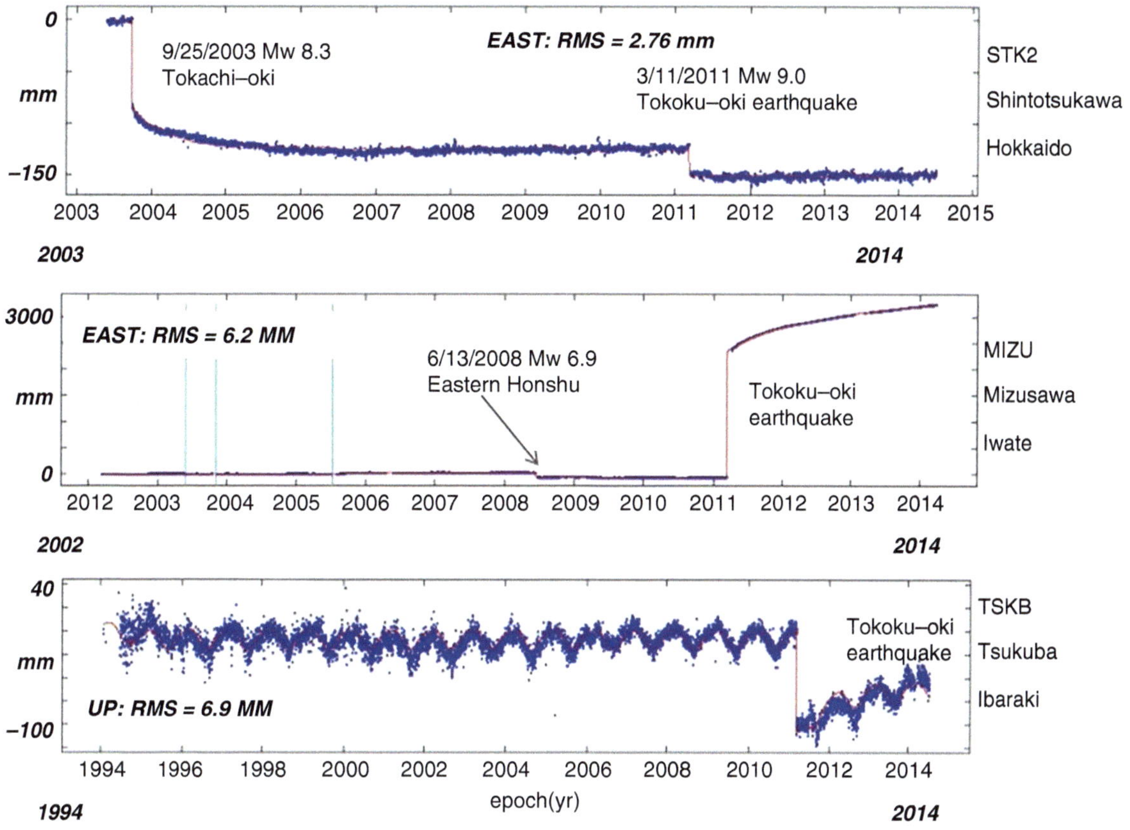

Fig. 4 Daily combined east and up component time series from three stations in Japan reveal aspects of the earthquake cycle (interseismic, coseismic, postseismic) for three significant subduction zone earthquakes

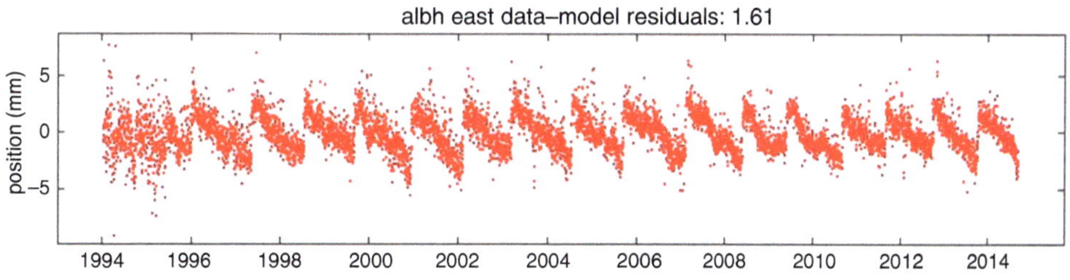

Fig. 5 SESES 20-year east residual time series for station ALBH in northwest Cascadia clearly shows the effect of ETS, with peak-to-peak amplitudes of less than 10 mm and periodicity of less than a year (17 events over two decades)

References

Argus DF, Fu Y, Landerer FW (2014) Seasonal variations in total water storage in California inferred from GPS observations of vertical land motion. Geophys Res Lett. doi:10.1002/2014GL059570

Bock Y, Melgar D, Crowell BW (2011) Real-time strong-motion broadband displacements from collocated GPS and accelerometers. Bull Seismol Soc Am 101:2904–2925. doi:10.1785/0120110007

Crowell BW, Bock Y, Melgar D (2012) Real-time inversion of GPS data for finite fault modeling and rapid hazard assessment. Geophys Res Lett 39, L09305. doi:10.1029/2012GL051318

Crowell BW, Melgar D, Bock Y, Haase JS, Geng J (2013) Earthquake magnitude scaling using seismogeodetic data. Geophys Res Lett 40:1–6. doi:10.1022/2003GL058391

Dong D, Fang P, Bock Y, Cheng MK, Miyazaki S (2002), Anatomy of apparent seasonal variations from GPS-derived site position time series. J Geophys Res. 107(B4):doi:10.1029/JB2001JB000573

Dong D, Fang P, Bock Y, Webb F, Prawirodirdjo L, Kedar S, Jamason P (2006) Spatiotemporal filtering using principal component analysis and Karhunen–Loeve expansion approaches for regional GPS network analysis. J Geophys Res 111, B03405. doi:10.1029/2005JB003806

Geng J, Bock Y, Melgar D, Crowell BW, Haase JS (2013) A new seismogeodetic approach applied to GPS and accelerometer observations of the 2012 Brawley seismic swarm: Implications for earthquake early warning. Geochem Geophys Geosyst 14. doi:10.1002/ggge.20144

Herring TA, King RW, McClusky SC (2010) Introduction to GAMIT/GLOBK release 10.4. Massachusetts Institute of Technology, Cambridge

Lohman RB, Murray JR (2013) The SCEC geodetic transient–detection validation exercise. Seismol Res Lett 84(3):419–425

Melgar D, Bock Y (2013) Near-field tsunami models with rapid earthquake source inversions from land and ocean based observations: the potential for forecast and warning. J Geophys Res 118. doi:10.1102/2013JB010506

Melgar D, Bock Y, Sanchez D, Crowell BW (2013a) On robust and reliable automated baseline corrections for strong motion seismology. J Geophys Res 118:1–11. doi:10.1002/jgrb.50135

Melgar D, Crowell BW, Bock Y, Haase JS (2013b) Rapid modeling of the 2011 Mw 9.0 Tohoku-oki earthquake with seismogeodesy. Geophys Res Lett 40:1–6. doi:10.1002/grl.50590

Nikolaidis R (2002) Observation of geodetic and seismic deformation with the global positioning system, PhD Thesis. University of California San Diego

Rogers G, Dragert HH (2003) Episodic tremor and slip on the Cascadia subduction zone: the chatter of silent slip. Science 300(5627):1942–1943

Williams SDP, Bock Y, Fang P, Jamason P, Nikolaidis RM, Prawirodirdjo L, Miller M, Johnson DJ (2004) Error analysis of continuous GPS position time series. J Geophys Res 109, B03412. doi:10.1029/2003JB0022741

Zhang J, Bock Y, Johnson H, Fang P, Genrich J, Williams S, Wdowinski S, Behr J (1997) Southern California Permanent GPS geodetic array: error analysis of daily position estimates and site velocities. J Geophys Res 102:18035–18055

Earthquake Cycles on the San Andreas Fault System in Northern California

M. Burak Yıkılmaz, Don L. Turcotte, Olga Beketova, Louise H. Kellogg, and John B. Rundle

Abstract

The 24 August, 2014, M=6.0 South Napa earthquake was the first earthquake to significantly affect the populous San Francisco Bay region since the 1989 Loma Prieta earthquake. The Napa earthquake has resulted in a new interest in earthquake risk on the San Andreas Fault (SAF) system north of San Francisco. The deformation in this region is dominantly right lateral shear between the rigid Pacific Plate and the rigid Sierra-Nevada-Central-Valley Plate. GPS observations are well approximated by a uniform shear strain across this 100 km wide zone. This zone is recognized to be a "slab window" with a relatively thin lithosphere. In this paper, we hypothesize that this lithosphere is composed of a 12 km thick brittle elastic upper lithosphere and a 15 km thick viscoplastic lower lithosphere. We attribute the observed near uniform surface strain to flow in the viscoplastic zone. Deformation in the brittle upper lithosphere results in seismicity. A substantial fraction, about 20 mm year^{-1}, of the 32 mm year^{-1} of the right lateral deformation takes place on the SAF. However, the SAF is not parallel to the motion of the bounding rigid plates. This geometrical incompatibility requires distributed deformation across the 100 km wide zone. A fraction of this distributed deformation takes place on three relatively well defined strike-slip fault zones to the east of the SAF as well as on other faults. We suggest that when a large earthquake occurs, the localized stress concentrations are relaxed by flow in the viscoplastic zone. We give a detailed description of this process during and following the 1906 earthquake and show the process is consistent with observations. We also relate our tectonic model to the present distribution of seismicity in our study area.

Keywords

Earthquakes • GPS • Northern California • Tectonics • Uniform shear

1 Introduction

The San Andreas Fault (SAF) is a primary boundary between the Pacific Plate (PP) and the North American Plate (NAP). However, the associated deformation occurs over much of the western United States. In this paper, we focus our attention on the SAF system in northern California. We consider this region for several reasons. The geometry is relatively simple since the zone of deformation is bounded on the west by the near rigid PP and on the east by the near rigid Sierra-Nevada Central-Valley Plate (SNCVP) (Fig. 1a). The deformation in this zone is transpression with some 32 mm year^{-1} of right lateral strike-slip displacement and about 3 mm year^{-1} of compression. There are many strike-slip faults in the zone with parallel ranges that absorb the compressional component. The strike-slip displacement is primarily accommodated on the SAF with 16–27 mm year^{-1} displacement.

M.B. Yıkılmaz (✉) • D.L. Turcotte • O. Beketova • L.H. Kellogg • J.B. Rundle
Department of Earth & Planetary Sciences, University of California Davis, One Shields Ave, Davis, CA 95616, USA
e-mail: mbyikilmaz@ucdavis.edu; dlturcotte@ucdavis.edu; obeketova@ucdavis.edu; kellogg@ucdavis.edu; jbrundle@ucdavis.edu

M. Hashimoto (ed.), *International Symposium on Geodesy for Earthquake and Natural Hazards (GENAH)*, International Association of Geodesy Symposia 145, DOI 10.1007/1345_2015_203

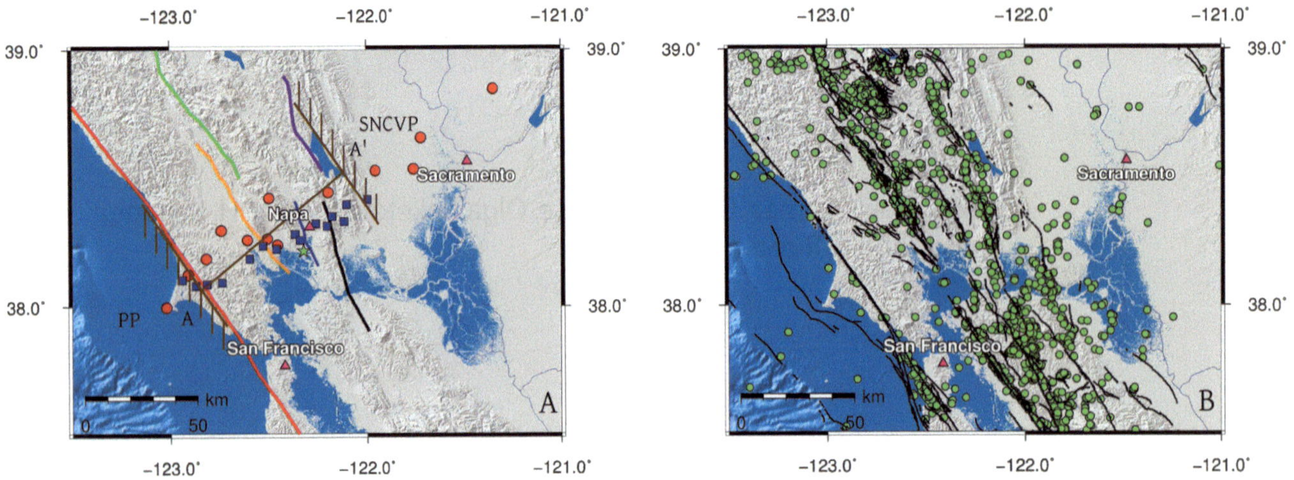

Fig. 1 (**a**) Illustration of the transect A-A' upon which we base our study of the seismic cycles on the San Andreas fault system. The 100 km wide zone of deformation is bounded by the rigid Pacific Plate (PP) and the rigid Sierra Nevada Central Valley Plate (SNCVP). The major faults in the zone of deformation are shown: San Andreas (*red*), Maacama (*green*), Rodgers Creek (*orange*), West Napa (*blue*), Bartlett Springs (*purple*), and Green Valley (*black*). Also shown are continuous GPS stations (*red circles*) and campaign GPS stations (*blue squares*). *Green star* along the West Napa fault indicates the epicenter of the August 24, 2014 earthquake. (**b**) Distribution of $M > 2$ earthquakes in the study region during the period 2000–2010

Strike-slip deformation is also attributed to three sub-parallel fault systems to the east of the San Andreas (Freymueller et al. 1999; D'Alessio et al. 2005). The first of these is the Rodgers Creek Fault (RCF). This is a well-defined fault with an 6–11 mm year^{-1} displacement. The RCF is the northern extension of the Hayward fault to the south. Further to the east, strike-slip deformation is documented on the West Napa (WNF) and the Concord-Green Valley (CGVF) faults. These are not well defined geologically and the displacements are poorly known. These faults are suggested to be the northern extension of the Calaveras Fault to the South. The last great earthquake in the study region was the 18 April, 1906 San Francisco earthquake on the SAF. Felt intensity studies indicated that this earthquake had a magnitude of about $M = 8.0$. Displacements averaging about 4 m extended over some 200 km including our study region. Triangulation data are available to assess both the coseismic deformation and the subsequent stress relaxation process (Thatcher 1974, 1975; Matthews and Segall 1993). The first moderate earthquake to occur in our study region since the 1906 earthquake was the 24 August, 2014, $M = 6.0$ South Napa earthquake. This earthquake occurred on a branch of the West Napa fault and had right-lateral, strike-slip surface deformation extending over some 15 km with a maximum offset of about 460 mm.

Strain rates across our zone of deformation are obtained using GPS data. We utilize observations obtained on the transect A-A' across the zone of deformation as illustrated in Fig. 1a. This transect is chosen because of the availability of relevant data in the region. Specifically the use of strain data associated with the 1906 coseismic deformation and subsequent strain relaxation as well as the current GPS strain field. This transect also includes the rupture of the 2014 South Napa earthquake.

The measured velocities at the stations shown in Fig. 1a are given in Fig. 2. These velocities are parallel to the trend of the SAF and are relative to the stable NAP. Included are both continuous GPS observations from the beginning of the records in the 2000s up to 1 January 2011 and campaign GPS observations during various intervals during the 1990s. We interpret this observed behavior to reflect a region of near uniform shear strain between the rigid PP and the rigid SNCVP. This region is a 100 km wide zone extending across the Coast Ranges from 20 km west of the SAF to 80 km east of the SAF. Based on the linear correlation given in Fig. 2, we take the velocity across the 100 km wide deformation zone to be 32 mm year^{-1} and the uniform shear strain to be $\dot{\epsilon} = 0.32\,\mu$strain year^{-1}. A number of authors have previously recognized the near linear strain field given by the GPS data (D'Alessio et al. 2005; Freymueller et al. 1999; Prescott et al. 2001; Smith and Sandwell 2003; Johnson and Segall 2004; Savage et al. 2004; Jolivet et al. 2009). Several of these authors have utilized block models for the earthquake cycle on independent faults to explain the strain field in this region. Various block models and a wide range of parameter values haven been used by these authors. For comparison with the linear strain field in Fig. 2, we derive a strain field expected for a block model. We consider a block model that extends a uniform vertical strike-slip fault to infinite depth. Beneath a locking depth d, a uniform slip velocity V_a is applied across the fault. The earthquake rupture occurs above the locking depth. The surface velocity distribution $V_s(x)$ during strain

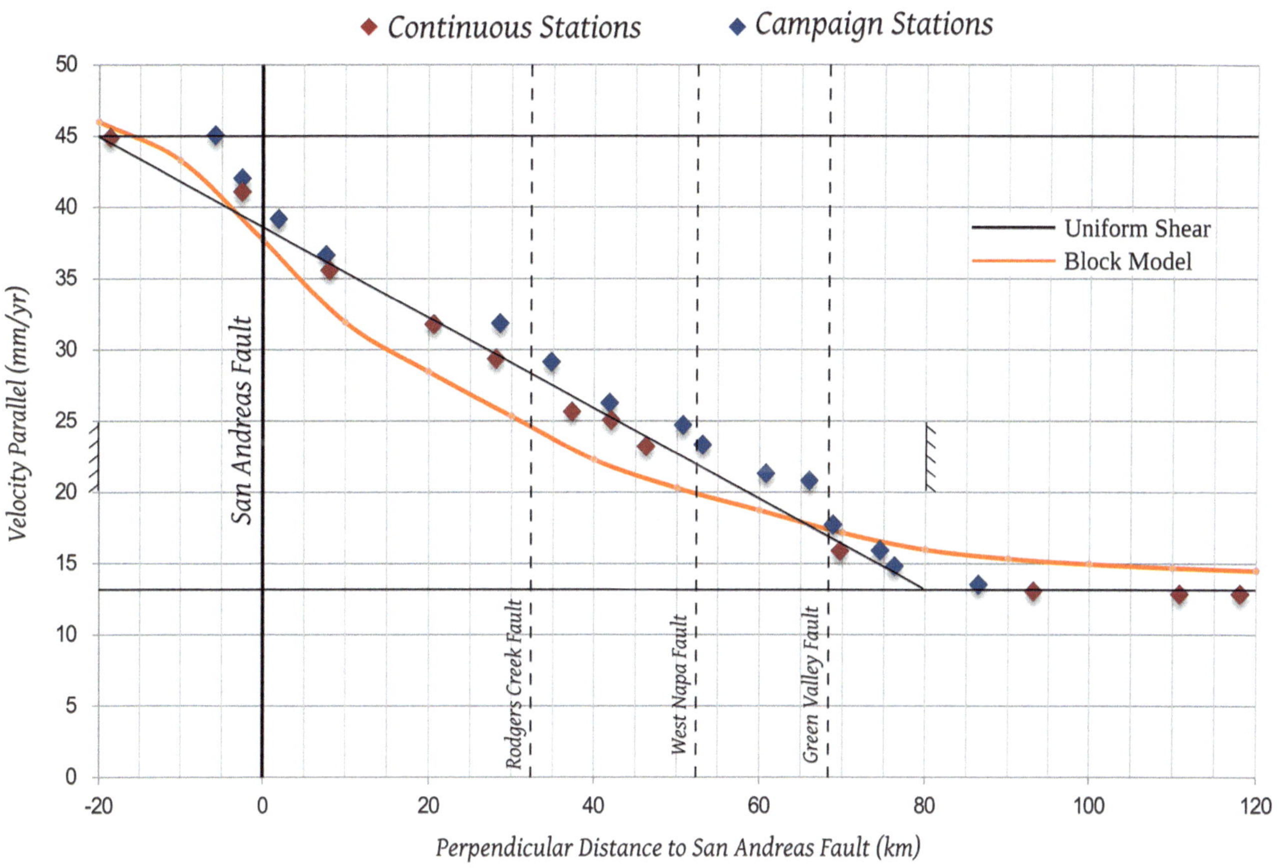

Fig. 2 Velocities of the GPS stations illustrated in Fig. 1. These velocities are relative to the stable North American Plate. We give the velocity components parallel to the SAF as a function of the distance from the SAF. Also shown is the linear correlation with the data that we use in this paper and the velocity distribution for our block model from Eq. (1)

accumulation is given by Freymueller et al. (1999):

$$V_s = V_a(\frac{1}{2} + \frac{1}{\pi}\tan^{-1}(\frac{x - x_f}{d})) \tag{1}$$

where x_f is the location of the fault and x-x_f is the distance normal to the fault. The surface velocity distribution depends on the prescribed slip velocity V_a and the locking depth d. The solution given in Eq. (1) is for an elastic half-space with uniform properties but does not depend on the elastic rigidity.

For our block model we consider four parallel strike-slip faults separated by three internal blocks with two boundary blocks. The partitioning of slip between these four faults depends on geological data. Detailed studies of this distribution of slip have been obtained recently in order to constrain the Uniform California Earthquake Rupture Forecast, version 3 (UCERF3). Estimated slip rates with bounds for the four faults that we study have been given by Dawson and Weldon (2013). These are, with their comments;

1. San Andreas (North Coast). Best estimate: 24 mm year^{-1}, bounds (16–27 mm year^{-1}). UCERF3 slip rate based on reported geologic rates.
2. Rodgers Creek. Best estimate: 9 mm year^{-1}, bounds (6–11 mm year^{-1}). UCERF3 slip rate based on reported geologic rate and rate on Hayward Fault. Note that slip rate is tapered in the area of overlap with the Maacama Fault so that the rate along this system is not doubly counted.
3. West Napa. Best estimate: 1 mm year^{-1}, bounds (1–5 mm year^{-1}). UCERF3 adopts the UCERF2 assigned rate because there is no direct slip rate available. Slip rate bounds expanded based on mapping by Clahan (2011) that suggests fault may be more active than previously thought.
4. Green Valley. Best estimate: 4 mm year^{-1}, bounds (2–9 mm year^{-1}). Rate based on reported geologic rate and USGS category, as well as the rate on the adjacent Concord Fault. Upper bound assigned based on upper bound on Concord Fault.

Assuming all deformation across the deformation zone occurs on the four faults (required by the block model), the total slip given by UCERF3 is 38 mm year^{-1}. This compares with our preferred value of 32 mm year^{-1} from the GPS data given in Fig. 2.

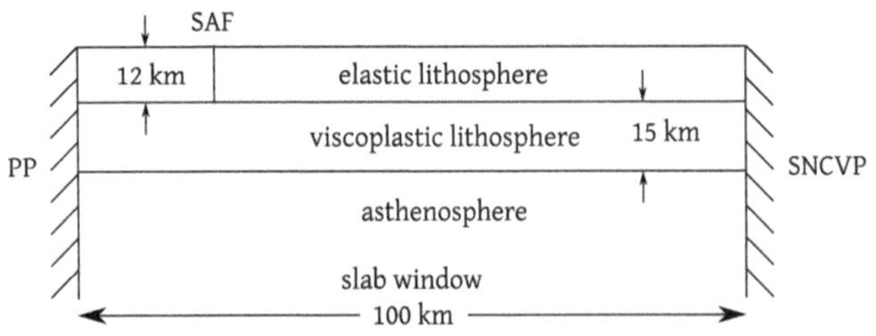

Fig. 3 Illustration of our model for the behavior of deformation in the slab window between the rigid PP and the rigid SNCVP (transect A-A' in Fig. 1)

In addition to the preferred slip velocities given above, we take the locking depth of each fault to be d = 12 km. This is typical of values selected by previous authors (i.e. Smith and Sandwell 2003) and is consistent with the maximum depth of seismicity. The San Andreas Fault is at $x_f = 0$ km, the Rodgers Creek Fault at $x_f = 32$ km, the West Napa Fault at $x_f = 52$ km, and the Green Valley fault at $x_f = 68$ km. The velocity field across the zone of deformation is obtained by solving Eq. (1) for each of the four faults and adding the results (appropriate for linear elasticity).

The velocity distribution obtained by our block model is given in Fig. 2. There are clearly significant difference between the block model results and the observed linear strain field. The primary cause for this difference is the dominant role played by the San Andreas Fault. A slip velocity of 24 mm year^{-1} out of a total slip of 38 mm year^{-1} (63% of the total). The San Andreas velocity field dominates the total strain field causing a significant deviation from the linear observed behavior. In order for the block model to be consistent with the observed near uniform strain, a significant reduction of the slip velocity on the San Andreas Fault would be required. Considering the uncertainties this is possible, but we consider it to be highly unlikely.

The block model described above implies that the strain accumulation and release cycles on the three major faults are only weakly coupled. The observed uniform strain field requires equal partitioning of stress accumulation to the three faults. The purpose of this paper is to propose a model for the region that explains the partitioning of coseismic slip on the three faults with uniform strain accumulation. Chéry (2008) also argued against the block model and concluded that the uniformity of the shear strain is evidence that the 100 km wide region between the two rigid plates is relatively soft.

2 Proposed Model

This paper utilizes the model illustrated in Fig. 3. We postulate a 12 km thick elastic lithosphere based on the depth of earthquakes in the deformation zone. In this brittle elastic lithosphere, deformation is dominated by displacements on faults. We further postulate a 15 km thick viscoplastic lower lithosphere (lower crust). The near-uniform deformation in this zone results in the near-uniform shear strain illustrated in Fig. 2. Beneath the lithosphere, the mantle asthenosphere is sufficiently weak to provide little resistance to the shearing motion. The thin brittle lithosphere responds to the near uniform shear deformation imposed by the bounding rigid plates and the near uniform shear strain in the viscoplastic lower lithosphere. Because of the geometrical incompatibility between the orientation of the SAF and the imposed deformation at the rigid plate boundaries, the brittle lithosphere is subjected to distributed internal deformation. Some of the deformation occurs on the three faults described above. When an earthquake occurs, like the 2014 South Napa earthquake, the localized coseismic deformation is redistributed by viscoplastic deformation of the lower crust.

The thin lithosphere and the deformable region beneath are attributed to the recent subduction of the Farallon Plate (FP) beneath the SNCVP in northern California. The separation of the FP from the PP left a gap in the lithosphere. This gap, known as a slab window, extends from the PP to the SNCVP as illustrated in Fig. 1. The GPS data given in Fig. 2 clearly illustrate the boundaries of the slab window and the near uniform strain accumulation to the West. The eastern boundary is at 80 km from the San Andreas Fault as shown. The SNCVP is basically rigid, implying a thick lithosphere. The eastern boundary is at the San Andreas Fault. Strain accumulation associated with earthquakes on the San Andreas Fault is predominantly to the east in the thinner lithosphere. Furlong and Schwartz (2004) have presented details of the slab window hypothesis.

The slab window is characterized by high heat flow and volcanism. Lachenbruch and Sass (1980) demonstrated high heat flow in the slab window region. The thin lithosphere hypothesis is further solidified by studies of the extensive recent volcanism throughout the slab window region (Dickinson 1997).Our objective is to present a model for the earthquake cycle in the study region. Our study of the earthquake cycle begins with the coseismic behavior associated with the 1906 San Francisco earthquake. Thatcher (1974, 1975) and Matthews and Segall (1993) used surface and geodetic triangulation data to quantify the surface rupture during the 1906 earthquake. Based on this work we take the coseismic surface displacement to be $\Delta w_{cs} = 4$ m and the depth of rupture to be 12 km. Consistent with these values and also with global compilations of stress drop in major earthquakes (Kanamori and Anderson 1975) we take the mean stress drop in the 1906 earthquake to be $\Delta\sigma_{cs} = 5$ MPa. These values are illustrated in Fig. 4.

We assume the shear stress across the 100 km wide window had increased to a uniform value $\Delta\sigma = 5$ MPa just prior to the 1906 earthquake. This assumption is based on the uniform increase in shear strain illustrated in Fig. 2 and

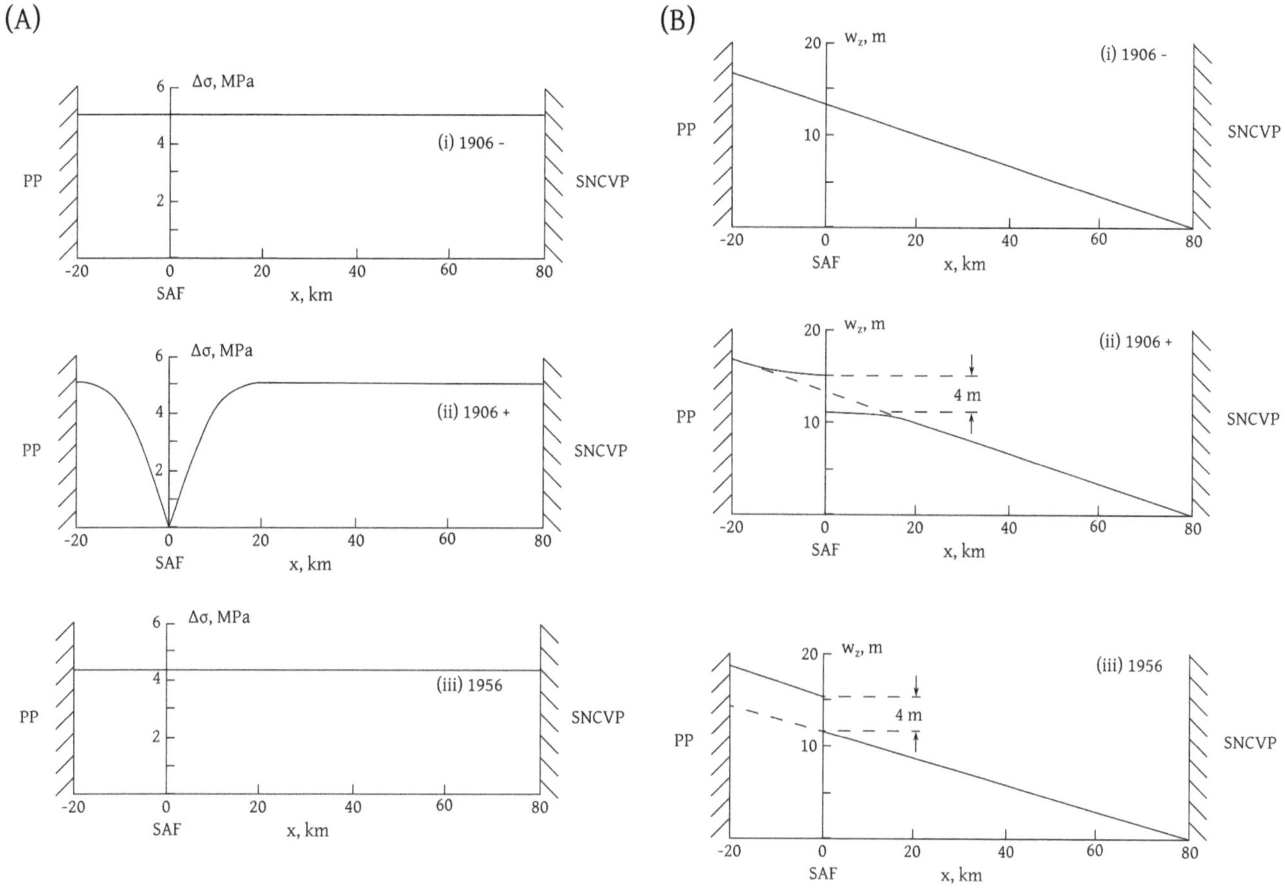

Fig. 4 An overview of our proposed earthquake cycle on the SAF system in Northern California. In (**a**), we show the dependence of the excess stress $\Delta\sigma$ during the cycle on the distance across the slab window x from the bounding PP on the southwest to the bounding SNCVP on the north east. In (**b**), we show the dependence of the surface displacement w during the cycle on the distance across the slab window x. The PP is moving at a constant velocity 32 mm $year^{-1}$ relative to the fixed SNCVP. The distribution of excess stress $\Delta\sigma$ and displacement w are given; (*i*) just before the 1906 earthquake, (*ii*) just after the 1906 earthquake, and (*iii*) in 1956 after post seismic relaxation

shown in Fig. 4a(i). The total shear stress σ would require the addition of a uniform background stress, which is not important for our analysis since we consider only the stress (and strain) variability during an earthquake cycle. Taking $\Delta\sigma = 5$ MPa and the shear modulus $G = 30$ GPa, the corresponding constant shear strain $\Delta\epsilon = \Delta\sigma / G = 1.67 * 10^{-4}$. This constant shear strain implies a linear increase in surface displacement w across the slab window. The total surface displacement is $w = 16.7$ m as illustrated in Fig. 4b(i). During the earthquake the mean stress on the fault drops from $\Delta\sigma = 5$ MPa to $\Delta\sigma = 0$ MPa [see Fig. 4a(ii)]. Geodetic studies before and after the earthquake restrict the depth of rupture to about 12 km (Thatcher 1974, 1975; Matthews and Segall 1993). Studies of the distribution of surface stress associated with a strike-slip fault rupture have been given by Turcotte and Schubert (2014). The region that has a significantly lower stress immediately after the earthquake and its aftershocks sequence has a width of about 30 km as illustrated in Fig. 4a(ii). After the earthquake there is a large variation of the stress in the lithosphere across the slab window (5 MPa). Because the upper lithosphere is thin, it is expected that flow in the viscoplastic lower lithosphere will lead to a relaxation of this stress difference (Yu et al. 1996; Pollitz et al. 2004, 2008; Pollitz and Nyst 2005; Pollitz and Schwartz 2008). Kenner and Segall (2000, 2003) studied the postseismic deformation following the 1906 earthquake. They utilized triangulation data from the Point Arenas and Point Reyes – Petaluma areas. They inferred that the effective relaxation time for this long-term postseismic deformation was 36 ± 16 years. They also noted that deep afterslip models on a vertical fault could not explain the concentration of this deformation on the northeast side of the SAF. This concentration of the post-seismic deformation to the northeast of the SAF is consistent with stress relaxation across the 100 km wide slab window. Based on the relaxation time given above, we assume that the stress relaxation was completed by 1956. This stress relaxation leads to a reloading of the shear stress on the SAF and a reduction in the shear stress elsewhere. This state of stress and strain is illustrated in Fig. 4a(iii), b(iii).

Since 1956 we assume a steady rate of uniform shear loading associated with the deformation given in Fig. 2. The current observation of near constant shear strain has important implications for the force balance and rigidity of the lithosphere in our study region. (1) The force transmitted through the lithosphere across the region is nearly constant. (2) The rigidity of the lithosphere in the study region is also nearly constant. If this were not the case, shear strain would be variable across the region, not consistent with observations. We attribute this rigidity to the viscoplastic flow in the lower lithosphere (lower crust). The associated deformation of the brittle, elastic upper lithosphere generates earthquakes on a wide range of scales including the 2014 South Napa Earthquake.

To summarize our model based on the profiles given in Fig. 4 we have made the following assumptions:

1. The stress and strain in Fig. 4a(i), b(i) are based on the uniform distribution of strain increase inferred from the GPS observations in Fig. 2.
2. The stress and strain in Fig. 4a(ii), b(ii) are based on an analytic solution for the change in stress and strain from the prescribed 1906 rupture.
3. The stress and strain in Fig. 4a(iii), b(iii) are based on an assumption of the relaxation of stress differences across the slab window by 1956.

3 Seismicity

We next consider the implications of the seismicity in our study region. In Fig. 1b we show that the distribution of M > 2 earthquakes in the study region during the period 2000–2010. This interval was chosen to correspond to the period during which the continuous GPS data were obtained. We also show the epicenter of the 2014 South Napa earthquake. This is the first moderate sized earthquake in the region since 1906. It is clear from the data in Fig. 1b that seismic activity is concentrated on and east of the Maacama-Rodgers Creek fault system with another concentration on the Bartlett Springs-Green Valley fault system. There is distributed seismicity between these two fault zones which including some activity on the West Napa fault system where the 2014 South Napa earthquake occurred.

The current seismicity is clearly concentrated in the eastern half of the deformation zone. We suggest that this is a consequence of the stress relaxation diffusing eastward from the western half of the zone following the 1906 earthquake. Combined with the coseismic stress reduction, this led to a net stress increase in the eastern half of the zone causing the increase in seismicity we see today.

4 Discussion

Geodetic observations and seismicity demonstrate conclusively that the PP and the SNCVP plate act as rigid plates bounding the zone of deformation associated with the San Andreas Fault in our study area. The deformation zone is transpressional with some 32 mm year^{-1} of right lateral strike-slip motion and some 3 mm year^{-1} of compressional motion. These values are quite well defined by GPS observations as we have shown. A substantial fraction of the strike-slip motion is taken up on the SAF. However, the preferred value of 24 mm year^{-1} on this fault has a large uncertainty, 16–27 mm year^{-1}, because the partition of displacements between the San Andreas Fault and the rest of the zone of deformation is not well constrained.

We have shown that the GPS observations across the deformation zone are well approximated by a uniform shear strain. We argue that this uniform shear strain is evidence that the lithosphere in this zone is thin. We propose a model in which this lithosphere has thickness of about 27 km compared with 50–100 km in the bounding rigid plates. We divide this lithosphere in an upper elastic lithosphere with a thickness of 12 km (the seismogenic zone) and lower viscoplastic lithosphere with a thickness of 15 km. The shear deformation in this viscoplastic zone gives the near uniform surface shear illustrated in Fig. 2. The thickness of the lithosphere approximates the thickness of the continental crust in the deformation zone. The mantle is a low viscosity asthenosphere which responds passively to the shear motion in the lithosphere. We have attributed the thin lithosphere to the slab-window hypothesis (Furlong and Schwartz 2004). These authors have presented extensive evidence for an anomalously thin lithosphere in our study region.

At the present time the brittle-elastic upper lithosphere also responds passively to the shear deformation with a near uniform increase in the shear stress. We suggest that this increase in shear stress triggers the background seismicity illustrated in Fig. 1b. We further suggest that this increase in shear stress also triggered the 2014 South Napa earthquake. Based on observations associated with the 1906 earthquake on the SAF, the subsequent stress relaxation, and the present uniform shear we hypothesize the following earthquake cycle for this region: (1) A large earthquake occurs on one of the major faults. The depth of rupture is about 12 km and the width of stress drop is about 30 km. (2) Stress relaxation occurs across the slab window. The stress in the lower lithosphere equilibrates, thereby reloading the fault that ruptured. (3) Uniform shear loading (as shown in Fig. 2) continues until there is again a large earthquake on one of the major faults.

Acknowledgements The GPS and the seismic data were obtained from UNAVCO and Advanced National Seismic System respectively.

References

Advanced National Seismic System (ANSS) (2012). http://www.ncedc.org/anss/catalog-search.html. Accessed 25 Aug 2012

Chéry J (2008) Geodetic strain across the San Andreas fault reflects elastic plate thickness variations (rather than fault slip rate). Earth Planet Sci Lett 269(3–4):351–364. doi:10.1016/j.epsl.2008.01.046

Clahan KB (2011) Paleoearthquake chronology along the Northern West Napa fault zone, Napa County, CA: U.S. Geological Survey Final Technical Report, Award Number 07HQGR0081, 6p

D'Alessio MA, Johanson IA, Bürgmann R, Schmidt DA, Murray MH (2005) Slicing up the San Francisco Bay area: block kinematics and fault slip rates from GPS-derived surface velocities. J Geophys Res 110(B6):B06403. doi:10.1029/2004JB003496

Dawson TE, Weldon RJ (2013) Uniform California earthquake rupture forecast, Version 3 (UCERF3) - The time-independent model, Appendix B, Geological slip rate data and geologic deformation model. USGS Open-File Report 2013-1165

Dickinson WR (1997) Tectonic implications of Cenozoic volcanism in coastal California. Geol Soc Am Bull 109(8):936–954. doi:10.1130/0016-7606(1997)109<0936:OTIOCV>2.3.CO;2

Freymueller JT, Murray MH, Segall P, Castillo D (1999) Kinematics of the Pacific North America plate boundary zone, northern California. J Geophys Res 104(B4):7419–7441. doi:10.1029/1998JB900118

Furlong KP, Schwartz SY (2004) Influence of the Mendocino triple junction on the tectonics of coastal California. Annu Rev Earth Planet Sci 32:403–433. doi: 10.1146/annurev.earth.32.101802.120252

Johnson KM, Segall P (2004) Viscoelastic earthquake cycle models with deep stress-driven creep along the San Andreas fault system. J Geophys Res 109(B10):B10403. doi: 10.1029/2004JB003096

Jolivet R, Bürgmann R, Houlié N (2009) Geodetic exploration of the elastic properties across and within the northern San Andreas fault zone. Earth Planet Sci Lett 288(1–2):126–131. doi:10.1016/j.epsl.2009.09.014

Kanamori H, Anderson DL (1975) Theoretical basis of some empirical relations in seismology. Bull Seismol Soc Am 65(5):1073–1095

Kenner SJ, Segall P (2000) Postseismic deformation following the 1906 San Francisco earthquake. J Geophys Res 105(B6):13195–13209. doi:10.1029/2000JB900076

Kenner SJ, Segall P (2003) Lower crustal structure in northern California: implications from strain rate variations following the 1906 San Francisco earthquake. J Geophys Res 108(B1):2011. doi:10.1029/2001JB000189

Lachenbruch AH, Sass JH (1980) Heat-flow and energetics of the San Andreas fault zone. J Geophys Res 85(NB11):6185–6222. doi:10.1029/JB085iB11p06185

Matthews MV, Segall P (1993) Estimation of depth-dependent fault slip from measured surface deformation with application to the 1906 San-Francisco earthquake. J Geophys Res 98(B7):12153–12163. doi:10.1029/93JB00440

Pollitz FF, Nyst M (2005) A physical model for strain accumulation in the San Francisco Bay region. Geophys J Int 160(1):302–317. doi:10.1111/j.1365-246X.2005.02433.x

Pollitz FF, Schwartz DP (2008) Probabilistic seismic hazard in the San Francisco Bay area based on a simplified viscoelastic cycle model of fault interactions. J Geophys Res 113(B5):B05409. doi:10.1029/2007JB005227

Pollitz FF, Bakun WH, Nyst M (2004) A physical model for strain accumulation in the San Francisco Bay region: stress evolution since 1838. J Geophys Res 109(B11):B11408. doi:10.1029/2004JB003003

Pollitz FF, McCrory P, Svarc J, Murray J (2008) Dislocation models of interseismic deformation in the western United States. J Geophys Res 113(B4):B04413. doi:10.1029/2007JB005174

Prescott WH, Savage JC, Svarc JL, Manaker D (2001) Deformation across the Pacific-North America plate boundary near San Francisco, California. J Geophys Res 106(B4):6673–6682. doi:10.1029/2000JB900397

Savage JC, Gan W, Prescott WH, Svarc JL (2004) Strain accumulation across the Coast Ranges at the latitude of San Francisco. 1994–2000. J Geophys Res 109(B3):B03413. doi:10.1029/2003JB002612

Smith B, Sandwell D (2003) Coulomb stress accumulation along the San Andreas Fault system. J. Geophys Res 108(B6):2296. doi:10.1029/2002JB002136

Thatcher W (1974) Strain release mechanism of 1906 San-Francisco earthquake. Science 184(4143):1283–1285. doi:10.1126/science.184.4143.1283

Thatcher W (1975) Strain accumulation and release mechanism of 1906 San-Francisco earthquake. J Geophys Res 80(35):4862–4872. doi:10.1029/JB080i035p04862.

Turcotte DL, Schubert G (2014) Faulting. In: Geodynamics. Cambridge University Press, New York, pp 361–364

UNAVCO, Plate Boundary Observatory (PBO) (2012). http://pbo.unavco.org/network. Accessed 25 Aug 2012

Yu TT, Rundle JB, Fernandez J (1996) Surface deformation due to a strike-slip fault in an elastic gravitational layer overlying a viscoelastic gravitational half-space. J Geophys Res 101(B2):3199–3214. doi:10.1029/95JB03118

Rheological Structure Beneath NE Japan Inferred from Coseismic Strain Anomalies Associated with the 2011 Tohoku-oki Earthquake (Mw9.0)

Yasuo Yabe, Jun Muto, Mako Ohzono, Yusaku Ohta, and Takeshi Iinuma

Abstract

Strain concentration zones (SCZs), in which the E–W contraction under a slow tectonic loading is larger than the surrounding area, in NE Japan have been attributed to low viscosity anomalies (LVAs) in their lower crust. The 2011 Tohoku-oki earthquake (Mw9.0) induced a stepwise stress change over NE Japan. The coseismic E–W extension in the SCZ along the Ou backbone Range (OBR) was smaller than theoretical one, whereas it was larger in the forearc SCZ (FSCZ). This suggests variation in rheological structure beneath the SCZs. We numerically evaluated responses of variety of rheological models to the tectonic slow loading and the coseismic instantaneous unloading. A model with a viscoelastic upper crust, which is caused by high temperature relating to magmatic processes along volcanic front, below the OBR can reproduce the observed deformations in the OBR. The surface deformations in the FSCZ were explained as enhanced deformations of thick, compliant sediment. The viscoelasticity in the upper crust was not allowed beneath the FSCZ. LVA in the lower crust was not essential to reproduce the preseismic and the coseismic deformation anomalies in the FSCZ. However, the postseismic deformation was strongly affected by the existence of LVA in the lower crust. Precise observations of the postseismic deformations should provide key clues to elucidate the rheological structure beneath the FSCZ.

Keywords

FEM modeling • Rheological structure beneath NE Japan • Seismic cycle • Strain concentration zones • The 2011 Tohoku-oki earthquake

1 Introduction

Based on the observed velocities of GPS sites, Miura et al. (2004) evaluated the residual strain rate over NE Japan for 1997–2001 by subtracting the effect of inter-plate coupling calculated from the back-slip model (Suwa et al. 2006). They showed that two strain (E–W contraction) concentration zones (SCZ) occur in NE Japan under a tectonic slow loading (Fig. 1a, c); one along the Ou backbone Range (OBR), another along a zone branching from the OBR to forearc side (forearc SCZ, FSCZ). Shallow seismicity is high in the SCZs and many large inland earthquakes occurred along them. Low seismic velocity and high attenuation anomaly in the lower crust beneath the OBR is known from seismic

Y. Yabe (✉) • Y. Ohta
Research Center for Prediction of Earthquakes and Volcanic Eruptions, Graduate School of Science, Tohoku University, Sendai, Miyagi 980-8578, Japan
e-mail: yabe@aob.gp.tohoku.ac.jp

J. Muto
Department of Earth Science, Tohoku University, Sendai, Miyagi 980-8578, Japan

M. Ohzono
Department of Earth and Environmental Sciences, Yamagata University, Yamagata, Yamagata 990-8560, Japan

T. Iinuma
Research and Development Center for Earthquake and Tsunami, Japan Agency for Marine-Earth Science and Technology, 3173-25, Showa-machi, Kanazawa-ku, Yokohama 236-0001, Japan

M. Hashimoto (ed.), *International Symposium on Geodesy for Earthquake and Natural Hazards (GENAH)*,
International Association of Geodesy Symposia 145, DOI 10.1007/1345_2015_135

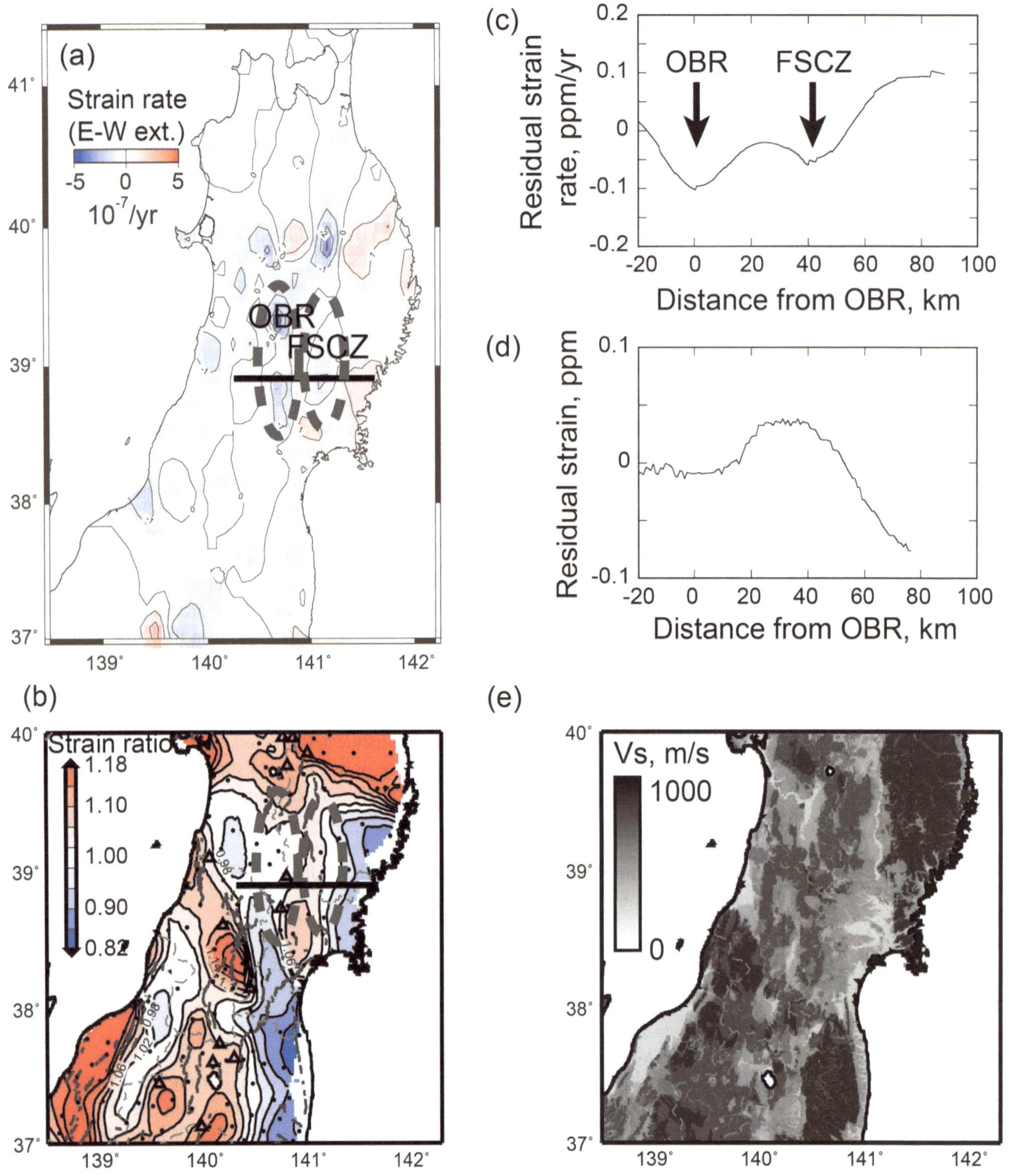

Fig. 1 (**a**) Distribution of the preseismic strain rate in NE Japan for 1997–2001 (after Miura et al. 2004). (**b**) Distribution of ratio of the observed coseismic strain to the theoretical one in NE Japan (after Ohzono et al. 2012a). *Dashed ellipses* in (**a**) and (**b**) enclose the OBR and the FSCZ. *Open triangles* in (**b**) denote Quaternary volcanoes. (**c**) Preseismic residual strain rate along the thick line in (**a**). Extension rate is taken as positive. (**d**) Coseismic residual strain along the thick line in (**b**). Extension is taken as positive. A static bias of 1 ppm is removed. (**e**) Distribution of the S wave velocity averaged over 30 m depth from the surface (Wakamatsu and Matsuoka 2013)

tomography (Tsumura et al. 2000; Nakajima et al. 2001a). Considering these, Hasegawa et al. (2005) proposed a conceptual rheological model for NE Japan, in which a low viscosity anomaly (LVA) is assumed in the lower crust beneath the OBR. A rheological structure based on the seismic, the thermal and the petrological structures also supports the LVA in the lower crust beneath the OBR (Muto 2011). Two-dimensional (2D) FEM models show that a LVA (Shibazaki et al. 2007, 2008) or a weak zone (Iio et al. 2004) in the lower crust is essential to the occurrence of the OBR.

Okada et al. (2010) found a similar seismic structure beneath the FSCZ. Omuralieva et al. (2012) showed that the cut-off depth of shallow seismicity (D90) in the FSCZ is shallower than that in the surrounding areas, suggesting a shallower depth of the brittle-ductile transition beneath the FSCZ. Therefore, a similar rheological model to the OBR should be applicable to the FSCZ.

On 11 March 2011, a great earthquake of Mw 9.0 (Tohoku-oki earthquake) occurred along the plate interface off the Pacific coast of NE Japan. Large coseismic deformations up to 5.3 m (e.g., Iinuma et al. 2011) were observed in NE Japan. Ohzono et al. (2012a) examined responses of the inland crust of NE Japan to the stepwise stress perturbation by the Tohoku-oki earthquake using a dense GPS network. They evaluated the coseismic strain field from the observed displacement field based on the method developed by Shen et al. (1996) and Sagiya et al. (2000). Theoretical strain field was calculated from a source model consisting of two rectangular faults on the plate interface that well reproduced the observed displacement field. A residual strain field that is a deviation of the observed coseismic strain field from the theoretical one shows such a regional variation that the observed E–W extension is smaller in the forearc and larger in the back-arc than the theoretical ones. This is clearer in ratios of the observed strain to the theoretical one (Fig. 1b), and may reflect variation in elastic moduli across NE Japan suggested by a seismic tomography (e.g., Nakajima et al. 2001b). Further, a local variation in the strain ratio shows a rather clear correspondence with the surface geology (Fig. 1e). Figure 1d shows a typical E–W profile of the residual strain. Here, a static bias of 1 ppm, which should be caused by a difference between the actual and the assumed elastic moduli of the crust, was removed. The residual E–W extension takes a local minimum in the OBR, while it is a local maximum in the FSCZ. Two SCZs showed different responses to the coseismic instantaneous unloading, though both of them behaved as the strain concentration zones under the slow loading before the Tohoku-oki earthquake. The coseismic instantaneous responses are mostly controlled by the elasticity of the crust and the uppermost mantle. In contrast, the slow deformations before the Tohoku-oki earthquake reflect their viscoelastic properties. Therefore, the difference in the rate-dependent responses between two SCZs should provide a clue to elucidate differences between their rheological structures.

Ohzono et al. (2012a) stated that the smaller extension in the OBR is consistent with the conceptual model proposed by Hasegawa et al. (2005). Because of faster strain energy dissipation in the low viscosity body beneath the OBR, the crust accumulates less strain energy than the surrounding area before the Tohoku-oki earthquake. As a result, amplitude of the elastic rebound in the OBR should be smaller than that in the neighboring areas. On the other hand, the larger deformations in the FSCZ under both of the preseismic (tectonic) slow loading and the coseismic instantaneous unloading are a typical response of an elastically compliant material, which enhances deformations irrelevant to loading rates. The FSCZ is covered by thick sediment (e.g. Ichiki et al. 1999; Kato, et al. 2004). A smaller stiffness of the thick sediment can be a cause for the observed large deformations in the FSCZ.

To test these suppositions on the preseismic and the coseismic strain anomalies in NE Japan, we examined surface deformation patterns of various rheological models under the slow loading and the instantaneous unloading using 2D viscoelastic FEM. The model predictions are compared with the observed pattern of deformations (Fig. 1c, d) to constrain rheological structure beneath NE Japan.

2 FEM Model

Topography, seismic structure and deformation distributions in and around NE Japan indicate that its crustal structure is relatively uniform along NE Japan arc (e.g. Nakajima et al. 2002; Miura et al. 2004; Suwa et al. 2006; Iinuma et al. 2011). We have constructed a simplified 2D FEM model in E–W cross-section to evaluate surface deformation of NE Japan under the plane strain condition. For simplicity, we adopted a flat surface model with a structural heterogeneity in a scale of the strain concentration zones (a few tens of km).

The linear Maxwell viscoelastic responses of models were solved using a commercial FEM code of MSC. Marc operated by Cyberscience Center, Tohoku University. In this FEM code, the Maxwell viscoelasticity is implemented by the following equations;

$$\dot{\overline{\varepsilon}}^c = \frac{\overline{\sigma}}{\eta} \tag{1}$$

$$\dot{\varepsilon}^c_{ij} = \dot{\overline{\varepsilon}}^c \left(\frac{\partial \overline{\sigma}}{\partial \sigma_{ij}} \right) \tag{2}$$

where $\overline{\varepsilon}^c$ and $\overline{\sigma}$ are the equivalent inelastic strain and the equivalent stress, respectively, and η is viscosity. $\varepsilon^c{}_{ij}$ and

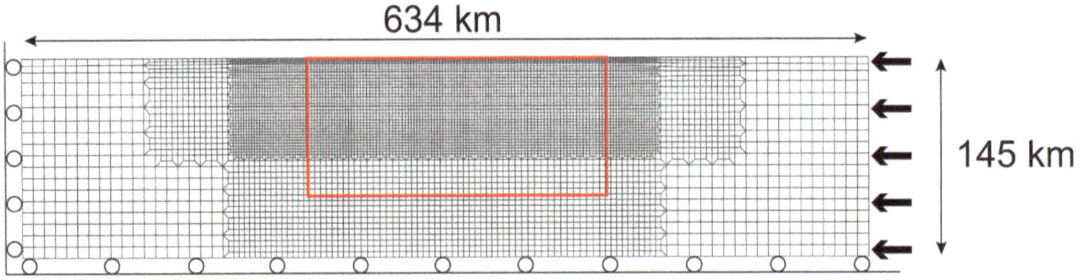

Fig. 2 FEM model meshes and boundary conditions. Left and bottom edges of the model are allowed to move only in vertical and horizontal directions, respectively. Time-dependent external load is applied to the right edges. Gravitational acceleration of 9.81 m/s^2 is applied to all elements. *Red box* indicates the region in which material distribution is shown in Fig. 3

σ_{ij} are each component of inelastic strain and stress, respectively.

Figure 2 shows the whole FEM mesh and boundary conditions. Horizontal and vertical extents of the model space are 634 and 145 km, respectively. We adopted 8-node quadratic elements that have the integration node on individual corners and edges of each rectangular element. Element size in the shallow central portion is 2–2.5 km in widths and heights, but the height of elements shallower than 4 km from the top surface is reduced to 1 km. The gravitational acceleration of 9.81 m/s^2 is applied to all elements.

We consider three layers model consisting of the upper crust, the lower crust and the mantle (Fig. 3). Thickness of the upper and the lower crust are 17 and 18 km, respectively. The OBR is located at the center of the model.

The relative plate motion rate between the Pacific plate and the overriding plate along the Japan Trench is about 8 cm/year (DeMets et al. 1994). The tsunami deposit surveys (Minoura et al. 2001; Sawai et al. 2008) indicate that earthquakes as gigantic as the Tohoku-Oki earthquake struck NE Japan every ~500 years on average for the past 3,000 years. Coseismic horizontal displacements of the Tohoku-oki earthquake were 31 and 60–70 m at sites 50 and 20–30 km, respectively, from the trench (Kido et al. 2011; Ito et al. 2011). Therefore, the interseismic loading of 8 cm/year and the coseismic unloading of 40 m every 500 years are applied to the east (right) edge of the model independently of depths. Duration of the 40-m unloading was 150 s (e.g. Ohta et al. 2012). To constrain translational motions of the model, only vertical and horizontal displacements are allowed for nodes on the west and the bottom edge, respectively.

Material properties assumed in this study are summarized in Table 1. They are based on the estimation by Muto (2011) and Ohzono et al. (2012b). The upper crust is usually treated as a pure elastic media. However, the low seismic velocity and the high attenuations beneath the OBR and the FSCZ suggest the occurrence of partial melting in the lower crust (e.g. Hasegawa et al. 2005; Muto 2011). Further, the heat flow is locally high along the OBR (Furukawa 1993; Tanaka and Ishikawa 2002). Considering these, temperature of the crust beneath the SCZs should be high, causing plastic deformation in some part of the brittle upper crust. To approximate rheology of the partially plastic upper crust, we assumed a viscoelasticity in the upper crust beneath the SCZs in some models.

To eliminate effects of the initial conditions, we adopted the following procedures: (1) Horizontal displacement of nodes on the east edge is prohibited and the model is relaxed for 5,000 years under the gravitational acceleration to achieve the equilibrium state. (2) The nodes on the east edge are slowly moved to west by a constant rate of 8 cm/year for 500 years and retracted to east by 40 m. By repeating this loading-unloading cycles a few times, load-deformation curves became almost identical among the cycles. (3) After 10 loading-unloading cycles, we evaluated crustal deformations due to the coseismic unloading followed by the slow loading of 8 cm/year.

3 Results

Suwa et al. (2006) estimated that the lower edge of slip deficit on the plate interface beneath NE Japan is about 100 km by using 3D displacement field measured by GPS network. They demonstrated that the horizontal displacement fields for slip deficit models with the lower edge of ~120 and ~200 km are very similar to that with the lower edge of 100 km. That is, the three models with slip deficit down to ~100, ~120 and ~200 km cannot be distinguished by the horizontal displacement field alone. However, the vertical displacement field was quite different among the three models. This indicates that the vertical displacement is sensitive to depth variation in interplate coupling and geometry of the plate interface, which are not considered in the present study, and that the horizontal displacement pattern is relatively insensitive to them. Therefore, we cannot compare the observed and the calculated pattern of vertical deformations. Only the patterns of horizontal surface deformations are discussed in the present study.

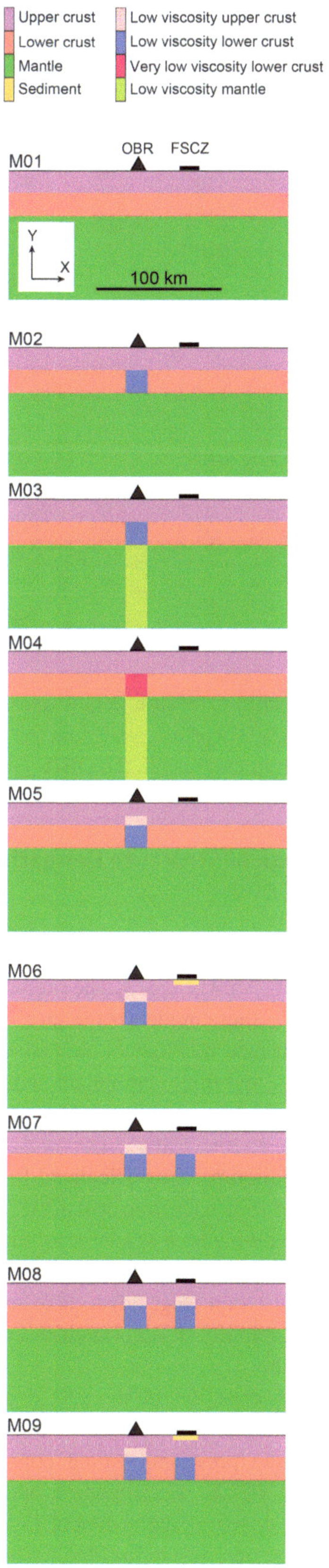

Fig. 3 Variation of rheological models used in this study. *Solid triangle* and *thick bar* attached on the top of each model indicate locations of the OBR and the FSCZ, respectively

Preliminary calculations indicate that rheological structures in the mantle negligibly affect the preseismic and the coseismic deformation pattern, probably because of a limited extent of region investigated in the present study. Therefore, we consider only limited variation in rheological anomalies in the mantle below.

3.1 Rheological Model Beneath the OBR

Since the crustal structure beneath the OBR is well studied, we first constrain rheological anomalies beneath the OBR by testing 4 models (M02 to M05 in Fig. 3). Surface deformation patterns of these models are shown in Fig. 4 and reproducibility with the observation is summarized in Table 2. We calculated the preseismic strain rate from deformations for 10 years just before the coseismic unloading. Since the E–W extension rate is taken as positive, the strain concentration zone, where the E–W contraction rate is larger than the surrounding area, is a local minimum in Fig. 4a. We call a local maximum of the strain rate in this figure the strain deficit zone, for convenience. Since the coseismic strain in our numerical model was predominated by almost uniform E–W extension, its anomaly shown in Fig. 4b was emphasized by subtracting the uniform E–W extension of 62.67 ppm calculated for the reference model M01 from the coseismic surface deformations of individual models. For the observed coseismic deformation, the reference strain field is heterogeneous and was calculated from the fault model on the plate interface. A positive strain anomaly means E–W extension lager than the reference value (62.67 ppm). Therefore, the extension deficit and the extension excess zone are a local minimum and a local maximum, respectively, in Fig. 4b. The reproducibility of each model was evaluated according to whether the preseismic strain concentration and the coseismic strain deficit occurred in the OBR. Amplitudes of strain anomalies were not considered in the evaluation, because our models are too simplified for quantitative comparison with the observation.

No observable coseismic anomaly in model M02 means that a LVA confined in the lower crust little affects the coseismic surface deformations. LVAs in both of the lower crust and the mantle (model M03) cause the preseismic strain deficit and the coseismic extension excess. When the viscosity of the lower crust beneath the OBR is extremely lowered (M04), the preseismic strain concentration can be reproduced, but the OBR behaves as the coseismic extension excess zone in contrast to the observation. Larger amplitude of the coseismic extension excess in M04 than in M03 suggests that the coseismic extension deficit cannot be reproduced by further reducing viscosity of the lower curst. Models M02 to M04 in which LVAs are introduced in the lower crust and the mantle cannot reproduce the coseismic

Table 1 Material properties

	Young's modulus (GPa)	Poisson's ratio	Density (kg/m^3)	Viscosity(Pa s)	Relaxation time (year)
Upper crust	80.4	0.27	2,700	∞	∞
Lower crust	106	0.29	2,900	10^{21}	771
Mantle	160	0.29	3,380	10^{19}	5
Sediment	60	0.27	2,000	10^{20}	134
Viscoelastic upper crust	80.4	0.29	2,700	10^{22}	10,169
Low viscosity lower crust	106	0.29	2,900	10^{19}	8
Very low viscosity lower crust	106	0.29	2,900	10^{16}	2.8 days
Low viscosity mantle	160	0.29	3,380	10^{16}	1.9 days

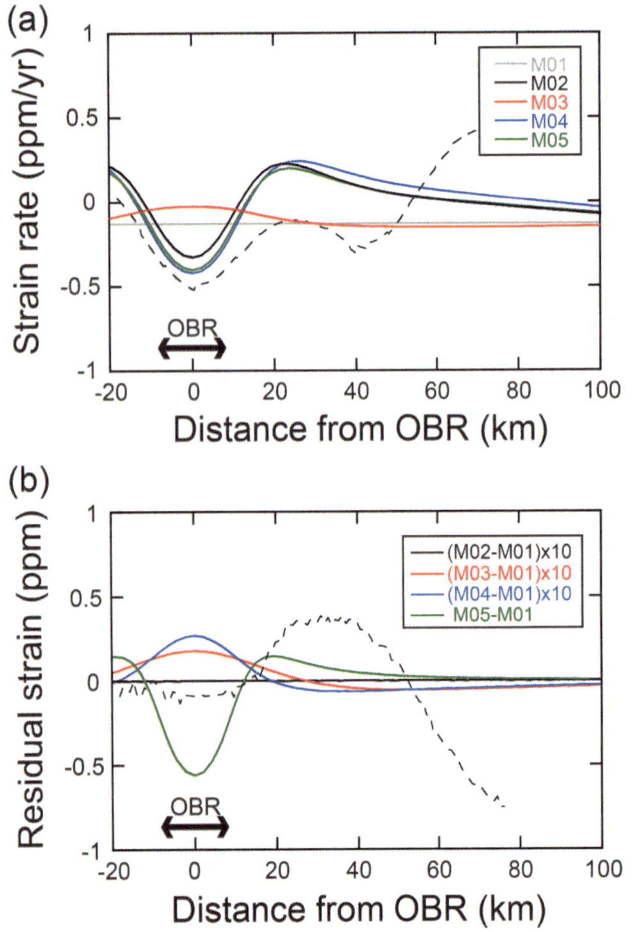

Fig. 4 Surface deformation associated with (**a**) the preseismic slow loading and (**b**) the coseismic instantaneous unloading for M02–M05. *Arrows* indicate the extent of the low viscosity anomaly beneath the OBR. E–W extension rate and E–W extension is taken as positive. *Dashed line* in (**a**) and (**b**) indicates the observed preseimic residual strain rate (Miura et al. (2004), vertically exaggerated by 5 times) and the coseismic residual strain (Ohzono et al. (2012a), static bias of 1 ppm removed and vertically exaggerated by 10 times), respectively

extension deficit, whereas the preseismic strain concentration can occur.

In model M05, a viscoelastic body is introduced in a lower half of the upper crust. This model successfully reproduces both of the coseismic extension deficit and the preseismic strain concentration. Therefore, we adopt the model M05 as the rheological structure beneath the OBR.

Table 2 Reproducibility of model prediction with the observations in the OBR

Model	Preseismic strain concentration	Coseismic extension deficit
M02	Yes	No
M03	No	No
M04	Yes	No
M05	Yes	Yes

3.2 Rheological Structure Beneath the FSCZ

To determine the rheological structure beneath the FSCZ, we tested models M06–M09 in Fig. 3. The rheological structure beneath the OBR in these models is the same as that in M05. In M06, no subsurface rheological anomalies are assumed beneath the FSCZ, but thick, compliant sediment covering the surface of the FSCZ is considered. M07–M09 have LVAs in the crust beneath the FSCZ. The preseismic and the coseismic deformation patterns calculated for these models are shown in Fig. 5. Since Ohzono et al. (2012a) obtained the coseismic deformation with a spatial resolution of 20 km, a boxcar window of 20 km in width is applied to the calculated coseismic residual strain (Fig. 5c) to take running average (Fig. 5b). Reproducibility of deformations by M06–M09 is summarized in Table 3.

The strain concentration in the FSCZ under the preseismic slow loading is clearly seen in M07 and M08. In M06 and M09, the preseismic strain concentration over the width of the FSCZ is recognized, when we ignore fluctuation at the edges of the sediment layer, which may be caused by numerical instability due to too coarse mesh to calculate surface deformation around a shallow, sharp material boundary. All of these four models reproduce the preseismic strain concentration in the FSCZ. Thus, all of

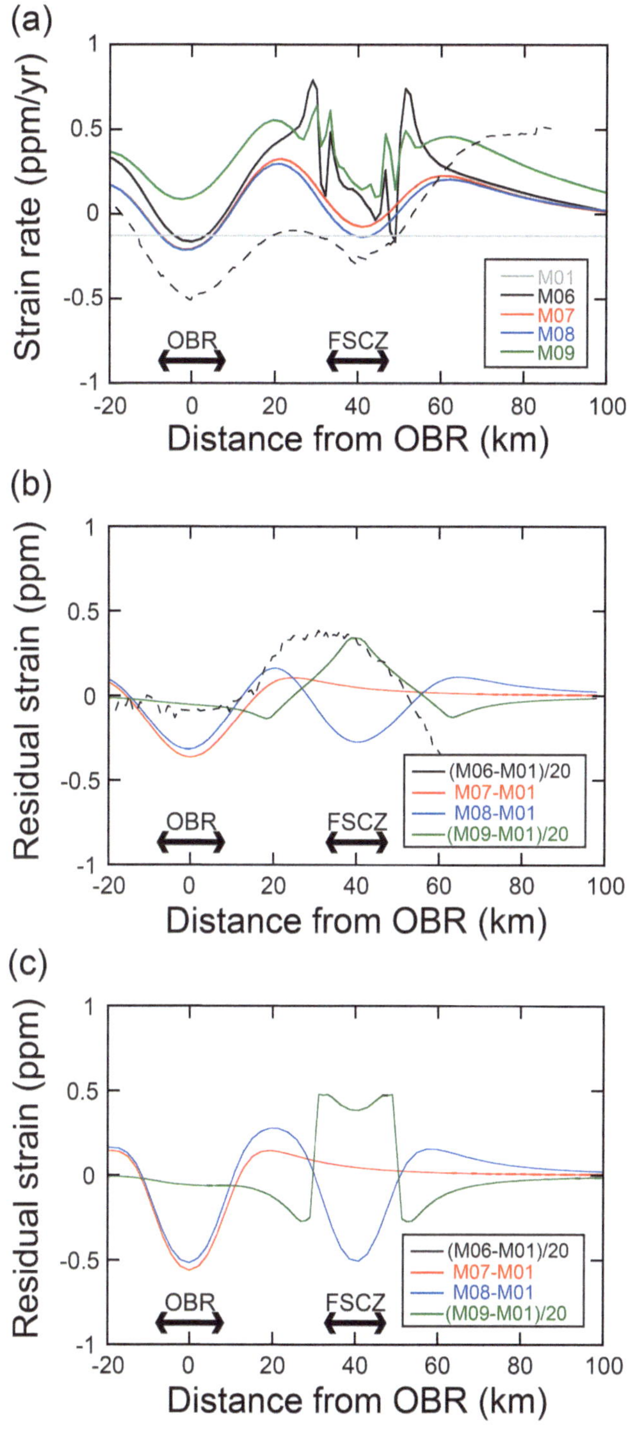

Fig. 5 (**a**) The preseismic surface deformation and (**b**) the smoothed and (**c**) the raw coseismic deformation for M06–M09. Since the coseismic deformations for M06 and M09 are identical to each other, the deformation pattern for M06 in (**b**) and (**c**) is hidden by that for M09. *Arrows* in (**b**) and (**c**) represent the extent of the rheological anomaly in the OBR and the FSCZ. E–W extension rate and E–W extension is taken as positive. *Dashed line* in (**a**) and (**b**) indicates the observed preseimic residual strain rate (Miura et al. (2004), vertically exaggerated by 5 times) and the coseismic residual strain (Ohzono et al. (2012a, b), static bias of 1 ppm removed and vertically exaggerated by 10 times), respectively

Table 3 Reproducibility of model prediction with the observations in the FSCZ

Model	Preseismic strain concentration	Coseismic extension excess
M06	Yes	Yes
M07	Yes	No
M08	Yes	No
M09	Yes	Yes

M06–M09 can reproduce the preseismic strain concentration, in spite of wide variety of the rheological anomalies beneath the FSCZ. This suggests that the preseismic deformation is insensitivity to the vertical extent of rheological anomaly.

When a viscoelasticity is assumed in a lower half of the upper crust (M08), the coseismic response of the FSCZ is the extension deficit opposite to the observation, meaning that the upper crust of the FSCZ must be elastic. In M07, the coseismic extension excess occurs over the FSCZ. However, it does not concentrate to the FSCZ, but distributes over a wide area east of the OBR, which is the same as that of M05. Considering the coseismic deformation of M02 in that LVA only in the lower crust does not affect the coseismic deformation pattern, the LVA in the lower crust beneath the FSCZ should play no role in the occurrence of the coseismic extension excess over the FSCZ of M07. The extension excess of M07 should occur to compensate the extension deficit in the OBR. M06 and M09 reproduce the coseismic extension excess in the FSCZ. Therefore, the observed preseismic and the coseismic behaviors at the FSCZ can be explained as enhanced deformations by thick, compliant sediment.

4 Discussion

Rheological structure of the crust beneath the OBR is constrained rather uniquely. On the other hand, only what we can say about rheological structure beneath the FSCZ is that a model having viscosity in the upper crust (M08) cannot reproduce the observations. It is evident that the FSCZ is covered by thick sediment (e.g. Ichiki et al. 1999; Kato et al. 2004). Therefore, models M06 and M09 that have sediment layer over the FSCZ should be more plausible than M07 for rheological structure of NE Japan. The preseismic and the coseismic deformations in the FSCZ are similar between M06 and M09. It is difficult to judge which model is a better approximation for the FSCZ.

Amplitude of the coseismic extension excess in the FSCZ of M06 and M09 is much larger than the amplitude of the coseismic extension deficit in the OBR. This should be caused by small Young's modulus and/or viscosity of the sediment. Two extreme cases are examined to evaluate which parameter is essential to the amplitude of the coseismic extension excess. When as large Young's modulus as the upper crust (80.4 GPa) is assumed for the sediment in model M09, the preseismic deformation is not changed from M09. However, the coseismic extension excess in the FSCZ disappeared. On the other hand, if the viscosity of the sediment is increased to the same value as that of the viscoelastic upper crust (10^{22} Pa s), the preseismic deformation in the FSCZ is not the strain concentration, but the strain deficit, while the amplitude of coseismic extension excess is not changed. These results suggest that the viscosity of the sediment controls the preseismic behavior of the FSCZ, while the coseismic behavior depends on the Young's modulus.

The brittle-ductile transition depth inferred by D90 (Omuralieva et al. 2012) beneath the FSCZ is estimated to be about 10 km, being shallower than the surrounding area. The precise seismic structure indicates a low velocity zone beneath the FSCZ (Okada et al. 2010). This low velocity zone is interpreted as a branch of the low velocity zone beneath the OBR, to where a large amount of fluid is supplied from a depth (Nakajima et al. 2001a; Hasegawa and Nakajima 2004; Mitsuhata et al. 2001). These lines of evidence suggest that M09 having a LVA in the lower crust beneath the FSCZ should be more appropriate than M06. Figure 6 shows the postseismic strain rate calculated from deformations for 10 years following the coseismic unloading. The FSCZ of M06 behaves as the strain concentration zone, whereas it is the strain deficit zone in M09. Detailed observations of the postseismic deformations in and around the FSCZ should provide an important clue to constrain the rheological structure beneath the FSCZ.

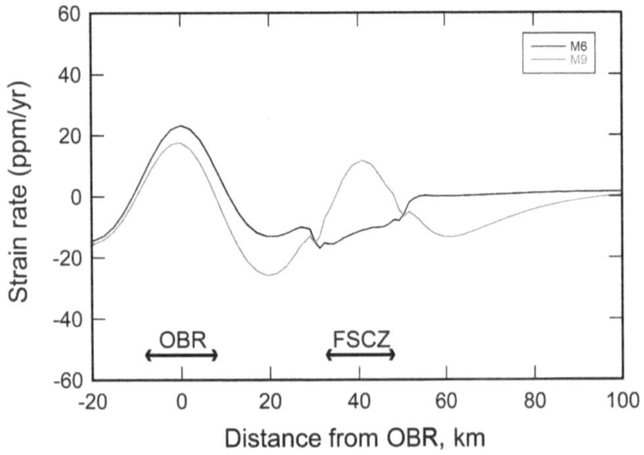

Fig. 6 Spatial variation in strain rate for 10 years following the coseismic unloading in M06 and M09. E–W extension rate is taken as positive

5 Conclusions

Rheological structure of NE Japan that reproduces the preseismic and the coseismic deformation was constrained by comparing the observations and the FEM simulations. The OBR behaves as the SCZ under the preseismic slow loading and responds as the extension deficit zone to the coseismic instantaneous unloading associated with the Tohoku-oki earthquake. The rheological structure having a LVA in the lower crust and a thinner elastic upper crust was the most suitable to reproduce the observed behavior of the OBR.

Occurrence of the coseismic extension excess in the FSCZ is explained as enhanced deformations by thick, compliant sediment, while the preseismic strain concentration was induced by the low viscosity of the sediment. Numerical simulations showed that the preseismic and the coseismic behavior of the FSCZ do not depend on rheology of the lower crust. This means that the surface deformation data available at the present are insensitive to the rheological structure at a depth beneath the FSCZ. However, the simulations demonstrate that the post-seismic deformation strongly depends on the rheological structure of the lower crust. Precise observation of the postseismic deformations in and around the FSCZ will give a clue to elucidate the rheological structure of the lower crust beneath the FSCZ.

Acknowledgements Critical comments by two anonymous reviewers were useful to improve quality of the manuscript. The authors thank Prof. S. Miura, Tohoku University, for giving the results of the preseismic deformation rate. This study was supported in part by a grant from the Ministry of Education, Culture, Sports, Science and Technology of Japan under the Research Program for Prediction of Earthquakes and Volcanic Eruptions and Grant-in-Aid for Scientific Research on Innovative Areas (KAKENHI No. 26109007) from the Ministry of Education, Culture, Sports, Science and Technology.

References

DeMets C, Gordon RG, Argus DF, Stein S (1994) Effect of recent revisions to the geomagnetic reversal time scale on estimates of current plate motion. Geophys Res Lett 21:2191–2194. doi:10.1029/94GL02118

Furukawa Y (1993) Depth of the decoupling plate interface and thermal structure under arcs. J Geophys Res 98:20005–20013. doi:10.1029/93JB02020

Hasegawa A, Nakajima J (2004) Geophysical constrains on slab subduction and arc magmatism. In: Sparks RSJ, Hawkesworth CJ (eds) State of the planet: frontiers and challenges in geophysics, AGU Geophysical Monograph, vol 19. IUGG, pp 81–94. doi:10.1029/150GM08

Hasegawa A, Nakajima J, Umino N, Miura S (2005) Deep structure of the northeastern Japan arc and its implications for crustal deformation and shallow seismic activity. Tectonopnyscs 403:59–75. doi:10.1016/j.tecto.2005.03.018

Ichiki M, Mishina M, Goto T, Oshiman N, Sumitomo N, Utada H (1999) Magnetotelluric investigations for the seismically active area in northern Miyagi prefecture, northeastern Japan. Earth Planets Space 51:351–361

Iinuma T, Ohzono M, Ohta Y, Miura S (2011) Coseismic slip distribution of the 2011 off the Pacific coast of Tohoku Earthquake (M 9.0) estimated based on GPS data—Was the asperity in Miyagi-oki ruptured? Earth Planets Space 63:643–648. doi:10.5047/eps.2011.06.013

Iio YT, Sagiya N, Umino T, Nishimura KT, Homma T (2004) A comprehensive model of the deformation process in the Nagamachi-Rifu Fault Zone. Earth Planets Space 53:1339–1345

Ito Y, Tsuji K, Osada Y, Kido M, Inazu D, Hayashi Y, Tsushima H, Hino R, Fujimoto H (2011) Frontal wedge deformation near the source region of the 2011 Tohoku-Oki earthquake. Geophys Res Lett 38:L00G05. doi:10.1029/2011GL048355

Kato N, Sato H, Imaizumi T, Ikeda Y, Okada S, Kagohara K,Kawahara T, Kasahara K (2004) Seismic reflection profiling across the source fault of the 2003 Nothern Miyagi earthquake (Mj6.4), NE Japan: basin inversion of Miocene back-arc rift. Earth Planets Space 56:1369–1374

Kido M, Osada Y, Fujimoto H, Hino R, Ito Y (2011) Trench-normal variation in observed seafloor displacements associated with the 2011 Tohoku-Oki earthquake. Geophys Res Lett 38:L24303. doi:10.1029/2011GL050057

Minoura K, Imamura F, Sugawara D, Kono Y, Iwashita T (2001) The 869 Jogan tsunami deposit and recurrence interval of large-scale tsunami on the Pacific coast of northeast Japan. J Natural Disaster Sci 23:83–88

Mitsuhata Y, Ogawa Y, Mishina M, Kono T, Yokokura T, Uchida T (2001) Electromagnetic heterogeneity of the seismogenic region of 1962 Northern Miyagi Earthquake, northeastern Japan. Geophys Res Lett 28:4371–4374. doi:10.1029/2001GL013079

Miura S, Sato T, Hasegawa A, Suwa Y, Tachibana K, Yui S (2004) Strain concentration zone along the volcanic front derived by GPS observations in NE Japan arc. Earth Planets Space 56:1347–1355

Muto J (2011) Rheological structure of northeast Japan lithosphere based on geophysical observation and rock mechanics. Tectonophysics 503:201–206. doi:10.1016/j.tecto.2011.03.002

Nakajima J, Matsuzawa T, Hasegawa A, Zhao D (2001a) Seismic imaging of arc magma and fluids under the central part of northeast Japan. Tectonophysics 341:1–17. doi:10.1016/S0040-1951(01)00181-0

Nakajima J, Matsuzawa T, Hasegawa A, Zhao D (2001b) Three-dimensional structure of Vp, Vs and Vp/Vs beneath the northeast Japan arc: implications for arc magmatism and fluids. J Geophys Res 106:21843–21857. doi:10.1029/2000JB000008

Nakajima J, Matsuzawa T, Hasegawa A (2002) Moho depth variation in the central part of northeastern Japan estimated from reflected and converted waves. Phys Earth Planet Inter 130:31–47. doi:10.1016/S0031-9201(01)00307-7

Ohta Y, Kobayashi T, Tsushima H, Miura S, Hino R, Takasu T, Fujimoto H, Iinuma T, Tachibana K, Demachi T, Sato T, Ohzono M, Umino N (2012) Quasi real-time fault model estimation for near-field tsunami forecasting based on RTK-GPS analysis: application to the 2011 Tohoku-Oki earthquake (Mw 9.0). J Geophys Res 117:B02311. doi:10.1029/2011JB008750

Ohzono M, Yabe Y, Iinuma T, Ohta Y, Miura S, Tachibana K, Sato T, Demachi T (2012a) Strain anomalies induced by the 2011 Tohoku Earthquake (Mw9.0) as observed by a dense GPS network in northeast Japan. Earth Planets Space. doi:10.5047/eps.2012.05.015

Ohzono M, Ohta Y, Iinuma T, Miura S, Muto J (2012b) Geodetic evidence of viscoelastic relaxation after the 2008 Iwate-Miyagi Nairiku earthquake. Earth Planets Space 64:759–764. doi:10.5047/eps.2012/04/001

Okada T, Umino N, Hasegawa A (2010) Deep structure of the Ou mountain range strain concentration zone and the focal area of the 2008 Iwate-Miyagi Nairiku earthquake. NE Japan – seismogenesis related with magma and crustal fluid. Earth Planets Space 62:347–352. doi:10.5047/eps.2009.11.005

Omuralieva AM, Hasegawa A, Matsuzawa T, Nakajima J, Okada T (2012) Lateral variation of the cutoff depth of shallow earthquakes beneath the Japan Islands and its implications for seismogenesis. Tectonophysics 518:93–105. doi:10.1016/j.tecto.2011.11.013

Sagiya T, Miyazaki S, Tada T (2000) Continuous GPS array and present-day crustal deformation of Japan. Pure Appl Geophys 157:2303–2322

Sawai Y, Fujii Y, Fujiwara O, Kamataki T, Komatsubara J, Okamura Y, Satake K, Shishikura M (2008) Marine incursions of the past 1500 years and evidence of tsunamis at Suijin-numa, a coastal lake facing the Japan Trench. Holocene 18:517–528. doi:10.1177/0959683608089206

Shen Z-K, Jackson DD, Ge BX (1996) Crustal deformation across and beyond the Los Angeles basin from geodetic measurements. J Geophys Res 101:27957–27980. doi:10.1029/2004JB003203

Shibazaki B, Garatani K, Okuda H (2007) Finite element analysis of crustal deformation in the Ou Backbone Range, northeast Japan, with non-linear visco-elasticity and plasticity: effects of non-uniform thermal structure. Earth Planets Space 59:499–512

Shibazaki B, Garatani K, Iwasaki T, Tanaka A, Iio Y (2008) Faulting processes controlled by the nonuniform thermal structure of the crust and uppermost mantle beneath the northeast Japanese island arc. J Geophys Res. doi:10.1029/2007JB005361

Suwa Y, Miura S, Hasegawa A, Sato T, Tachibana K (2006) Interplate coupling beneath NE Japan inferred from three-dimensional displacement field. J Geophys Res. doi:10.1029/204JB003203

Tanaka A, Ishikawa Y (2002) Temperature distribution and focal depth in the crust of the northeastern Japan. Earth Planets Space 54:1109–1113

Tsumura N, Matsumoto S, Horiuchi S, Hasegawa A (2000) Three-dimensional attenuation structure beneath the northeast Japan arc estimated from spectra of small earthquakes. Tectonophysics 319:241–260. doi:10.1016/S0040-1951(99)00296-6

Wakamatsu K, Matsuoka M (2013) Nationwide 7.5-arc-second Japan engineering geomorphologic classification map and Vs30 zoning. J Disaster Res 8:904–911

Paradoxical Vertical Crustal Movement Along the Pacific Coast of Northeast Japan

Takeshi Sagiya

Abstract

Rapid subsidence had been observed along the Pacific coast of northeast Japan and the 2011 M_w9.0 Tohoku-oki earthquake caused additional subsidence over 1 m. On the other hand, geomorphological evidence shows the same area is slightly uplifting in a long-term. In order to interpret this paradoxical vertical movement, we construct a simple model of earthquake deformation cycle. Viscous relaxation in the asthenosphere can significantly affect crustal deformation pattern over an earthquake cycle. When earthquake recurrence interval is longer than the asthenospheric relaxation time, temporal variation in the interseismic deformation pattern becomes significant. Spatially heterogeneous behavior in earthquake recurrence may be responsible for the enigmatic vertical movement in northeast Japan.

Keywords

2011 Tohoku-oki earthquake • Crustal deformation cycle • Northeast Japan

1 Introduction

Vertical crustal movement along the Pacific coast of northeast Japan has been controversial for a long time. Geomorphologists found Stage 5e marine terraces at an elevation of 20–60 m, implying a slight long-term uplift at the rate of less than 0.5 mm/year over 125,000 years (Koike and Machida 2001). On the other hand, leveling surveys and tidal records showed a rapid coastal subsidence as fast as 5 mm/year during the latter half of the twentieth century (El-Fiky and Kato 1999). Ikeda (2003) postulated that a giant megathrust earthquake associated with the subduction of the Pacific plate may occur and recover the subsidence.

On March 11, 2011, the M_w9.0 Tohoku-oki earthquake occurred and the Pacific coast of northeast Japan had a coseismic subsidence up to 1.1 m (Ozawa et al. 2011). After the main shock, significant postseismic uplift has been observed. However, the postseismic uplift for 3 years from the main shock amounts to only a few tens of centimeters and it is uncertain if the coseismic as well as interseismic subsidence will be recovered (Fig. 1). So the question about the long-term and short-term vertical movements remains unresolved. We have to wait for at least several decades to answer the question through geodetic observations.

Crustal deformation around the subduction zone is an important manifestation of mechanical interaction at the plate interface. There have been various studies to estimate interplate slip deficit based on geodetic observations. As one of such studies, Suwa et al. (2006) analyzed three-dimensional GPS velocity data during the interseismic period to find the slip deficit existing down to a depth of about 100 km, while other studies based on horizontal data only reported much shallower locked zone (down to a depth of 50–60 km) (Nishimura et al. 2004; Hashimoto et al. 2009). Thus there exist large uncertainties regarding seismic potential evaluation along the Japan trench originated from vertical crustal movement.

Coseismic fault rupture of the 2011 earthquake occurred shallower than the depth of 40 km and afterslip is estimated

T. Sagiya (✉)
Disaster Mitigation Research Center, Nagoya University, Furo-cho, Chikusa-ku, Nagoya, Japan
e-mail: sagiya@nagoya-u.jp

M. Hashimoto (ed.), *International Symposium on Geodesy for Earthquake and Natural Hazards (GENAH)*, International Association of Geodesy Symposia 145, DOI 10.1007/1345_2015_189

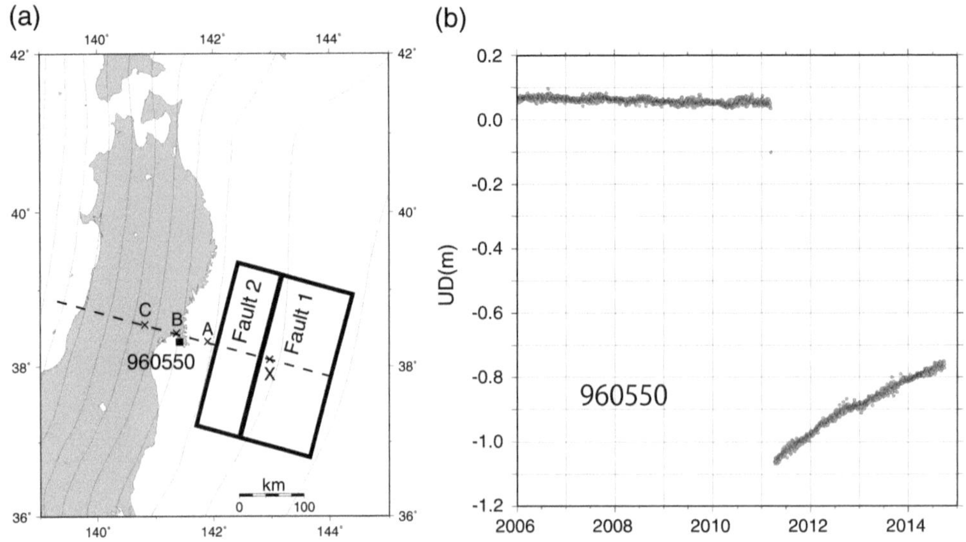

Fig. 1 (**a**) Map of northeast Japan. Configuration of the subducted Pacific slab is denoted by *dashed contour lines* for every 20 km depth (Kita et al. 2010; Nakajima and Hasegawa 2006). *Rectangles* are the location of model faults. *Thick dashed line* is the projection line. A, B, C, and X denote the location of displacement calculation. (**b**) Vertical displacement data at GEONET station 960550

at the downward extension of the coseismic rupture (Ozawa et al. 2011). How we can reconcile all these coseismic and interseismic observations is an essential question to understand physical mechanism of megathrust earthquakes at subduction zones. Vertical movement data is an important key to tackle such a problem.

In this study, we present a simplified kinematic model of plate subduction to interpret the paradoxical vertical movements along the Pacific coast of northeast Japan, in which we consider an earthquake cycle with viscoelastic response of the asthenosphere and co-existence of the magnitude-9 and magnitude-8 class earthquakes at the same plate boundary. There are many studies to understand earthquake cycle at subduction zones with sophisticated numerical approaches and realistic structural model. Compared with such studies, our simplified model enables us to identify responsible sources for characteristic aspects in deformation patterns easily.

2 An Earthquake Cycle Model

2.1 Formulation

In order to discuss time-dependent vertical crustal movements along the Pacific coast of northeast Japan, we construct a simple kinematic model of an earthquake deformation cycle. For our kinematic modeling, we apply an approach by Savage and Prescott (1978). That is, time evolution of the plate boundary fault slip is divided into the steady part Δu_s and the perturbation part Δu_p. When the relative plate velocity is v, the steady part is represented by a linearly increasing slip,

$$\Delta u_s(t) = vt \tag{1}$$

If we assume that interseismic locking is perfect and the whole slip deficit is released by regular earthquakes with a recurrence interval of T, the perturbation term becomes purely cyclic and can be described as follows.

$$\Delta u_p(t) = -vt + vT \sum_{k=1}^{\infty} H\,(t - kT) \tag{2}$$

Hashimoto et al. (2004) calculated long-term contribution of steady plate subduction without interplate locking around Japan. They demonstrated that subduction of the Pacific and the Philippine Sea plates yields long-term deformation pattern of the Japanese islands, which resembles long-term vertical deformation shown by the free-air gravity anomaly. Their model predicted that the Pacific coast of northeast Japan has an uplift at about 1 mm/year. So the geological long-term uplift (0.5 mm/year) may be attributed to the steady plate subduction. However, the model of Hashimoto et al. was simple and other factors such as intraplate faulting can affect long-term deformation. Thus we just point out that steady plate subduction as a candidate mechanism of the long-term vertical movement and will not discuss it in the following. Observed interseismic subsidence rate is much larger than the long-term one. We deal with only the cyclic part of the crustal deformation associated with time-dependent plate interaction.

Surface deformation caused by the cyclic part $d(t)$ can be described by the following formula.

$$d(t) = \int_{-\infty}^{t} \Delta \dot{u}_p(\tau) G(t - \tau) d\tau \tag{3}$$

Here, $G(t)$ is the viscoelastic displacement response to a unit fault slip. From Eqs. (2) and (3), the surface displacement due

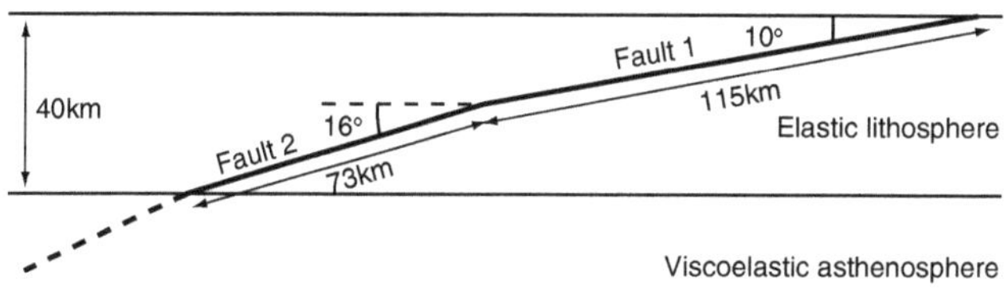

Fig. 2 Model configuration

to a periodic earthquake cycle starting from $t = 0$ is written as follows.

$$d(t) = -v \int_0^t G(t-\tau)d\tau + vT \sum_{k=1}^{n} G(t-kT) \quad (4)$$

Here, n is the number of earthquakes until the time t. If t is much larger than the asthenospheric relaxation time, we can regard the response function $G(t)$ becomes constant $G(\infty)$ for $t > t_0$, and Eq. (4) can be rewritten as follows.

$$\begin{aligned} d(t) &= -v \int_0^{t-t_0} G(t-\tau)d\tau - v \int_{t-t_0}^{t} G(t-\tau)d\tau \\ &\quad + vT \sum_{k=1}^{n} G(t-kT) \\ &= -vG(\infty)t + vT \sum_{k=1}^{n} G(t-kT) + const. \quad (5) \end{aligned}$$

Then we can calculate cyclic crustal displacement pattern by using completely relaxed response $G(\infty)$ and a finite summation of viscoelastic response function $G(t)$.

2.2 Model Setup

In our model, the seismogenic part of the subducting Pacific plate interface is represented by two rectangular faults embedded in the 40 km thick elastic lithosphere overlying a viscoelastic asthenosphere. The shallow fault covers the depth range of 0–20 km with a dip angle of 10°, while the deeper one covers the 20–40 km depth with a dip angle of 16° (Fig. 2). Both faults are 250 km long along the Japan trench. Plate subduction rate is 0.1 m/year in the perpendicular direction to the faults' strike so that only a dip slip component is considered. We only model faulting in the elastic lithosphere since the dislocation in the asthenosphere does not have a permanent contribution to the surface displacement after many cycles of earthquake occurrence (Matsu'ura and Sato 1989). We assume Maxwell viscosity of 1.0×10^{19} Pa · s for the asthenosphere. The corresponding asthenospheric relaxation time is 4.37 years but the effective relaxation time

Table 1 Structure model

Thickness(km)	V_p(km/s)	V_s(km/s)	ρ(kg/m^3)	η(Pa · s)
40	6.70	3.87	2,900	∞
∞	8.00	4.62	3,400	1.0×10^{19}

is larger by an order of magnitude due to interaction with the elastic lithosphere. Structural parameters are summarized in Table 1. Viscoelastic deformation response to a unit fault slip of each rectangular fault is calculated using PSGRN/PSCMP code (Wang et al. 2006).

As already mentioned, we assume perfect locking during interseismic periods and accumulated slip-deficit is fully released by earthquakes. Then variations in earthquake size is described by different recurrence intervals. That is, a recurrence interval of 50 years corresponds to coseismic slip of 5 m (magnitude-8 class), and a recurrence interval of 500 years corresponds to 50 m of coseismic slip (magnitude-9 class).

We consider such a simple earthquake cycle model to investigate consequences of an earthquake cycle with a viscoelastic relaxation effects. In addition, by introducing two faults in the shallow and the deeper parts, we can investigate the effects of heterogeneous earthquake recurrence intervals on the plate interface. This second effect may be important considering there apparently exist earthquake recurrences with different intervals and different earthquake sizes along the Japan trench.

We calculate two cases of long (500 years) and short (50 years) recurrence intervals for the shallow and the deeper faults. The calculated deformation shows regular behavior after 1,000 years or two cycles of the longer recurrence interval. We evaluate crustal deformation after 9,000 years from the beginning with a time interval of 5 years.

3 Results and Discussion

In the following, we show calculated displacement time series at three points (A, B, and C) shown in Fig. 1. These points are located at 200 km, 250 km, and 300 km from the Japan trench or the shallower edge of the modeled shallow fault along the profile line running across the center of the two faults. Figure 3 shows calculated deformation responses to an earthquake cycle for each fault. The shallow fault (Fault 1) does not contribute to the coastal vertical motion if the recurrence interval is 50 years (Fig. 3a). On the other hand, in the case of long (500 years) recurrence interval, postseismic uplift up to 2 m occurs, for 50–200 years depending on the location, and an interseismic subsidence follows as the contribution of the shallow fault. In this case, interseismic subsidence rate caused by the shallow fault is

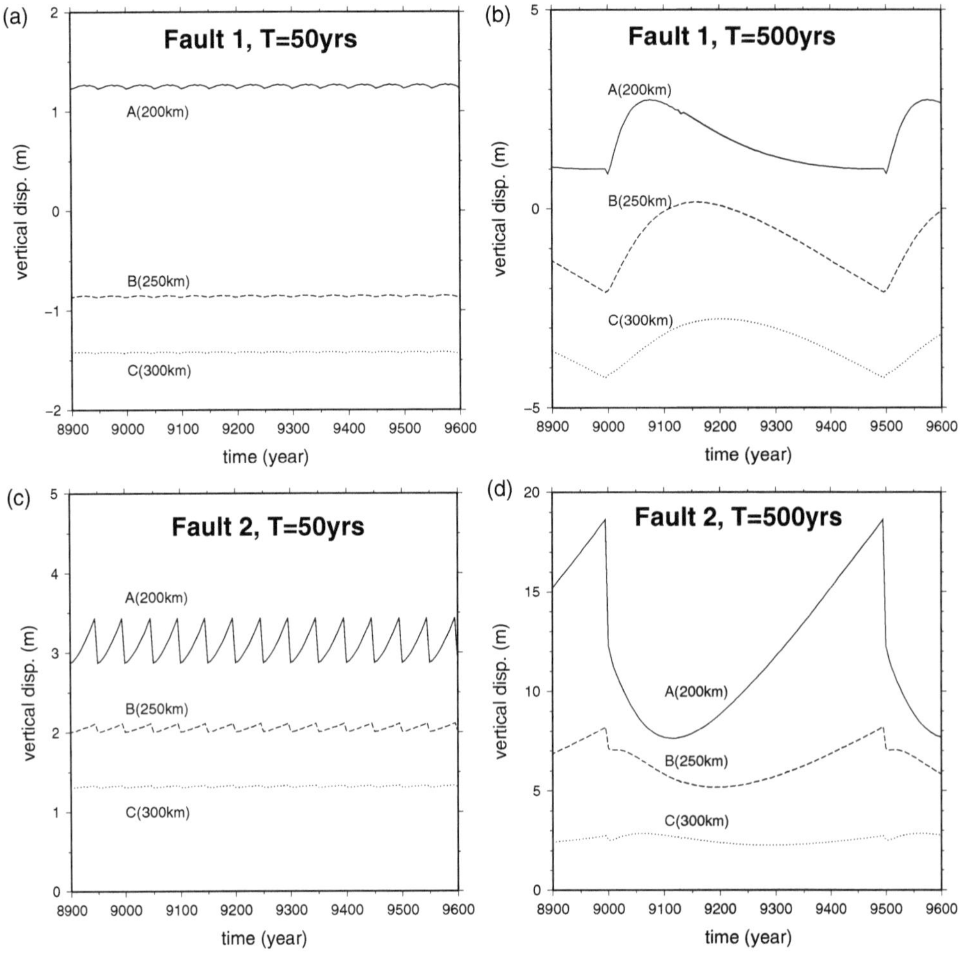

Fig. 3 Calculated vertical displacement for each fault. Two cases of recurrence intervals (50 and 500 years) are shown for each fault. *Curves* show results for different locations around the coastal region shown in Fig. 1

around 5 mm/year (Fig. 3b), which is close to the observed value during the latter half of the twentieth century.

The deeper fault (Fault 2) located close to the coastal area has significant effects on coastal crustal movement all through the earthquake cycle. Resultant displacement becomes too large in the case of the long (500 years) recurrent interval (Fig. 3d). However, in the case of the short recurrence interval, magnitude of interseismic uplift and coseismic subsidence is less than 1 m (Fig. 3c).

In our model, total surface displacement is represented by a sum of contributions from the two faults. In this calculation, we assume that less frequent shallow fault always ruptures with the deep fault. The calculation result is shown in Fig. 4. In order to reproduce characteristic features of the observed crustal deformation, that is, rapid interseismic subsidence (5 mm/year), coseismic subsidence, and postseismic uplift, we need to combine the case of the long recurrence interval at the shallow fault and the short recurrence interval at the deep fault. As long as we consider in the current framework, this is the only possible combination to explain the observed pattern. This combination is also consistent with the seismicity. At the plate interface off Miyagi Prefecture, before the 2011 earthquake, M7- or M8-class earthquakes have repeatedly occurred in 1897, 1933, 1936, 1937, 1978, and 2005 (Headquaters for Earthquake Research Promotion 2003). Though these events are not considered to be truly characteristic (Kanamori et al. 2006), the deeper part of the plate boundary had repeated large earthquakes. In the shallow portion of the plate boundary, earthquakes with estimated magnitude of 7.5–8.0 occurred in 1793, 1897, and 1915 (Headquaters for Earthquake Research Promotion 2003). But it is possible that a large portion had remained unruptured for a long time until the 2011 Tohoku-oki earthquake. Such observation is in general consistent with a long recurrence interval of the shallow fault in our model.

In Fig. 4, calculation result for point B is representative of the coastal displacement. It shows interseismic as well as coseismic subsidence followed by a postseismic uplift. Although the coseismic subsidence at point B is smaller than the observation, it can be increased by minor changes to the model. One possibility is to assume variable coseismic slip for the deep fault. If we assume the deep fault to carry its slip deficit over ordinary earthquake cycles and release all the accumulated slip deficit when the shallow fault ruptures, we can increase coseismic subsidence as shown by B' in Fig. 4. Coseismic slip analyses of the 2011 earthquake showed that

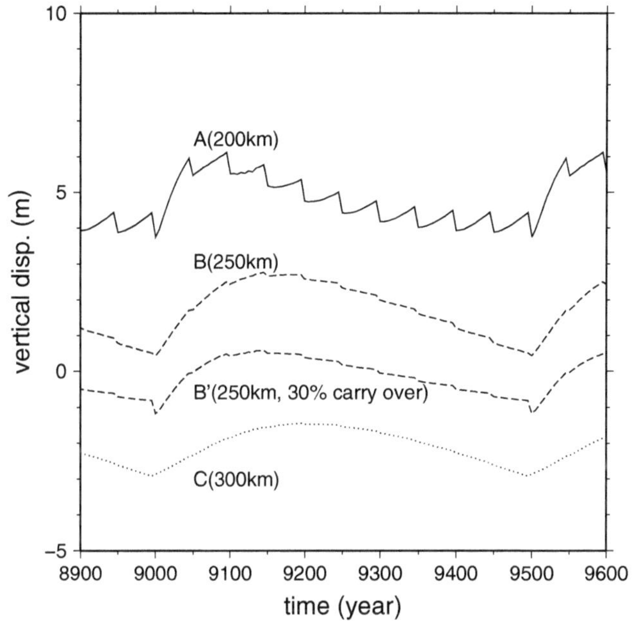

Fig. 4 Calculated vertical displacement time series. B' denotes the case of 30% carry over of slip deficit at fault 2

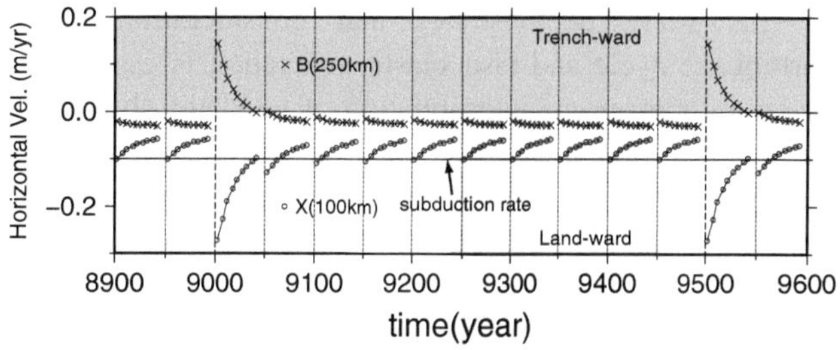

Fig. 5 Calculated temporal changes of interseismic horizontal displacement rate at points X and B

the source area of the 1978 Miyagi-oki earthquake had a much larger fault slip in 2011. So modeling earthquakes with a short recurrence interval with variable coseismic slips is a reasonable assumption. In Fig. 4, it should be also noted that temporal change pattern of the vertical displacement is very sensitive to the location. Point A, located 50 km offshore, has totally different temporal change pattern from that of point B.

After the 2011 Tohoku-oki earthquake, significant postseismic signals have been detected by various geodetic observations such as satellite gravity, on-land GPS, and seafloor GPS-acoustic measurements. Viscoelastic relaxation is considered to be the most important origin of the postseismic deformation (Han et al. 2014; Sun et al. 2014; Watanabe et al. 2014). The viscoelastic cycle model in this study predicts horizontal crustal deformation pattern, which shows significant postseismic landward displacements just above the shallow fault (X in Fig. 5). Since this study applies various simplifying assumptions, a quantitative comparison with the observation data is beyond the scope of this paper. But it should be noted that even such a simplified model can reproduce a basic feature of the postseismic deformation, implying that viscoelastic relaxation has been playing an intrinsic role in the postseismic process.

This study demonstrates that viscoelastic relaxation in the asthenosphere causes significant temporal variation in crustal deformation pattern during the interseismic time period. This effect is of particular importance when an earthquake recurrence interval is longer than the asthenospheric relaxation time. The coastal area of northeast Japan is located above the down-dip edge of the interplate locked zone. Elastic dislocation models predict interseismic uplift due to interplate locking. But our model suggests that long-lasting interplate coupling in the shallow part may result in coastal subsidence. Thus, if we observe interseismic subsidence at the down-dip edge of the locked zone, the shallow part of the plate boundary may have been locked for a long time. In such a situation, a high potential for a big tsunami is implied. Actually, there is such a situation along the Pacific coast of eastern Hokkaido, northern Japan (Murakami and Ozawa 2004).

There are other factors not considered in this study but may affect the crustal deformation pattern significantly. First, structural heterogeneity can change viscoelastic relaxation pattern. Yoshioka and Suzuki (1999) showed that postseismic relaxation pattern is largely affected by the subducting slab. For more realistic modeling, we need to consider heterogenous structure by using a numerical approach such as the finite element method.

Another factor to be considered is spatio-temporal change of the plate boundary slip. Our assumption that interplate locking is perfect and the accumulated slip deficit is totally released by the earthquakes is apparently too simple. Numerical modeling of earthquake cycles based on fault constitutive law suggests there occur aseismic fault slips before and after the earthquake (Tse and Rice 1986). There are also plenty of observational evidence of aseismic slow slip events on the plate boundary (Schwartz and Rokosky 2007). El-Fiky and Kato (1999), through analysis of leveling and triangulation data, estimated about 64% of the plate motion was accommodated with an aseismic slip. On the other hand, recent studies with GPS data analysis have shown that interplate locking ratio was larger, close to full, at least in the deeper part of the off Miyagi region (Suwa et al. 2006; Nishimura et al. 2004; Hashimoto et al. 2009; Mazzaotti et al. 2000). So the simplified assumption of full locking in this study can be considered as a possible end member. Also, it should be noted that a case of partial but steady locking can be considered by simply multiplying the coupling ratio to the calculated displacement results shown above since our calculation purely depend on the perturbation term of the plate boundary slip.

This study demonstrates that consideration of an earthquake cycle and viscoelastic relaxation is essential to make an appropriate interpretation of geodetic observation data. The model predicts that significant deformation pattern change will occur over decades to centuries. The prediction should be verified through comparison with actual observation in the future, and such effort is indispensable in improving our understanding about geodynamic processes.

Acknowledgements Critical comments from three anonymous reviewers were helpful in improving the manuscript. This work was supported by JSPS KAKENHI Grant Number 22540437 and 25282111.

References

El-Fiky GS, Kato T (1999) Interplate coupling in the Tohoku district from geodetic data inversion. J Geophys Res 104:20361–20377

Han S-C, Sauber J, Pollitz F (2014) Broadscale postseismic gravity change following the 2011 Tohoku-Oki earthquake and implication for deformation by viscoelastic relaxation and afterslip. Geophys Res Lett 41. doi:10.1002/2014GL060905

Hashimoto C, Fukui K, Matsu'ura M (2004) 3-D modeling of plate interfaces and numerical simulation of long-term crustal deformation in and around Japan. Pure Appl Geophys 161:2053–2068

Hashimoto C, Noda A, Sagiya T, Matsu'ura M (2009) Interplate seismogenic zones along the Kuril-Japan trench inferred from GPS data inversion. Nat Geosci 2:141–144

Headquaters for Earthquake Research Promotion (2003) Evaluations of occurrence potentials of earthquakes off Sanriku-Boso area (version 2). http://www.jishin.go.jp/main/chousa/11nov_sanriku/sanriku_boso_4_hyoka.pdf

Ikeda Y (2003) Discrepancy between geologic and geodetic strain rates (in Japanese). Gekkan Chikyu 25:125–129

Kanamori H, Miyazawa M, Mori J (2006) Investigation of the earthquake sequence off Miyagi prefecture with historical seismographs. Earth Planets Space 58:1533–1541

Kita S, Okada T, Hasegawa A, Nakajima J, Matsuzawa T (2010) Anomalous deepening of a seismic belt in the upper plane of the upper-plane of the double seismic zone in the Pacific slab beneath the Hokkaido corner: possible evidence for thermal shielding caused by sub ducted forearm crust materials. Earth Planet Sci Lett 290:415–426

Koike K, Machida H (2001) Atlas of quaternary marine terraces in the Japanese islands (in Japanese). University of Tokyo Press, Tokyo, 105 pp.

Matsu'ura M, Sato T (1989) A dislocation model for the earthquake cycle at convergent plate boundaries. Geophys J Int 96:23–32

Mazzaotti S, Le Pichon X, Henry P, Miyazaki S (2000) Full interseismic locking of the Nankai and Japan-west Kurile subduction zones: an analysis of uniform elastic strain accumulation in Japan constrained by permanent GPS. J Geophys Res 105:13159–13177

Murakami M, Ozawa S (2004) Mapped vertical deformation field of Japan derived from continuous GPS measurements and its tectonic implications (in Japanese with English abstract). Zisin 57:209–231

Nakajima J, Hasegawa A (2006) Anomalous low-velocity zone and linear alignment of seismicity along it in the sub ducted Pacific slab beneath Kanto, Japan: reactivation of sub ducted fracture zone? Geophys Res Lett 33. doi:10.1029/2006GL026773

Nishimura T, Hirasawa T, Miyazaki S, Sagiya T, Tada T, Miura S, Tanaka K (2004) Temporal change of interpolate coupling in northeastern Japan during 1995–2002 estimated from continuous GPS observations. Geophys J Int 157:901–916

Ozawa S, Nishimura T, Suito H, Kobayashi T, Tobita M, Imakiire T (2011) Coseismic and postseismic slip of the 2011 magnitude-9 Tohoku-Oki earthquake. Nature 475:273–277

Savage JC, Prescott WH (1978) Asthenosphere readjustment and the earthquake cycle. J Geophys Res 83:3369–3376

Schwartz SY, Rokosky JM (2007) Slow slip events and seismic tremor at circum-Pacific subduction zones. Rev Geophys 45. doi:10.1029/2006RG000208

Sun T, Wang K, Iinuma T, Hino R, He J, Fujimoto H, Kido M, Osada Y, Miura S, Ohta Y, Hu Y (2014) Prevalence of viscoelastic relaxation after the 2011 Tohoku-oki earthquake. Nature 514:84–87. doi:10.1038/nature13778

Suwa Y, Miura S, Hasegawa A, Sato T, Tachibana K (2006) Interplate coupling beneath NE Japan inferred from three-dimensional displacement field. J Geophys Res 111. doi:10.1029/2004JB003203

Tse ST, Rice JR (1986) Crustal earthquake instability in relation to the depth variation of frictional slip properties. J Geophys Res 91:9452–9472

Wang R, Lorenzo-Martin F, Roth F (2006) PSGRN/PSCMP–a new code for calculating co- and post-seismic deformation, geoid and gravity changes based on the viscoelastic-gravitational dislocation theory. Comput Geosci 32:527–541

Watanabe S, Sato M, Fujita M, Ishikawa T, Yokota Y, Ujihara N, Asada A (2014) Evidence of viscoelastic deformation following the 2011 Tohoku-Oki earthquake revealed from seafloor geodetic observation. Geophys Res Lett. doi:10.1002/2014GL061134

Yoshioka S, Suzuki H (1999) Effects of three-dimensional inhomogeneous viscoelastic structures on post-seismic surface deformations associated with the great 1946 Nankaido earthquake. Pure Appl Geophys 154:307–328

Forecasting Earthquakes with the Virtual Quake Simulator: Regional and Fault-Partitioned Catalogs

Mark R. Yoder, Kasey W. Schultz, Eric M. Heien, John B. Rundle, Donald L. Turcotte, Jay W. Parker, and Andrea Donnellan

Abstract

We introduce a framework for forecasting earthquakes using Virtual Quake (VQ), the generalized successor to the perhaps better known Virtual California earthquake simulator. We briefly introduce the VQ simulator, including its availability to research organizations and statistics relevant to earthquake forecasting applications. We discuss contemporary, regional type, forecasts and also show that forecasts can be significantly improved by partitioning catalogs along fault sections.

Keywords

Earthquake • Earthquake forecasting • Earthquake predictability • Earthquake simulator • Virtual California • Virtual Quake

1 Introduction

We discuss the role of the Virtual Quake (VQ) earthquake simulator in future Earth science investigations. VQ, and simulators in general, facilitate a broad range of studies, including surface deformation and gravity change modeling (Schultz et al. 2015). In this manuscript we discuss earthquake forecasting and predictability applications, focusing specifically on the El Mayor-Cucapah (EMC) region of southern California and northern Baja California Norte, Mexico (see Figs. 1 and 2). We discuss two primary benefits of simulated data: (1) the ability to produce extended catalogs from which to calculate robust earthquake probabilities, and (2) data partitioning: the capacity to separate and group earthquakes into individual rupture sources. We introduce a forecasting framework based on the recurrence intervals of large $m > 7$ earthquakes, for which sufficient data to directly calculate robust statistics exist only in synthetic catalogs. We further show evidence that significant information gain, with respect to calculating earthquake probabilities, can be achieved by partitioning the catalog by fault sections – a task that is quite difficult to accomplish with real seismicity, but that is trivial in a simulated environment.

M.R. Yoder (✉) • K.W. Schultz
Department of Physics, University of California Davis, Davis, CA, USA
e-mail: mryoder@ucdavis.edu

E.M. Heien • D.L. Turcotte
Department of Earth and Planetary Sciences, University of California Davis, Davis, CA, USA

J.B. Rundle
Department of Physics, University of California Davis, Davis, CA, USA

Department of Earth and Planetary Sciences, University of California Davis, Davis, CA, USA

Santa Fe Institute, Santa Fe, NM, USA

J.W. Parker • A. Donnellan
NASA Jet Propulsion Laboratory (JPL), Pasadena, CA, USA

2 Virtual Quake: A Technical Overview

Virtual Quake (VQ) is a boundary element, fault type earthquake simulator based on slider-block dynamics and Green's function stress transfer. VQ is a generalization of the perhaps better known Virtual California (VC) earthquake simulator, originally developed by Rundle (1988), in which the fault parameterization has been generalized and simplified

M. Hashimoto (ed.), *International Symposium on Geodesy for Earthquake and Natural Hazards (GENAH)*, International Association of Geodesy Symposia 145, DOI 10.1007/1345_2015_198

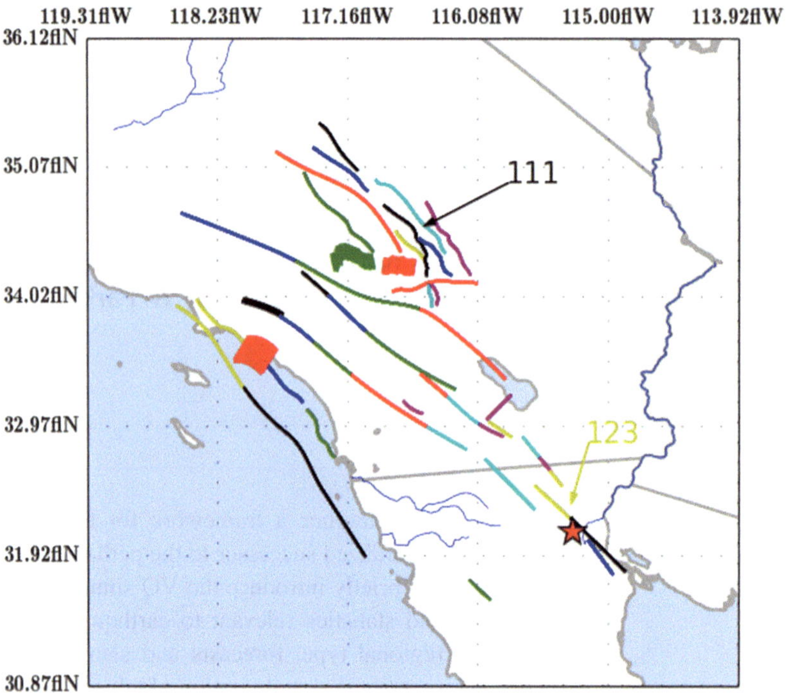

Fig. 1 Map of virtual El Mayor-Cucapah (EMC) fault sections. Individual fault sections are separated by *color*; clusters of *parallel lines* indicate dipping faults. Fault Sects. 111 and 123 (see Sect. 4 and Figs. 5, 6, 7 and 8) are indicated by *arrows* and labels of the *same color*

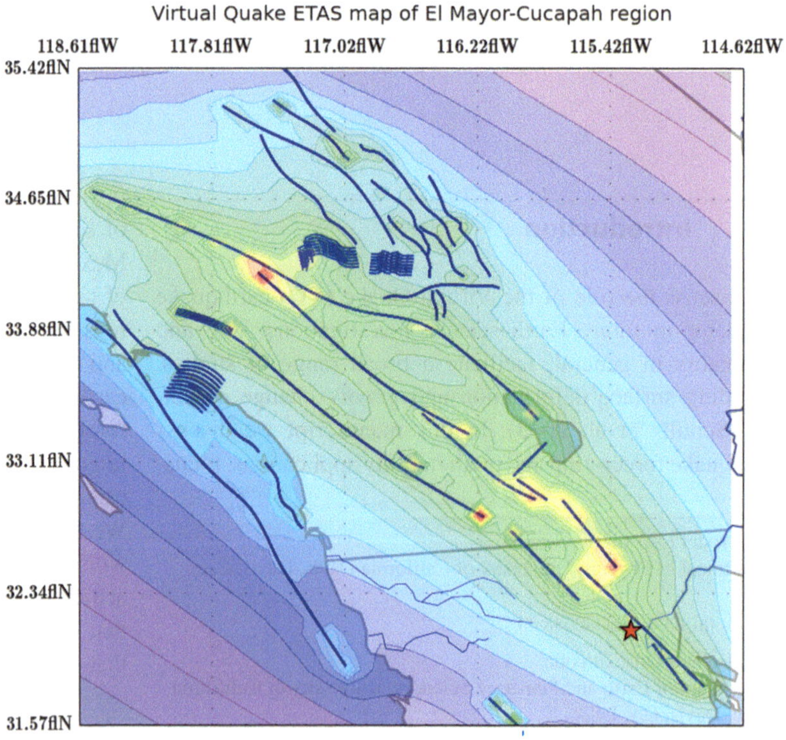

Fig. 2 Fault traces and historical (simulated) seismicity for the El Mayor-Cucapah (EMC) region. Fault traces are shown as *blue lines*; the EMC epicenter is shown by the *red/black star*. All earthquakes in the simulation occur on faults; off-fault seismicity rate fields (*colored* contours) are generated using the spatial component of an ETAS model (Yoder et al. 2015a; Glasscoe et al. 2014; Yoder 2013) based on the VQ (on-fault) catalog. As usual, "warm" ("cool") *colors* indicate regions with high (low) seismic activity

to facilitate modeling of any arbitrary fault system (Sachs et al. 2012; Heien and Sachs 2012; Rundle et al. 2005; Heien et al. 2014). The current release of VQ is written in C++; analytical tools are written primarily in Python and as Python-C++ "extensions" (Python.org 2014; swig.org 2014). VQ is an Open Source project; community development and participation is encouraged; current and past releases of VQ can be downloaded via the Computational Infrastructure for Geodynamics (CIG) or via various development "Forks" from the GitHub version control and repository system (CIG 2015a,b).

The VQ simulator consists of three major components: a fault model, a set of quasi-static elastic interactions (Green's functions), and an event model. The topology and regional significance is defined by the fault model, for example, "Virtual California" is an implementation of VQ with a California fault model. The fault model, itself, incorporates a three level hierarchy:

1. **Faults** are surfaces on which earthquakes occur
2. **Fault sections** are regions along a fault with similar parameterization (slip rate, fault strength, etc.)
3. **Fault elements** are discrete components from which faults are constructed. Slip, stress, and other physics are calculated at this level.

Each fault is meshed from a trace – its general path through the Earth's crust, a set of strength and slip rate parameters, and observed dip and rake angles along the trace. A fault surface is constructed by discretizing the trace into a lattice (of fault elements); faults and fault sections of the EMC region are shown in Fig. 1. In the simulation, fault elements accumulate stress gradually according to local fault parameterization; earthquakes (element slip events) occur when stress exceeds a failure threshold. When a fault element fails, rather than slip "forward," in the direction of stress, failed elements slip backwards returning to their equilibrium position. This "backslip" model, in addition to being a programmatic simplification, eliminates fault evolution and facilitates the calculation of extended earthquake catalogs representative of current earthquake hazard.

Our analysis focuses on a 50,000 year VQ catalog based on the *allcal2* fault model (SCEC 2011; Baral 2012). The spatial distribution of (simulated) earthquakes in the catalog is shown in Fig. 2; the merits and limitations of the VQ simulator, this data set, and the *allcal2* fault model are discussed in some detail by Tullis et al. (2012a,b) and Sachs et al. (2012). One salient conclusion that can be drawn from these analyses is that, while admittedly imperfect, the simulators in the study – including VQ and this catalog specifically, are suitable tools for studying earthquake probabilities and various other aspects of earthquake related physics. See also Yoder et al. (2015b) and Schultz et al. (2015) (this issue) for further discussion and examples of VQ based Earth science investigations.

3 Extended Catalogs: Regional Forecasts

Perhaps the most obvious advantage of simulator based studies, as opposed to using observed earthquake catalogs, is the quantity of available data. For example, according to the Advanced National Seismic System (ANSS) search engine (NCEDC 2014), just three $m > 7$ earthquakes have occurred in the EMC region of southern California and northern Baja California Norte (see Fig. 2) over the past 100 years. These are specifically the Landers (28 June 1992, $m = 7.3$), Hector Mine (16 October 1999, $m = 7.1$), and of course the El Mayor-Cucapah (10 April 2010, $m = 7.2$) events. In practice, this small sample size suggests a rate of, applying Poisson statistics, $3 \pm \sqrt{3}$ (with a coefficient of variation $c_v \approx 0.58$) earthquakes every ≈ 18 years (the interval over which the most recent $m > 7$ events occurred) or ≈ 100 years (or more – the duration of the available catalog). Foregoing, for the time being, an extended discussion of how to properly quantify the mean recurrence interval (and uncertainty) of these three events, we suggest that the available data – by themselves, are simply not sufficient in quantity to calculate robust rates and probabilities of $m > 7$ earthquakes in the region. In contrast, our synthetic catalog contains $2{,}207 \pm 47$ events and $c_v \approx 0.02$ for the same geographical region. These extended catalogs facilitate statistical modeling of regional seismicity that is simply not possible using current observation based seismographic data. Again, Tullis et al. (2012a,b) discuss the validity of the catalog and simulator in some detail.

The VQ quantities of primary interest for forecasting are the recurrence intervals Δt_r between large $m \geq m_0$ earthquakes and the "waiting times" Δt_w until the next large earthquake after some time t_0 since the most recent large event, $\Delta t_r = t_0 + \Delta t_w$. For this manuscript, we have chosen (somewhat arbitrarily) $m_0 = 7$, and for brevity – unless otherwise specified, we refer to $m > m_0$ and $m > 7$ interchangeably (Yoder et al. 2015b). Following Rundle et al. (2005) and Shcherbakov et al. (2005), we model recurrence interval probabilities (RI) as a conditional Weibull distribution,

$$P(\Delta t_r; t_0) = 1 - exp\left[\left(\frac{t_0}{\tau}\right)^{\beta} - \left(\frac{\Delta t_r}{\tau}\right)^{\beta}\right], \quad (1)$$

where, β and τ are fitting parameters, and $P(\Delta t_r, t_0)$ represents the probability that an interval $\Delta t_r'$ is greater than t_0 and less than Δt_r. In Fig. 3, we show RI distributions, and their fits to Eq. (1), for $t_0 = \{0,\ 11.13,\ 22.27,\ 33.40,\ 44.53,$ and $55.66\}$ years, or $t_0 = \langle \Delta t_r \rangle_{EMC} \times \{0,\ 0.5,\ 1,\ 1.5,\ 2,$ and $2.5\}$ – where $\langle \Delta t_r \rangle_{EMC}$ is the mean interoccurrence interval of $m > 7$ earthquakes in the (simulated) EMC region.

Figure 3 shows, in addition to (1) the observed distributions (solid lines), (2) the expected distribution based on the best fit parameters for $t_0 = 0$ – keeping these parameters fixed and changing only t_0 in Eq. (1), and (3) a direct fit to Eq. (1) for each value of t_0 – noting that these individual plots may not be easily distinguished because they plot closely on top of one another. These fits produce values of $\langle \tau \rangle_{t_0} = 25.3 \pm 9.0$ and $\langle \beta \rangle_{t_0} = 1.17 \pm 0.24$, which is close to the Poisson value of $\beta = 1$. These fits suggests reasonable agreement between the expected (theory) distribution and observed (simulated) values – and we will further qualify this statement in Sect. 4.

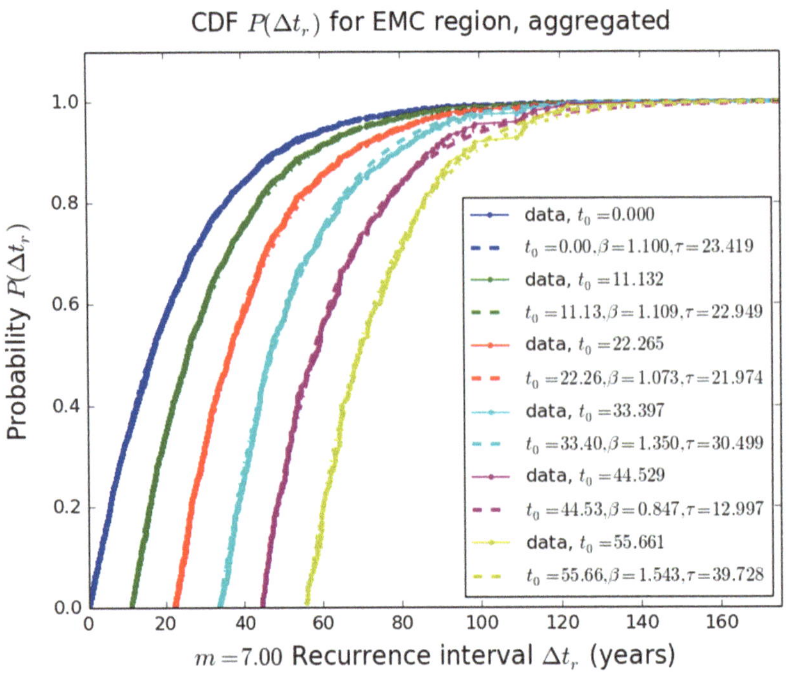

Fig. 3 EMC regional recurrence interval probability distributions: the probability distribution of the recurrence intervals for $m > 7$ earthquakes given the observation that some time t_0 has already passed since the previous $m > 7$ event; the x axis indicates the total interval $\Delta t_r = t_0 + \Delta t_w$ between $m > 7$ earthquakes. The values of t_0 correspond to 0, 0.5, 1, 1.5, 2, and 2.5 times the mean interoccurrence interval, $\langle \Delta t_r \rangle \approx 22.7$ years. *Each color* represents the distribution for a value of t_0; the *solid line* indicates the observed (in the simulation) distribution; a *dotted line* indicates the best fit to a Weibull distribution for that curve; there is also a *dashed line* that indicates the waiting time distributions based on the t_0 fit, and applying Eq. (1). The best fit parameters are shown in the caption; their mean values are $\langle \beta \rangle = 1.17 \pm 0.24$ and $\langle \tau \rangle = 25.3 \pm 9.0$. These best fit parameters suggest that the EMC regional seismicity is in good agreement with the random Weibull theory Eq. (1), and is in fact nearly Poissonian $\beta \approx 1$

Inverting Eq. (1), the expected waiting time intervals (EWT) are

$$\langle \Delta t_w(t_0; P) \rangle = \tau \left[\left(\frac{t_0}{\tau} \right)^{\beta} - \ln(1 - P) \right]^{\frac{1}{\beta}} - t_0. \quad (2)$$

We show EWT for the EMC region in Fig. 4. Note that Figs. 3 and 4 show that RI and EWT for the EMC region, in aggregate, are approximately Poissonian, $\beta \approx 1$.

4 Data Partitioning: Faultwise Forecasts

Perhaps more subtle, but no less important than the quantity of data, the ability to precisely partition those data constitutes a significant advantage to working with synthetic catalogs. While quite challenging (and important) for real earthquakes in observed catalogs (Yoder and Rundle 2015; Yoder 2014), this problem – of defining spatio-temporal extents and magnitude domains to resolve individual earthquake sequences and facilitate information gain, is easy and straightforward in the VQ simulated environment.

In order to isolate individual earthquake source processes, keeping in mind the fault model hierarchy discussed in Sect. 2, we partition our catalog by fault section. Specifically, each fault section sub-catalog contains all events where fault elements on that section slipped – noting that (1) the epicenter (initial slip) or center of mass of the event might have occurred on a different fault section, and (2) large events might exist in multiple sub-catalogs (span multiple sections). For all working catalogs, in cases where multiple section sub-catalogs are combined, each event is unique (multi-section duplicates are removed).

Figures 5, 6, 7 and 8 show RI and EWT distributions for two fault sections in the EMC region (Sects. 111 and 123; see Fig. 1 for their geographical locations). Preliminary analysis suggests two classes of fault sections: (1) those for which the random Weibull theory Eq. (1) *over-predicts* large earthquakes (Sect. 111) and (2) sections where Eq. (1) *under-predicts* large events (Sect. 123). Figures 6 and 8 further imply that on some faults, large earthquakes are most probable (EWT are *shortest*) immediately after a large earthquake – implying activation type triggering. On other faults, the probability is lowest (EWT are longest) immediately following a large earthquake and then systematically

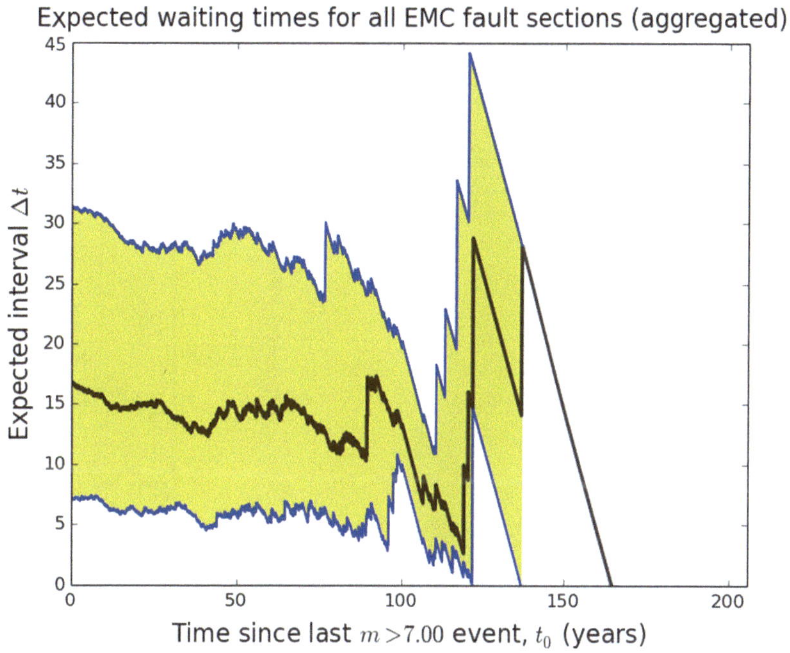

Fig. 4 Expected waiting times $\Delta t_w = \Delta t_r - t_0$ until the next $m > 7$ earthquake, in the EMC region, at a time $t = t_0$ since the previous $m > 7$ event. The x axis represents the time t_0 since the previous $m > 7$ event in the catalog; the y axis shows Δt_w. The *black line* shows the 50th percentile probability; the upper and lower bounds of the *yellow* area show the 75th and 25th probabilities, respectively

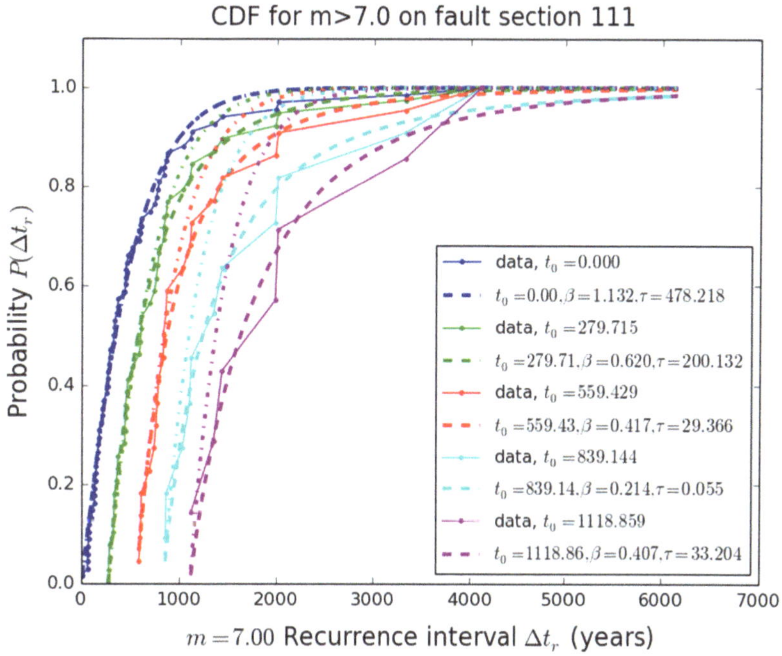

Fig. 5 RI probability distributions for fault Sect. 111, similar to Fig. 3. The *dot-dash curves* represent the expected conditional Weibull distribution based on the $t_0 = 0$ fit (first, *blue curve*), Eq. (1). The *point* and *solid lines* indicate data (simulated earthquakes), and the *dashed lines* indicate the "local" best fit – the best fit of that data set (for a given t_0 to Eq. (1). Again, the values of t_0 shown correspond to 0, 0.5, 1.0, 1.5, 2.0, and 2.5 times the mean interoccurrence interval $\langle \Delta t_r \rangle$ for $m > 7$ earthquakes on this fault section. Note that for these earthquakes, the default $t_0 = 0$ Weibull theory systematically **over-predicts** events for $t_0 > 0$. For this fault section, $t_0 = 0$ is nearly Poissonian $\beta = 1.13$ and $\beta < 1$ for $t_0 > 0$

increases – consistent with classic elastic-rebound theory or quiescent type earthquake triggering (Rundle et al. 2011; Yoder et al. 2012). Yoder et al. (2015b) discuss additional examples of this distinction and show RI and EWT distributions for all fault sections in the EMC region.

Yoder et al. (2015b) further show that interval statistics of faultwise sub-catalogs, as defined above, exhibit measurable deterministic behavior. Specifically, they show that the rate of smaller $m < 7$ (simulated) earthquakes increases precursory to large $m > 7$ earthquakes. This is similar to a scaling based framework, for real earthquake sequences, developed by Yoder (2011); Yoder and Rundle (2015), with the notable distinction that the synthetic catalog contains many more testable cases (large earthquakes) than are presently available in observed data. In general, Figs. 5, 6, 7 and 8 suggest that simulators, because they facilitate sophisticated data partitioning that is difficult, if not impossible, using observed data and extended catalogs not available from current

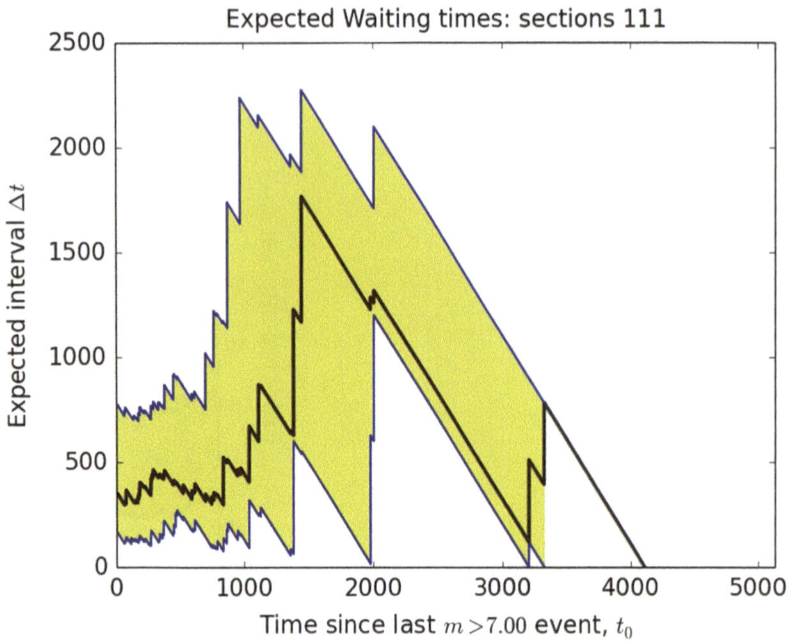

Fig. 6 Expected waiting times $\langle \Delta t_w \rangle$ for fault Sect. 111, similar to Fig. 4. Note that, for this fault section, the shortest expected interval Δt_w until the next $m > 7$ earthquake occurs immediately following the previous $m > 7$ event

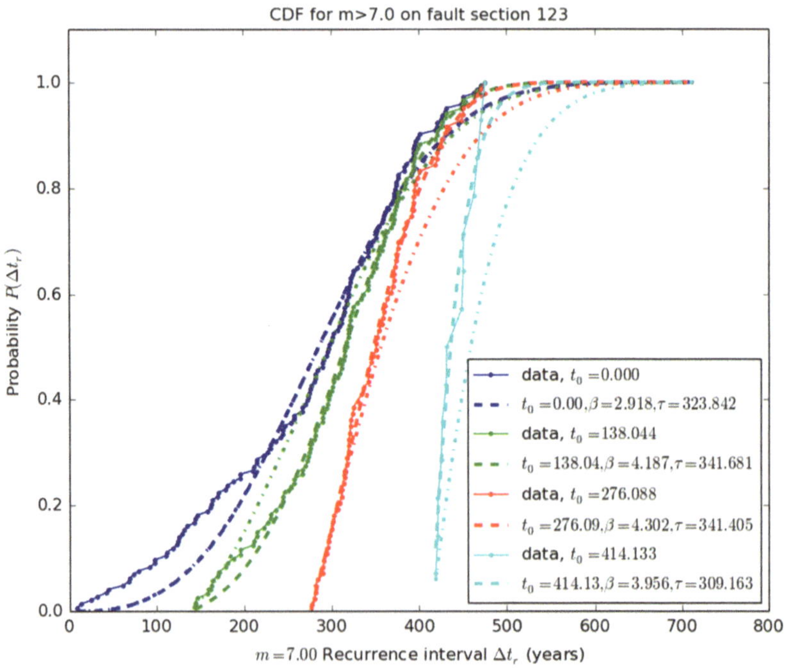

Fig. 7 RI probability distributions, Eq. (1) for events on fault Sect. 123, similar to Fig. 5. In this case, however, the $t_0 = 0$ Weibull theory systematically **under-predicts** $m > 7$ events for $t > t_0$ and $\beta \gg 1$ for all times t_0

seismographic databases, will play an important role in future studies to understand earthquake predictability and other characteristics of large earthquakes.

5 Discussion and Conclusions

Given the geological time scales that characterize large earthquake recurrence (on the order of $\Delta t > 1{,}000$ years), as compared to the length of the current seismographic record ($\Delta t \approx 100$ years – arguably $10 < \Delta t < 20$ years for high quality data), earthquake simulators will invariably play an important role in the future of earthquake forecasting and other areas of solid-earth science. In addition to producing extended catalogs, simulators also facilitate sophisticated catalog partitioning, and other techniques, that promise to facilitate significantly improved earthquake probability calculations. In this paper, we specifically discuss the benefits of catalog partitioning and show that faults fall into two categories – those for which Weibull theory

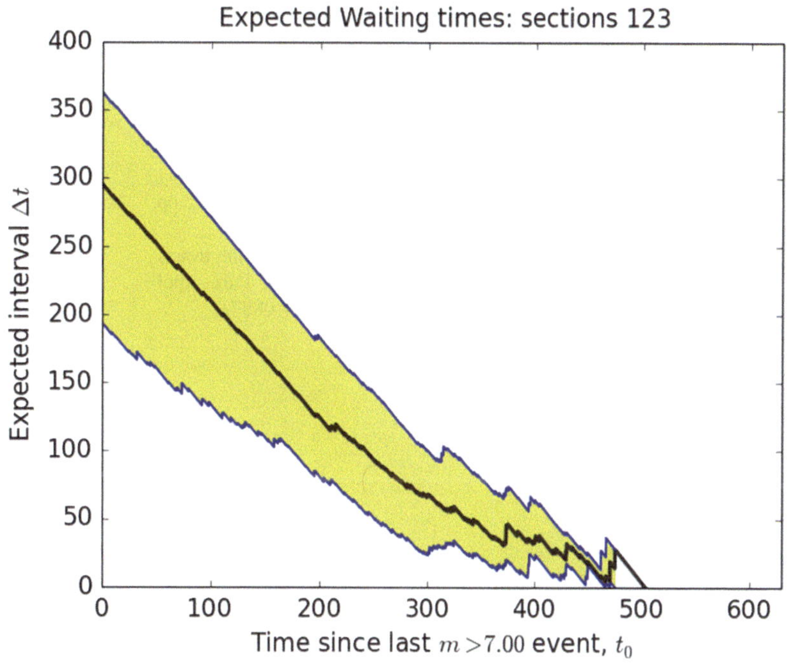

Fig. 8 Expected waiting times $\langle \Delta t_w \rangle$ for fault Sect. 123. Note that these waiting time distributions are consistent with classic elastic-rebound type models, in which the probability of failure (an earthquake) increases (or equivalently, the expected waiting time decreases) monotonically with increasing time t_0, presumably as a fault gradually accumulates stress

Eq. (1) under-predicts, and those for which it over-predicts, the rate of large earthquakes. Based on this, in addition to using simulators to improve time-independent earthquake probability calculations, it may be possible to develop time dependent earthquake forecasts based on the distinction between activation and quiescence type triggered faults.

Acknowledgements This research is supported by the NASA Earth and Space Science fellowship number NNX11AL92H, NASA grant NNX08AF69G, and JPL Subcontract 1291967.

References

Baral M (2012) Data transfer file formats for earthquake simulators. Seismol Res Lett 83(6):991–993

CIG (2015a) Computational infrastructure for geodynamics: virtual quake. http://geodynamics.org/cig/software/vq, (last viewed, Jan 2015)

CIG (2015b) Virtual quake: Github. https://github.com/geodynamics/vq

Glasscoe M, Wang J, Pierce M, Yoder M, Parker J, Burl M, Stough T, Granat R, Donnellan A, Rundle J, Ma Y, Bawden G (2014) E-decider: using earth science data and modeling tools to develop decision support for earthquake disaster response. Pure Appl Geophys (Published online), Topical Volume on Multihazard Simulation and Cyberinfrastructure

Heien EM, Sachs MK (2012) Understanding long term earthquake behavior through simulation. Comput Sci Eng 14(5):10–15

Heien EM, Sachs MK, Schultz KW, Turcotte DL, Rundle JB (2014) Cig virtual quake manual v1.1. http://geodynamics.org/cig/software/github/vq/v1.1.0/vq_manual_1.1.0.pdf

NCEDC (2014) Advanced national seismic system (anss) catalog search (ncedc). doi:10.7932/NCEDC. http://www.ncedc.org/anss/catalog-search.html

Pythonorg (2014) Extending Python with C or C++. https://docs.python.org/2/extending/extending.html, (viewed in 2014)

Rundle JB (1988) A physical model for earthquakes. 2. Application to southern California. J Geophys Res Solid Earth 93(B6): 6255–6274

Rundle J, Rundle P, Donnelan A, Turcotte D, Shcherbakov R, Li P, Malamud B, Grant LB, Fox G, Mcleod D, Yakovlev G, Parker J, Klein W, Tiampo K (2005) A simulation-based approach to forecasting the next great san francisco earthquake. Proc Natl Acad Sci 102(43):15,363–15,367. doi:10.1073pnas.0507528102

Rundle JB, Holliday JR, Yoder MR, Sachs MK, Donnellan A, Turcotte DL, Tiampo KF, Klein W, Kellogg LH (2011) Earthquake precursors: activation or quiescence. Geophys J Int 187(1):225–236

Sachs MK, Heien EM, Turcotte DL, Yikilmaz M, Rundle JB, Kellogg LH (2012) Virtual California earthquake simulator. Seismol Rev Lett 83(6):959–963

SCEC (2011) http://scec.usc.edu/research/eqsims/documentation.html

Schultz KW, Sachs MK, Heien EM, Yoder MR, Rundle JB, Turcotte DL, Donnellan A (2015) Virtual quake: statistics, co-seismic deformations and gravity changes for driven earthquake fault systems. In: Proceedings of the International Association of Geodesy Symposia (available online)

Shcherbakov R, Yakovlev G, Turcotte DL, Rundle JB (2005) Model for the distribution of aftershock interoccurrence times. Phys Rev Let 95:218501. doi:10.1103/PhysRevLett.95.218501

Swigorg (2014) Swig software development tool. http://www.swig.org/Doc1.3/Python.html. http://www.swig.org/Doc1.3/Python.html (viewed 2014)

Tullis TE, Richards-Dinger K, Barall M, Dieterich JH, Field EH, Heien EM, Kellogg LH, Pollitz FF, Rundle JB, Sachs MK, Turcotte DL, Ward SN, Yikilmaz MB (2012a) A Comparison among observations and earthquake simulator results for the allcal2 California fault model. Seismol Res Lett 83(6):994–1006. doi:10.1785/0220120094

Tullis TE, Richards-Dinger K, Barall M, Dieterich JH, Field EH, Heien EM, Kellogg LH, Pollitz FF, Rundle JB, Sachs MK, Turcotte DL, Ward SN, Yikilmaz MB (2012b) Generic earthquake simulator. Seismol Res Lett 83(6):959–963. doi:10.1785/0220120093

Yoder MR (2011) Record-breaking earthquake precursors. Ph.D. thesis, University of California Davis
Yoder MR (2013) ETAS and Record-breaking Hazard Maps. http://gw11.quarry.iu.teragrid.org/myoder/ecatwww/edhazmapthin.html
Yoder MR (2014) Earthquakes: risk factors, seismic effects, and economic consequences. Nova Science Publishers, Inc., New York. Chapter 9, 1/F and the Earthquake Problem: Earthquake Forecasting and a Framework for Predictability – Past, Present and Future, pp 195–227
Yoder MR, Rundle J (2015) Record-breaking intervals: detecting trends in the incidence of self-similar earthquake sequences. Pure Appl Geophys 172(8):2215–2235. doi:10.1007/s00024-014-0887-7
Yoder MR, Turcotte DL, Rundle JB (2012) Extreme events and natural hazards: the complexity perspective. Geophysical Monograph Series, vol 196, AGU, Washington, DC. chap Earthquakes: complexity and extreme events, pp 17–26. doi:10.1029/2011GM001071
Yoder MR, Rundle JB, Glasscoe MT (2015a) Near-field ETAS constraints and applications to seismic hazard assessment. Pure Appl Geophys 172(8):2277–2293. doi:10.1007/s00024-014-0785-z
Yoder MR, Schultz KW, Heien EM, Rundle JB, Turcotte DL, Parker JW, Donnellan A (2015b) The virtual quake earthquake simulator: a simulation based forecast of the el mayor-cucapah region and evidence of predictability in simulated earthquake sequences (in production). doi:10.1093/gji/ggv320

Part III

Near Real-Time Warning

Development and Assessment of Real-Time Fault Model Estimation Routines in the GEONET Real-Time Processing System

Satoshi Kawamoto, Kohei Miyagawa, Toshihiro Yahagi, Masaru Todoriki, Takuya Nishimura, Yusaku Ohta, Ryota Hino, and Satoshi Miura

Abstract

The recent development of Global Navigation Satellite Systems (GNSS) and communication infrastructure provides real-time displacement data, enabling the real-time estimation of coseismic fault models for large earthquakes and the potential improvement of tsunami warning systems. In this paper, we present two real-time coseismic fault model estimation procedures implemented in the GEONET real-time processing system: a single rectangular fault model and a slip distribution model on the subducting plate interface. We evaluate the fault modeling procedures for two past large interplate earthquakes: the 2003 Tokachi-oki earthquake and the 2011 Tohoku earthquake. Furthermore, we also evaluate a potential large interplate earthquake along the Nankai trough based on simulated GPS time series. We obtained reasonable coseismic simple rectangular fault models for the past two large earthquakes. In contrast, the simulated interplate event along the Nankai trough clearly indicates the great advantage of estimation of the coseismic slip distribution on the plate interface. These results clearly show that reliable estimation of the coseismic fault model and/or slip distribution on the subducting plate interface will be possible in Japan based on the real-time GNSS data. This result will contribute to a more reliable tsunami early warning system.

Keywords

Earthquake early warning • Fault model inversion • GNSS • GPS • Real-time kinematic (RTK) positioning

1 Introduction

A fault model for a large earthquake, estimated from geodetic measurements, provides a moment magnitude (*Mw*), rupture length and width that are free from the saturation problem found for seismic data (Blewitt et al. 2006; Allen et al. 2009). The obtained fault parameters including true *Mw* constrain the size of a potential subsequent tsunami, being, therefore, very important for disaster mitigation, e.g., tsunami warnings. Since the modeled results must be obtained rapidly, a possible implementation combines the use of real-time Global Navigation Satellite System (GNSS) positioning with an automated estimation of a finite fault model (Blewitt et al. 2009; Ohta et al. 2012). For the 2011 Tohoku earthquake

S. Kawamoto (✉) • K. Miyagawa
Geospatial Information Authority of Japan, 1 Kitasato, Ibaraki, Japan
e-mail: kawamoto-s96tf@mlit.go.jp

T. Yahagi
Cabinet Secretariat, 2-4-12 Nagatacho, Tokyo, Japan

M. Todoriki
Center for Integrated Disaster Information Research, Interfaculty Initiative in Information Studies, The University of Tokyo, 1-1-1 Yayoi, Tokyo, Japan

T. Nishimura
Disaster Prevention Research Institute Kyoto University, Gokasho, Kyoto, Japan

Y. Ohta • R. Hino • S. Miura
Graduate School of Science, Tohoku University, 6-3 Aramaki Aza-Aoba, Miyagi, Japan

M. Hashimoto (ed.), *International Symposium on Geodesy for Earthquake and Natural Hazards (GENAH)*,
International Association of Geodesy Symposia 145, DOI 10.1007/1345_2015_49

(*Mw* 9.0), the tsunami warning was issued based on a saturated initial magnitude by a seismic-based Earthquake Early Warning (EEW) and the predicted tsunami heights were underestimated (Ozaki 2011). Therefore, as a response to the catastrophic tsunami caused by the Tohoku earthquake, the Japan Meteorological Agency (JMA) has been developing a more robust tsunami warning system. However, GNSS observations have not yet been integrated in this system because the reliability for GNSS-based tsunami warning in the operational phase must be carefully assessed.

The Geospatial Information Authority of Japan (GSI) and Tohoku University have jointly developed a real-time analysis system within Japan's national GNSS network, the GEONET (Yamagiwa et al. 2006): the Real-time GEONET Analysis system for Rapid Deformation monitoring (REGARD), to model an earthquake fault in real-time. The goal of this collaboration project was to provide reliable size estimation for large earthquakes to be used later in tsunami warning. The prototype system of REGARD involves the processing of continuous real-time streams from 146 GEONET GNSS stations, using the RTKLIB 2.4.1 program package (Takasu 2011), and the automatic modeling of a large earthquake event, detected either by the RAPiD algorithm (Ohta et al. 2012) or the EEW from the JMA, assuming the source geometry as a single rectangular fault model. In addition to the single rectangular fault modeling routine, we developed another routine to model the heterogeneous slip distribution on a curved fault surface, e.g., the Nankai Trough, in which a single rectangular fault routine may poorly reproduce the fault surface because the strike direction of the fault is not uniform along the trench (e.g., Furumura et al. 2011).

In the present study, we describe how the two earthquake fault modeling procedures are implemented in our REGARD system and assess their performance by analyzing their suitability in reproducing past large earthquakes and a simulated large interplate earthquake in the Nankai Trough.

2 Method

2.1 Automated Earthquake Fault Modeling Routines for REGARD

Currently, the first tsunami warning is issued 3 min after the origin of an earthquake in Japan. Thus, the inversion of the fault model must be quick and automated based on the real-time GNSS data. Therefore, earthquake fault modeling routines are designed to be completed within 10–30 s for each inversion analysis and to be repeated for a fixed period of time. In a sequence of inversions for an event, the latest static ground displacements were extracted from the GNSS time-series up to the present time. The two basic routine designs involve: (1) a nonlinear inversion of a single rectangular fault with a loosely constrained a priori model, and (2) a linear inversion of a slip distribution model on the assumed subducting plate interface.

2.1.1 Single Rectangular Fault Modeling Routine

The single rectangular fault modeling routine estimates 12 parameters: nine fault parameters described by $\mathbf{m_f}$ (latitude, longitude, depth, length, width, strike, dip, and rake angles, and amount of slip) and a translation parameter $\mathbf{T}$ that accounts for common mode noise in the east, north, and vertical components. The Green's function was calculated by a dislocation model in a homogeneous half-space (Okada 1992). To assume a normal prior distribution of the model parameters described by $\mathbf{m_f}$, we used the nonlinear optimization method proposed by Matsu'ura and Hasegawa (1987).

The parameters were set a priori in two steps: first, the initial location of the fault was set from the epicenter, identified by the EEW, or, if that information was not available, set to the site location where the largest displacement was detected; second, the initial strike, dip, and rake angles were taken from a pre-determined moment tensor database (Fig. 1a), which lists the typical mechanisms of the earthquakes occurring in each 0.5×0.5 degree grid. The standard deviations of the model parameters were set to be $\mathbf{m_s}$ = [0.02L (in deg.), 0.02L (in deg.), 5 (km), 2L (m), 2L (m), 10 (deg.), 10 (deg.), 10 (deg.), 10 (m)], where the order of the parameters is the same as $\mathbf{m_f}$ and L is the initial fault length. Here, the initial fault length L was determined from the moment identified by EEW with a scaling law: $W = L/2$ and $s = W/10{,}000$, where W is the fault width and s is the amount of slip. Finally, the 12 parameters were optimized starting with the initial parameters described above. A sequence of this inversion procedure was carried out for each event. Immediately after the end of an inversion, the next inversion was repeated updating the displacements with the latest data and the initial parameters with the best fault parameters found for the previous inversions for the same event.

2.1.2 Fault Slip Distribution Modeling Routine

The second fault modeling routine computes the slip distribution on the plate interfaces along the Japan Trench, the Nankai Trough, the Ryukyu Trench, and the Sagami Trough (Fig. 1b). We interpolated the plate interface models (Nakamura and Kaneshiro 2000; Baba et al. 2002; Nakajima and Hasegawa 2006, 2007; Hirose et al. 2008; Kita et al. 2010; Cabinet Office 2013) and discretized the curvature of the plate interfaces using triangular meshes. The Green's function was calculated with a triangular dislocation element (Meade 2007). The epicenter was determined in the same way as described in Sect. 2.1.1 and the plate interface model closest to the epicenter was chosen for the inversion analysis.

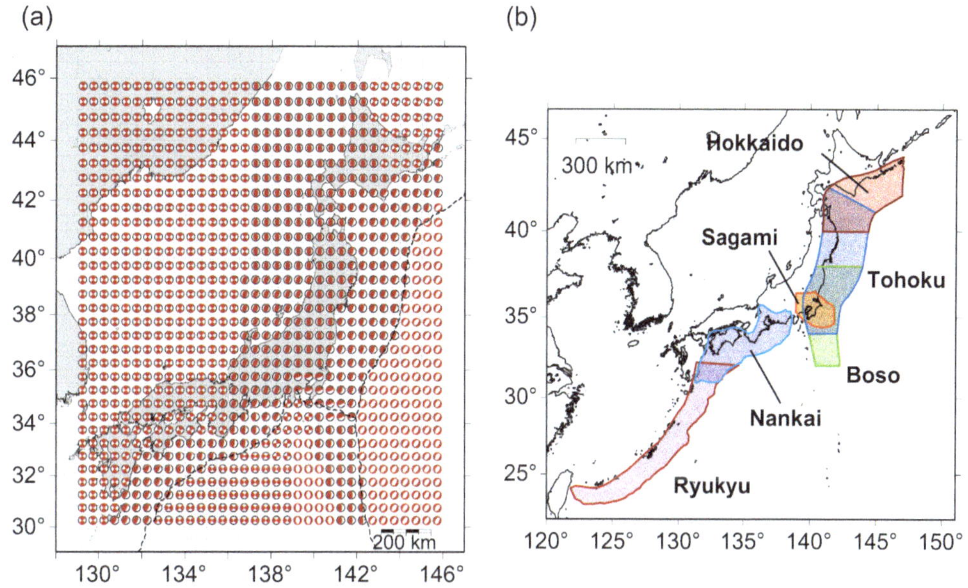

Fig. 1 (**a**) Pre-determined moment tensor database. The closest moment tensor was used to provide the initial strike, dip, and rake angles for the single rectangular fault model inversions. (**b**) Pre-determined triangular elements mapped on the plate interfaces

The slip distribution was smoothed by a second-order Tikhonov regularization (e.g., Aster et al. 2012) with the linear least squares problem solved with a minimization of the L2 norm:

$$min.\|\mathbf{Gm}-\mathbf{d}\|+\alpha\|\mathbf{Lm}\|,$$

where **G** is the Green's function, **m** is the fault slip, **d** is the static ground displacement, **L** is the roughness matrix, and α is the hyper parameter that controls the smoothness of the slip.

Since α differs for each case, a globally optimized α was approximated in advance. We assumed forward models with a *Mw* of 8.5 for the areas where large earthquakes repeatedly occurred (Fig. 2a–c), and calculated the static displacements due to the simulated slip distributions. The static displacements were recovered for varying values of α, and the best value was determined using a L-curve criterion (Hansen 1992), in which the best α corresponds to the solution at the corner of the L-curve (Fig. 2d–i) between the residual norm and the solution seminorm.

2.2 Posterior Analyses of Past Earthquakes

To test both fault modeling routines described in previous section, we used GEONET data for the 2003 Tokachi-oki earthquake (*Mw* 8.3) and the 2011 Tohoku earthquake (*Mw* 9.0). A simulated dynamic displacement dataset for a worst-case scenario for the Nankai Trough earthquake (Todoriki et al. 2013, Fig. 3) was also used. The Nankai Trough earthquake was based on the model of the 1707 Hoei earthquake (Furumura et al. 2011), consisting of ten fault segments along ~600 km from Suruga Bay to Hyuga-nada with input slips for each segment of 5.6, 7.0, 5.6, 9.2, and 9.2 m from east to west, respectively.

The previously dumped real-time 1 Hz GNSS streaming data was used for the REGARD system based on the post-processed kinematic baseline analysis. Therefore the reproduced real-time data reflected all missing observations. The time series were estimated by the RTKLIB 2.4.1 software, which offers kinematic baseline solutions. The second half-predicted orbit from the International GNSS Service (IGS) Ultra-Rapid orbit (Kouba 2009) was used. The tropospheric delay and the ionospheric effects were estimated epoch-by-epoch. We applied a 20 s-moving average filter to the 1 Hz site position time-series to smooth dynamic components because the Green's functions are valid only for static components. After the application of the moving average filter, the static displacements for the inversions were obtained by subtracting the reference position obtained 5 min before the event was detected. The *Mw* was estimated with a rigidity of 30 GPa.

3 Results

3.1 2003 Tokachi-oki Earthquake (*Mw* 8.3)

For the 2003 Tokachi-oki earthquake, the maximum coseismic slip was estimated to be ~50 km east from Cape Erimo, in the 4–9 m range with a *Mw* = 8.0–8.3 (Yagi 2004; Miyazaki et al. 2004; Ozawa et al. 2004; Yamanaka and Kikuchi 2003; Crowell et al. 2012; Global Centroid Moment Tensor (GCMT)).

The *Mw* estimates for time = 3 min showed a *Mw* = 7.94 for the single rectangular fault model and a *Mw* = 8.19 for the slip distribution model (Fig. 4a, S. 1). Both models

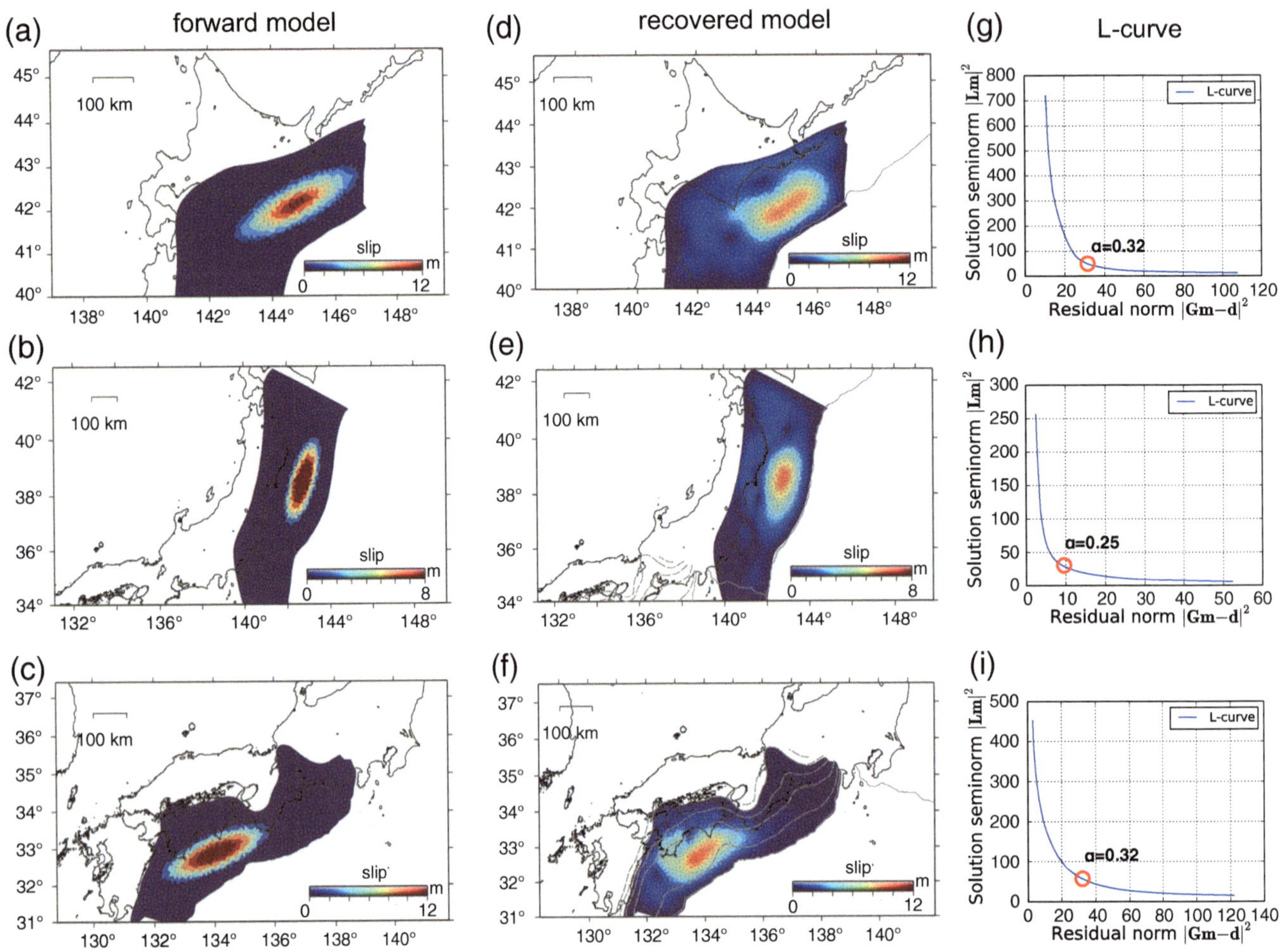

Fig. 2 Forward models for the approximation of the smoothing parameters (**a**–**c**), recovered models with the best smoothing parameters (**d**–**f**), and best values for the smoothing parameters (**g**–**i**) found at the corner of the L-curve

showed a good variance reduction (VR) above 90 % and a stabilization of the time variation of the *Mw* estimates ~120 s after the origin time. The average computation times were ~12 s for the single rectangular model and less than 10 s for the slip distribution model, which is an acceptable latency for operational use. Our slip distribution model was slightly better than the single rectangular fault model in VR by 3.0 %, and agreed well with Crowell et al. (2012) results (*Mw* = 8.23), who also uses a slip distribution model. However, the single rectangular fault model was located 5 km shallower than the actual plate interface and caused the underestimation of *Mw* = 7.94.

3.2 2011 Tohoku Earthquake (*Mw* 9.0)

The 2011 Tohoku earthquake caused a fault rupture over 300 km along the Japan Trench with the main rupture occurring ~150 km away from the coast line (e.g., Nishimura et al. 2011; Ozawa et al. 2011, 2012; Simons et al. 2011; Melgar et al. 2013; Melgar and Bock 2013). The estimates of *Mw* range from 8.8 to 9.0 and the maximum slip is estimated roughly in 25 m from onshore observations (Ozawa et al. 2011; Wright et al. 2012; Melgar et al. 2013) or over 50 m by adding sea-floor measurements (Simons et al. 2011; Ozawa et al. 2012). There is an asperity with an anomalously large slip of up to 50 m near the trench (Ide et al. 2011; Maercklin et al. 2012).

Our single rectangular fault and the slip distribution models showed significantly high VR of 98.8 and 99.1 % and *Mw* of 8.75 and 8.83 at 180 s, respectively (Fig. 4.b, S 2). The *Mw* estimates increased until ~130 s because the rupture time of the major slip area was over 100 s (Ide et al. 2011). The slip distribution model estimated a fault model well in agreement with Ozawa et al. (2011), who used only onshore GPS data from the GEONET. However, our model failed to recover the large slip near the trench. This could be due to the lack of offshore data. The estimated single rectangular fault model was located ~30 km closer to the coastline compared with the heterogeneous slip estimation. Furthermore, the

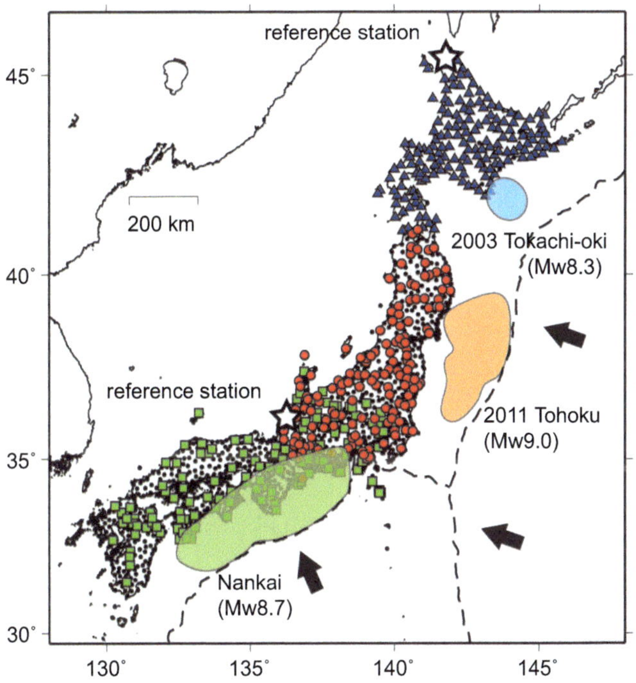

Fig. 3 Global Navigation Satellite System (GNSS) stations used in the posterior analyses to test the performance of the inversion routines in the REGARD. The *blue triangles*, *red circles*, and *green squares* denote the stations for the 2003 Tokachi-oki (*Mw* 8.3), the 2011 Tohoku (*Mw* 9.0), and the simulated Nankai (*Mw* 8.7) earthquakes, respectively. The stars identify the reference stations

estimated *Mw* was underestimated. This may be due to the heterogeneous fault rupture and geometry that cannot be approximated well by a single rectangular fault.

3.3 Simulated Nankai Earthquake (*Mw* 8.7)

The slip distribution model showed a VR over 96 % and stable *Mw* estimates after ~160 s (Fig. 4.c, S 3). The estimated *Mw* agreed well with the assumed *Mw* of 8.68. However, the single rectangular fault model clearly failed to fit the displacements, with a *Mw* of 8.5–9.8 and a significantly low VR because the complex plate geometry could not reproduce the simple rectangular fault model.

Although the average simulated final slip was 7.4 m long and the rupture 150 km wide, the slip distribution model showed larger slips up to ~15.0 m and a rupture only ~100 km wide. The underestimation in the rupture area may be compensated by the over-shrink of the total slip, implying that the fault ruptures near the trench could not be modeled from only land-based GEONET data. However, the true moment and rupture length were recovered well and consistent with the true model.

3.4 Instability of the *Mw* Estimates

The errors in *Mw* estimates are related to the background noise level of the real-time GNSS positioning. To check the instability of the inversion, we extracted a displacement field in the same way as our tests from the real-time positioning time-series on 26 January 2015 when no large event occurred. The RMSs of the extracted displacements were 0.026 m, 0.037 m and 0.040 m for east, north, and south components respectively, and spatially correlated (S. 4). Then we estimated a slip distribution model for the extracted displacements. The estimated moment was 4.62×10^{22} Nm (*Mw* 7.71) with a VR of 55.6 % (S. 4), corresponding to $Mw = 8.00 + 0.09$ or $Mw = 8.50 + 0.02$.

The missing observation caused a cycle slip and became another noise source. During the 2003 Tokachi-oki earthquake, stations “0789”, “0889”, and “0122” were disconnected for about 2 min. Station “0889” was the closest to the coastline and the estimated positions changed 0.4671 m, 0.2019 m, and 0.1005 m for the north, south, and upward directions respectively after the missing observation (S. 5). This jump was caused by the reset of the fixed integer ambiguity. S. 6 shows the inversion results derived from the displacement field, including the position jumps for stations “0789”, “0889”, and “0122”. The VR of the slip distribution model dropped by 8.3 % and the estimated *Mw* increased from 8.19 to 8.22. On the contrary, there was little effect on the result of the single rectangular fault model. From these results, we conclude that the slip distribution model is more sensitive to the missing observation, which is, however, not significant if the stations are densely distributed (~20 km each for the GEONET) and the earthquake is large enough, i.e. $Mw > 8.0$.

4 Discussions and Conclusion

In this study, we tested two fault model inversion routines for the REGARD: a single rectangular fault model and a slip distribution model, using 1 Hz displacement data from the 2003 Tokachi-oki earthquake (*Mw* 8.3), the 2011 Tohoku earthquake (*Mw* 9.0), and a Nankai Trough earthquake simulated using the 1707 Hoei type rupture source (*Mw* 8.7). Both routines gave fault models and unsaturated *Mw* estimates with a high VR for the 2003 Tokachi-oki earthquake and the 2011 Tohoku earthquake after 3 min. The single rectangular fault model, however, poorly estimated the assumed Nankai Trough earthquake. In contrast, the slip distribution modeling routine provided reasonable *Mw* estimates, being then likely effective in the *Mw* estimation of a future Nankai earthquake. The underestimation of the tsunami heights for the 2011

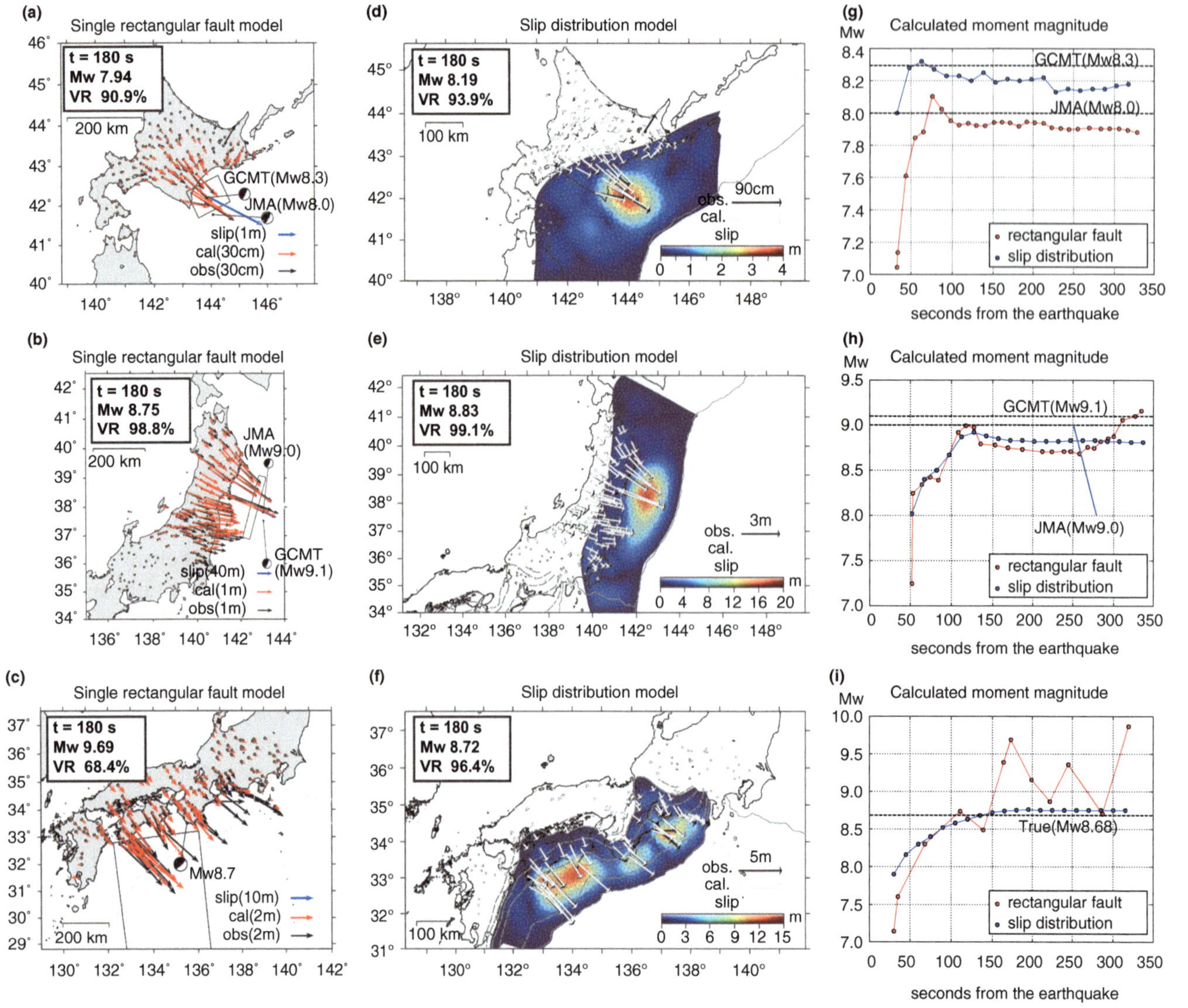

Fig. 4 Comparison between the results from the single rectangular fault model (*left*; **a**–**c**), the slip distribution model (*center*; **d**–**f**), and the time variation of the moment magnitude estimates (*right*; **g**–**i**) for (**a**) the 2003 Tokachi-oki earthquake (*Mw* 8.3), (**b**) the 2011 Tohoku earthquake (*Mw* 9.0), and (**c**) the simulated Nankai Trough earthquake (*Mw* 8.7). The results are shown for time = 180 s after the start of each earthquake and include the Japan Meteorological Agency (JMA) and the Global Centroid Moment Tensor (GCMT) project data represented by the CMT (*left*) and the dotted lines (*right*). **S.1** Comparison between the results from the single rectangular fault model (*upper*) and the slip distribution model (*bottom*) for the 2003 Tokachi-oki earthquake (*Mw* 8.3) every 15 s. **S.2** Comparison between the results from the single rectangular fault model (*upper*) and the slip distribution model (*bottom*) for the 2011 Tohoku earthquake (*Mw* 9.0) every 15 s. **S.3** Comparison between the results from the single rectangular fault model (*upper*) and the slip distribution model (*bottom*) for the simulated Nankai earthquake (*Mw* 8.7) every 15 s. **S.4** Inversion results for the background noises extracted from the real-time positioning time-series on 26 January 2015. **S.5** Time-series of the station "0889" during a missing observation occurred after the 2003 Tokachi-oki earthquake. The missing observation continued for 125 s and caused a cycle slip which reset the fixed integer ambiguity. **S6** Inversion results derived from the displacement field, including the position jumps for stations "0789", "0889", and "0122" where cycle slips occurred

Tohoku earthquake was due to a saturated initial magnitude of 7.9 determined by seismic-based EEW (Ozaki 2011). On the other hand, REGARD provided an unsaturated *Mw* of 8.83 at the time 3 min after the earthquake. Therefore, unsaturated *Mw* estimates provided by REGARD should significantly improve the seismic-based initial magnitude estimation and tsunami warning for large earthquakes with $Mw > 8.0$. For estimating the initial magnitude, the single rectangular fault model was too simple for a large earthquake. However, it was much better than the EEW for the 2011 Tohoku earthquake. Furthermore, it is important to note that the single rectangular fault model estimation is still useful for non-interplate events such as an intraslab earthquake (e.g., Ohta et al. 2011).

Several inversion schemes for inverting real-time GPS data have already been proposed (e.g., Nishimura et al. 2010;

Crowell et al. 2012; Melgar et al. 2012; Ohta et al. 2012; Wright et al. 2012; Minson et al. 2014). Our approaches were customized to model interplate earthquakes with a constraint on the fault geometry similar to previous studies (Crowell et al. 2012; Wright et al. 2012). Thus, earthquakes generated by other mechanisms can be modeled more robustly using inversion schemes with less constraints regarding the fault geometry (e.g., Melgar et al. 2012; Minson et al. 2014). The simple determination of the initial location for our single rectangular fault model could be improved by a hypocenter determination using the displacements of four or more stations (Crowell et al. 2009) even if no EEW was available. Moreover, the pre-determined moment tensor database could be modified in real-time by a CMT solution from the displacements (Melgar et al. 2012, 2013).

The instability of the slip distribution model due to noises on the real-time positioning was not significant on the *Mw* estimates for large earthquakes $Mw > 8.0$. However, the improvement on the positioning time-series was important to extend the lower limit of the targeted earthquake size. The extracted displacements from the RTK positioning were spatially correlated because of the common noise of the reference station. The implementation of precise point positioning (PPP) should reduce this kind of noise and its precision does not depend on the baseline length. The precision of the kinematic PPP was around 0.01 and 0.02 m for the horizontal and vertical components, respectively (Psimoulis et al. 2014), which is clearly better than the current noise level of our real-time RTK positioning with long baselines over 100 km. However, real-time PPP is still in the research phase (Huber et al. 2014), so the RTK is the most suitable for the current real-time warning system.

Our scheme should be particularly efficient for regions for which a detailed knowledge of the plate interfaces exists and with a repeated occurrence of large earthquakes, such as Japan. Furthermore, the resulting seafloor deformation based on the estimated coseismic fault model and/or slip distribution could be directly used for tsunami forecasting. To cover the next potential earthquake targets along the Nankai Trough and the Ryukyu Trench, we will extend the network of REGARD to cover the whole area of Japan. Thus, we conclude that unsaturated *Mw* estimates provided by REGARD should improve the tsunami warning in Japan. Currently, the real-time GNSS stations are limited to land, thus, there is no sensitivity to large coseismic slip in the offshore region. Melgar and Bock (2013b) demonstrated that the resolution of a shallow slip can be improved by a joint inversion using offshore shallow wave gauge data. Furthermore, the lack of station coverage might also be a problem in the modeling of earthquakes along the Ryukyu Trench, which could be improved with quasi-real time sea floor geodetic measurements such as GPS-acoustic techniques (Imano et al. 2015).

Acknowledgements We would like to thank Yehuda Bock and two anonymous reviewers for their detailed and constructive reviews. We thank the International GNSS Service (IGS) for providing the products used in this study, the Japan Meteorological Agency (JMA) and the Global CMT (GCMT) project (Dziewonski et al. 1981; Ekström et al. 2012) for providing the moment tensor solutions.

References

Allen RM, Gasparini P, Kamigaichi O, Boese M (2009) The status of earthquake early warning around the world; an introductory overview. Seismol Res Lett 80(5):682–693

Aster RC, Borchers B, Thurber CH (2012) Parameter estimation and inverse problems, 2nd edn. Academic, Amsterdam

Baba T, Tanioka Y, Cummins PR, Uhira K (2002) The slip distribution of the 1946 Nankai earthquake estimated from tsunami inversion using a new plate model. Phys Earth Planet Int 132:59–73. doi:10.1016/S0031-9201(02)00044-4

Blewitt G, Kreemer C, Hammond WC, Plag HP, Stein S, Okal E (2006) Rapid determination of earthquake magnitude using GPS for tsunami warning systems. Geophys Res Lett 33:L11309. doi:10.1029/2006GL026145

Blewitt G, Hammond WC, Kreemer C, Plag HP, Stein S, Okal E (2009) GPS for real-time earthquake source determination and tsunami warning systems. J Geod 83(3–4):335–343. doi:10.1007/s00190-008-0262-5

Cabinet Office, Government of Japan (2013) Report on the fault models of M7 class earthquake beneath the capital and M8 class earthquake on Sagami Trough, and seismic intensity distribution and tsunami height (in Japanese). http://www.bousai.go.jp/kaigirep/chuobou/senmon/shutochokkajishinmodel. Accessed 9 Sept 2014

Crowell BW, Bock Y, Squibb MB (2009) Demonstration of earthquake early warning using total displacement waveforms from real-time GPS network. Bull Seismol Soc Am 80(5):772–782. doi:10.1785/gssrl.80.5.772

Crowell BW, Bock Y, Melgar D (2012) Real-time inversion of GPS data for finite fault modeling and rapid hazard assessment. Geophys Res Lett 39(9):L09305. doi:10.1029/2012GL051318

Dziewonski AM, Chou TA, Woodhouse JH (1981) Determination of earthquake source parameters from waveform data for studies of global and regional seismicity. J Geophys Res 86:2825–2852. doi:10.1029/JB086iB04p02825

Ekström G, Nettles M, Dziewonski AM (2012) The global CMT project 2004–2010: centroid-moment tensors for 13,017 earthquakes. Phys Earth Planet Int 200–201:1–9. doi:10.1016/j.pepi.2012.04.002

Furumura T, Imai K, Maeda T (2011) A revised tsunami source model for the 1707 Hoei earthquake and simulation of tsunami inundation of Ryujin Lake, Kyushu. Jpn J Geophys Res 116:B02308. doi:10.1029/2010JB007918

Hansen PC (1992) Analysis of discrete ill-posed problems by means of the L-curve. SIAM Rev 34(4):561–580

Hirose F, Nakajima J, Hasegawa A (2008) Three-dimensional seismic velocity structure and configuration of the Philippine Sea slab in southwestern Japan estimated by double-difference tomography. J Geophys Res 113:B09315. doi:10.1029/2007JB005274

Huber K, Hinterberger F, Lesjak R, Weber R (2014) Real-time PPP with ambiguity resolution – determination and application of uncalibrated phase delays. In: Proceedings of the 27th international technical meeting of the ION Satellite Division, ION GNSS, pp 976–985

Ide S, Baltay A, Beroza GC (2011) Shallow dynamic overshoot and energetic deep rupture in the 2011 Mw 9.0 Tohoku-Oki earthquake. Science 332(6036):1426–1429. doi:10.1126/science.1207020

Imano M, Kido M, Ohta Y, Fukuda T, Ochi H, Takahashi N (2015) GPS/acoustic measurement using a multi-purpose moored buoy system. International Association of Geodesy Symposia, submitted, this issue

Kita S, Okada T, Hasegawa A, Nakajima J, Matsuzawa T (2010) Anomalous deepening of a seismic belt in the upper-plane of the double seismic zone in the Pacific slab beneath the Hokkaido corner: possible evidence for thermal shielding caused by subducted forearc crust materials. Earth Planet Sci Lett 290:415–426. doi:10.1016/j.epsl.2009.12.038

Kouba J (2009) A guide to using International GNSS Service (IGS) products. http://igscb.jpl.nasa.gov/components/usage.html. Accessed 9 Sept 2014

Maercklin N, Festa G, Colombelli S, Zollo A (2012) Twin ruptures grew to build up the giant 2011 Tohoku, Japan, earthquake. Sci Rep 2:709. doi:10.1038/srep00709

Matsu'ura M, Hasegawa Y (1987) A maximum likelihood approach to nonlinear inversion under constraints. Phys Earth Planet Int 47:179–187. doi:10.1016/0031-9201(87)90076-8

Meade BJ (2007) Algorithms for the calculation of exact displacements, strains, and stresses for triangular dislocation elements in a uniform elastic half space. Comput Geosci 33(8):1064–1075. doi:10.1016/j.cageo.2006.12.003

Melgar D, Bock Y (2013) Near-field tsunami models with rapid earthquake source inversions from land- and ocean-based observations: the potential for forecast and warning. J Geophys Res Solid Earth 118:1–17. doi:10.1002/2013JB010506

Melgar D, Bock Y, Crowell BW (2012) Real-time centroid moment tensor determination for large earthquakes from local and regional displacement records. Geophys J Int 188(2):703–718. doi:10.1111/j.1365-246X.2011.05297.x

Melgar D, Crowell BW, Bock Y, Haase JS (2013) Rapid modeling of the 2011 Mw 9.0 Tohoku-oki earthquake with seismogeodesy. Geophys Res Lett 40:2963–2968. doi:10.1002/grl.50590

Minson SE, Murray JR, Langbein JO, Gomberg JS (2014) Real-time inversions for finite fault slip models and rupture geometry based on high-rate GPS data. J Geophys Res Solid Earth 119(4):3201–3231. doi:10.1002/2013JB010622

Miyazaki S, Larson KM, Choi K, Hikima K, Koketsu K, Bodin P, Haase J, Emore G, Yamagiwa A (2004) Modeling the rupture process of the 2003 September 25 Tokachi-Oki (Hokkaido) earthquake using 1-Hz GPS data. Geophys Res Lett 31:L21603. doi:10.1029/2004GL021457

Nakajima J, Hasegawa A (2006) Anomalous low-velocity zone and linear alignment of seismicity along it in the subducted Pacific slab beneath Kanto, Japan: reactivation of subducted fracture zone? Geophys Res Lett 33:L16309. doi:10.1029/2006GL026773

Nakajima J, Hasegawa A (2007) Subduction of the Philippine Sea plate beneath southwestern Japan: slab geometry and its relationship to arc magmatism. J Geophys Res 112:B08306. doi:10.1029/2006JB004770

Nakamura M, Kaneshiro S (2000) Determination of subducted Philippine Sea Plate in the Nansei islands deduced from hypocenter data. Bull Fac Sci Univ Ryukyus 70:73–82 (in Japanese with English abstract)

Nishimura T, Imakiire T, Tobita M (2010) Development on the rapid estimation for an earthquake fault model using real-time 1-second sampling GPS data. J Geospat Inform Authority Jap 120:63–73 (in Japanese)

Nishimura T, Munekane H, Yarai H (2011) The 2011 off the pacific coast of Tohoku earthquake and its aftershocks observed by GEONET. Earth Planet Space 63:631–636

Ohta Y, Miura S, Ohzono M, Kita S, Iinuma T, Demachi T, Tachibana K, Nakayama T, Hirahara S, Suzuki S, Sato T, Uchida N, Hasegawa A, Umino N (2011) Large intraslab earthquake (2011 April 7 M7.1) after the 2011 off the Pacific coast of Tohoku earthquake (M9.0): coseismic fault model based on the dense GPS network data. Earth Planet Space 63(12):1207–1211. doi:10.5047/eps.2011.07.016

Ohta Y, Kobayashi T, Tsushima H, Miura S, Hino R, Takasu T, Fujimoto H, Iinuma T, Tachibana K, Demachi T, Sato T, Ohzono M, Umino N (2012) Quasi real-time fault model estimation for near-field tsunami forecasting based on RTK-GPS analysis: application to the 2011 Tohoku-Oki earthquake (Mw 9.0). J Geophys Res Solid Earth 117(B2):B02311. doi:10.1029/2011JB008750

Okada Y (1992) Internal deformation due to shear and tensile faults in a half-space. Bull Seismol Soc Am 82(2):1018–1040

Ozaki T (2011) Outline of the 2011 off the Pacific coast of Tohoku Earthquake (Mw 9.0) -tsunami warnings/advisories and observations. Earth Planet Space 63:827–830

Ozawa S, Kaidzu M, Murakami M, Imakiire T, Hatanaka Y (2004) Coseismic and postseismic crustal deformation after the Mw 8 Tokachi-oki earthquake in Japan. Earth Planet Space 56:675–680

Ozawa S, Nishimura T, Suito H, Kobayashi T, Tobita M, Imakiire T (2011) Coseismic and postseismic slip of the 2011 magnitude-9 Tohoku-Oki earthquake. Nature 475(7356):373–376. doi:10.1038/nature10227

Ozawa S, Nishimura T, Munekane H, Suito H, Kobayashi T, Tobita M, Imakiire T (2012) Preceding, coseismic, and postseismic slips of the 2011 Tohoku earthquake, Japan. J Geophys Res Solid Earth 117(B7), B07404. doi:10.1029/2011JB009120

Psimoulis PA, Houlié N, Michel C, Meindl M, Rothacher M (2014) Long-period surface motion of the multipatch Mw9.0 Tohoku-Oki earthquake. Geophys J Int 199:968–980

Simons M, Minson SE, Sladen A, Ortega F, Jiang J, Owen SE, Meng L, Ampuero JP, Wei S, Chu R, Helmberger DV, Kanamori H, Hetland E, Moore AW, Webb FH (2011) The 2011 magnitude 9.0 Tohoku-Oki earthquake: mosaicking the megathrust from seconds to centuries. Science 332(6036):1421–1425. doi:10.1126/science.1206731

Takasu T (2011) RTKLIB: an open source program package for GNSS positioning. http://www.rtklib.com. Accessed 9 Sept 2014

Todoriki M, Furumura T, Maeda T (2013) 3D FDM simulation of seismic wave propagation for Nankai trough earthquake: effects of topography and seawater. AGU Fall Meeting, 9–13 December, San Francisco, USA

Wright TJ, Houlié N, Hildyard M, Iwabuchi T (2012) Real-time, reliable magnitudes for large earthquakes from 1 Hz GPS precise point positioning: the 2011 Tohoku-Oki (Japan) earthquake. Geophys Res Lett 39(12):L12302. doi:10.1029/2012GL051894

Yagi Y (2004) Source rupture process of the 2003 Tokachi-oki earthquake determined by joint inversion of teleseismic body wave and strong ground motion data. Earth Planet Space 56:311–316

Yamagiwa A, Hatanaka Y, Yutsudo T, Miyahara B (2006) Real-time capability of GEONET system and its application to crust monitoring. Bull Geospat Inform Authority Jap 53:27–33

Yamanaka Y, Kikuchi M (2003) Source process of recurrent Tokachi-oki earthquake on September 26, 2003, inferred from teleseismic body waves. Earth Planet Space 55:e21–e24

Buoy Platform Development for Observation of Tsunami and Crustal Deformation

Narumi Takahashi, Yasuhisa Ishihara, Tatsuya Fukuda, Hiroshi Ochi, Jun'ichiro Tahara, Takami Mori, Mitsuyasu Deguchi, Motoyuki Kido, Yusaku Ohta, Ryota Hino, Katsuhiko Mutoh, Gosei Hashimoto, Osamu Motohashi, and Yoshiyuki Kaneda

Abstract

We constructed a buoy system for real-time observations of tsunamis and crustal deformation in collaboration with the Japan Agency for Marine-Earth Science and Technology, Tohoku University, and the Japan Aerospace Exploration Agency. The most important characteristics of our system are resistance to the strong sea currents in the large-earthquake rupture zone around Japan (e.g., the Kuroshio maximum speed >5 knots), and the capability to transmit data in real-time. Our system has four units: (1) a buoy station with a GPS/Acoustic station serving as a central base, (2) a wire-end station (WES) 1,000 m below the sea surface that serves as a staging base, (3) a pressure seafloor unit (PSU) comprising a pressure sensor, and (4) six GPS/Acoustic transponders to measure crustal deformation. The pressure data used to detect tsunamis and the vertical component of crustal deformation are sent to the land station via the wire-end and buoy stations at intervals of 1 h in normal mode and 15 s in tsunami mode. The data measured between the buoy and six transponders are also sent to the land station at 1-week intervals. The Iridium satellite is used for data transmission of all data to land station. The dynamic range for pressure observations is +/− 8 m with a fine resolution of 2 mm, and the accuracy of the crustal deformation measurements is less than 1 m. We tuned the

N. Takahashi (✉)
CEAT, Japan Agency for Marine-Earth Science and Technology, 3173-25 Showa-machi, Kanazawa-ku, Yokohama 236-0001, Japan
e-mail: narumi@jamstec.go.jp

Y. Kaneda
Nagoya University, Furo-cho, Chikusa-ku 464-8601 Nagoya, Japan

Y. Ishihara • T. Fukuda • H. Ochi • T. Mori • M. Deguchi
MARITEC, Japan Agency for Marine-Earth Science and Technology, 2-15 Natsushima-cho, Yokosuka 237-0061, Japan

J. Tahara
Tokyo University of Marine Science and Technology, 2-1-6, Etchujima, Koto-ku, 135-0044, Tokyo, Japan

M. Kido • R. Hino
IRIDeS Bureau, Tohoku University, 6-6-4 Aoba, Sendai 980-8579, Japan

Y. Ohta
RCPEV, Graduate School of Science, Tohoku University, Aoba, Sendai 980-8579, Japan

K. Mutoh • O. Motohashi
Satellite Navigation Office, Japan Aerospace Exploration Agency, 2-1-1 Sengen, Tsukuba 305-8505, Japan

G. Hashimoto
Satellite Applications and Promotion Center, Japan Aerospace Exploration Agency, 2-1-1 Sengen, Tsukuba 305-8505, Japan

M. Hashimoto (ed.), *International Symposium on Geodesy for Earthquake and Natural Hazards (GENAH)*,
International Association of Geodesy Symposia 145, DOI 10.1007/1345_2015_114

system for an observation period of 5 months and carried out a sea trial. The length of the observation period influences the total system due to the weight of the battery. We rearranged the geometry of the total system to new one with heavier weight and a lot of batteries on the buoy considering long period observation and decided upon a slack ratio of 1.6. In addition, it is important for a long observation period to minimize electrical consumption. We used double pulses for acoustic data transmission between the PSU and WES. The time difference between two pulses indicates the observed pressure value. For the PSU, we designed a tsunami mode on the basis of data from the tsunami generated by the 2011 earthquake off Tohoku, which were recorded by cabled network system data and offline bottom pressure data. The results confirmed that a tsunami can be detected even if the first tsunami signals include strong-motion signals. In this case, the tsunami was detected 10–20 s after the first seismic arrival. During sea trials, we successfully tested the tsunami mode we designed. We succeeded real-time observation of pressure and crustal deformation using buoy system in strong sea current speed area for 5 months. However, there are some issues to be resolved at this moment. For acoustic data transmission, 1 ms step difference of the detection of acoustic signals at the WES, wrong detection of the multiple phases are issues to be resolved. We will consider assigned mapping of transmitted data to the time difference of the double pulses and take measures on the PSU and WES. In addition, we consider strategy to reduce slack ratio in the future. For data transmission from the WES to the buoy station, we experienced electrical unhealthy of the wire rope due to damages by the fisheries activities and the torsion brought by rotation of the buoy. We consider the countermeasure to reduce the rotation.

Keywords
Buoy • Crustal deformation • Real-time monitoring • Tsunami

1 Introduction

Many large earthquakes have repeatedly occurred around Japan. Examples include the 2011 off Tohoku earthquake, the 1944 Tonankai earthquake, and the 1946 Nankai earthquake. These large earthquakes have generally located in the subduction zone under the Japanese coast. The Japan Agency for Marine-Earth Science and Technology (JAMSTEC) has constructed a seafloor network cable system named the Dense Oceanfloor Network system for Earthquakes and Tsunamis (DONET) in the Tonankai and Nankai earthquake rupture zones (Kaneda 2010; Kawaguchi et al. 2008). The DONET in the Tonankai area has been operating since 2011. The system has 20 stations with many types of sensors including strong-motion sensors, broadband seismometers, and pressure sensors. It is designed to provide early detection of earthquakes and tsunamis. However, the costs for construction and implementation of such large systems are high. Simulation studies of crustal displacement have indicated that changes of slip velocity are precursors of large earthquakes (see, for example, Hyodo and Hori 2010). Our focus is on the observation of tsunami and crustal deformation; to achieve this, we have developed an easy buoy system in collaboration with JAMSTEC, Tohoku University, and the Japan Aerospace Exploration Agency (JAXA).

Buoy systems are already implemented around the world (e.g., Bernard et al. 2014; Pandoe et al. 2012; Aasen et al. 2007), the best-known being the DART system developed by NOAA (e.g., Bernard and Meinig (2011)). However, because subduction zones are sometimes in areas where there are high-speed sea currents, we adopted slack mooring instead of the strain mooring used by DART, thus allowing our system to operate in high-speed currents.

The concept of our new buoy system has already been reported by Takahashi et al. (2014). Here we describe the results from recent sea trials and issues to be resolved in the near future.

2 Geometry of the System

Our system for real-time observation of tsunami and crustal deformation consists of four main units: a buoy station including a GPS/A unit, a wire-end station (WES), a pressure seafloor unit (PSU) with an attached pressure sensor, and six seafloor transponders (Takahashi et al. 2014, Fig. 1). The buoy station plays a role of the control center for the entire station to take timing of starting signals for many. And it collects many data, which are pressure sensor data from the WES, measured data from the GPS/A station, and

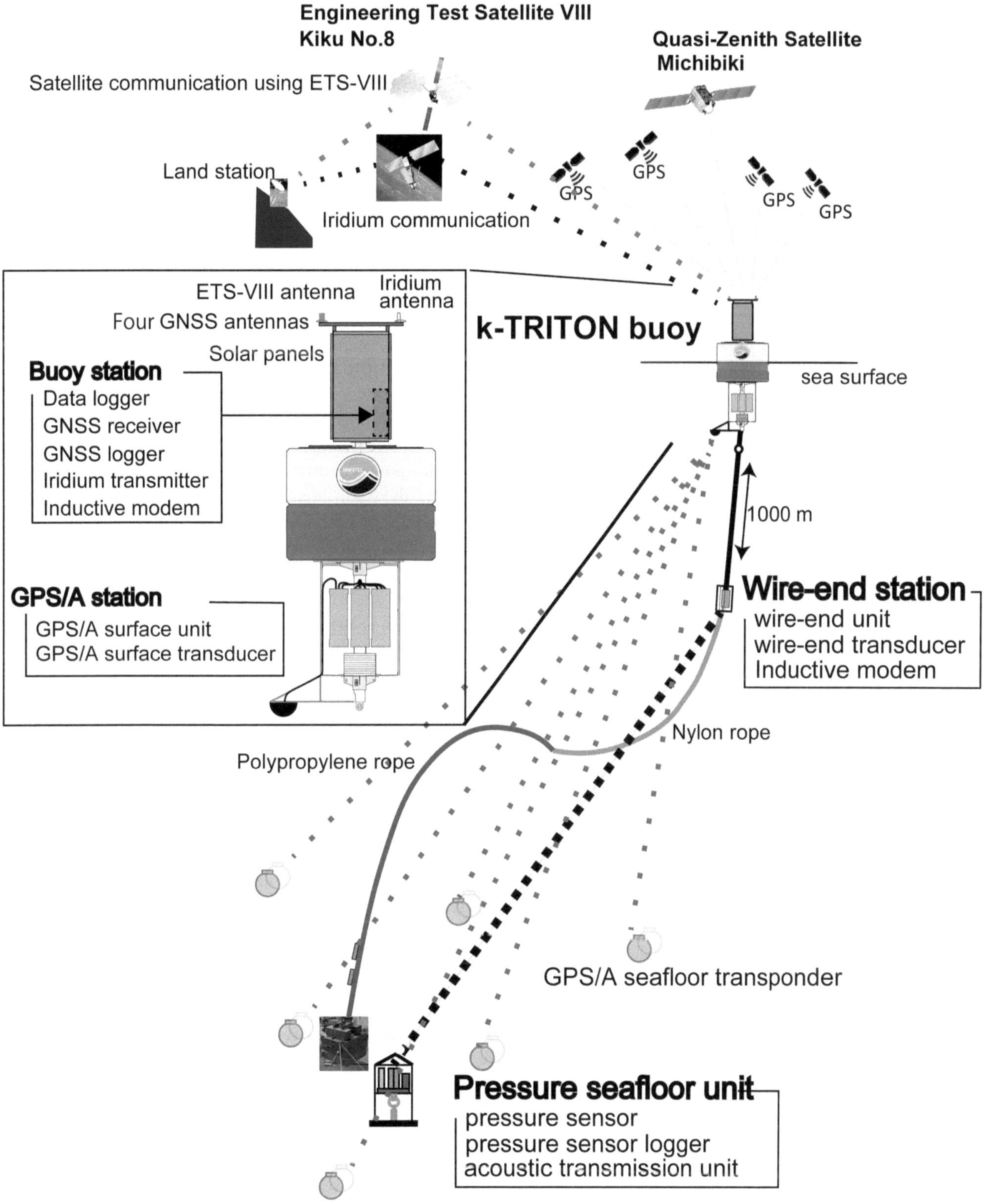

Fig. 1 Schematic figure of the total system

navigation data from the Global Navigation Satellite System (GNSS) antenna. The buoy station also stores these data on the on-board logger and sends them to land by using iridium data transmission. It receives instructions from the land station (for example, to switch to tsunami mode) and sends acoustic signals to the PSU. It also controls the timing of the management of these data. The WES serves as the base for management of the pressure data. It converts the acoustic signals from the PSU to digital data and attaches a time stamp. Moreover, it determines whether the PSU is in tsunami mode. The PSU serves as the base of observation of the pressure data.

Our buoy system is set up to accommodate the environment of the deployment location and the length of the observation period. The overall geometry of the system is determined by the total balance of the horizontal forces and vertical forces. For the sea trial, we selected same location to first sea trial in the Nankai Trough at a water depth of 2,983 m, where there are very high-speed sea currents (Takahashi et al. 2014). Our trial was conducted over a 6-month period and the total weight of the buoy station become heavier than the first trials. Therefore, the anchor weight made heavy 6.8 tons, but other parameters are same to the first one. Conditions were as follows: sea current of 5.5 knots, and wind speed of 10 m/s with direction similar to that of the sea current. Our estimation of the buoy draft under the above conditions was 1.29 m using the slack ratio of 1.6.

We determined that the maximum pressure variation that could be recorded in response to a tsunami and the vertical component of crustal deformation was 8 m and that its resolution was less than 5 mm (Takahashi et al. 2014). These parameters are important for deciding the data-sampling interval in tsunami mode. Considering real implementation for disaster prevention, we selected an interval of 15 s for data sampling. For observation of the horizontal component of crustal deformation, we used an acoustic method to measure the distance between the buoy and the seafloor transponders. We also collected GNSS data from the four attached antennas (Fig. 1) to estimate the position, tilt and rotation of the buoy. The sampling interval for observation of crustal deformation was 1 week. Kido et al. (2007) describe the method in detail; we used six seafloor transponders to maintain multiple paths between the buoy and these transponders. Further details are presented by Imano et al. in this issue Imano et al. (2015).

3 Action for Long Period Implementation

For long-term observation, it is important to minimize power consumption because there is limited space to accommodate batteries, and because the total weight of the system affects the required length of ropes and the weight of the anchor. If we have to use large buoy to accommodate much battery and to have buoyancy, it brings severe influences on the slack ratio. Because the resistance for the high speed sea current and winds increase. If slack ratio is enlarged, the success rate of the acoustic data transmission may decrease. It is indispensable to save battery for this buoy system and we considered that the effective methods are use of double pulses for acoustic data transmission and introduction of tsunami mode on the PSU. Here, we describe data transmission and tsunami mode setting in regards to the issue of saving space.

3.1 Minimizing Power Consumption

Instead of coding the digital data to express observed pressure values, we adopted the double pulse method in order to lower power consumption. However, the success of the acoustic data transmission depends on only two pulses of double pulses. Therefore, we adopted charp signals for the double pulses and their correlations. In this system, we assign the values to time step within the double pulses with 1 ms step in this sea trial. The minimum time difference of the double pulse is 201 ms and it means that transmitted pressure data just corresponds to the base water depth of the PSU. The maximum time difference is 8,201 ms and it means that the observed pressure is 8 m. In the time length of 8,000 ms with an interval of 1 ms, each pressure value (difference from the base pressure corresponding to base water depth) is assigned to plus and minus alternation with a step of 2 mm. The error after conversion of the double pulse value was less than 0.7 ms (Takahashi et al. 2014), but the error depends on the length of the double pulse and tended to increase at the large one. Therefore, we allotted to the time differences in order from small absolute ones. We estimated that the power consumption was lowered approximately 20–30% by using double pulses.

3.2 Introduction of Tsunami Mode

The setting of the tsunami mode on the PSU is also affected by the need to achieve a balance between power consumption and effectiveness for disaster prevention. In tsunami mode, we set the data-sampling interval to 15 s and the data were sent to the land station in real-time. In normal mode, the observed data were sent to the WES at 15-min intervals and the collected data on the WES were sent to the buoy station at 1-h intervals. We used the ratio of twice the short-time average (STA) calculation to twice the long-time average (LTA) as the trigger for tsunami mode. The STA and LTA have a main role of reduction of short period signals like

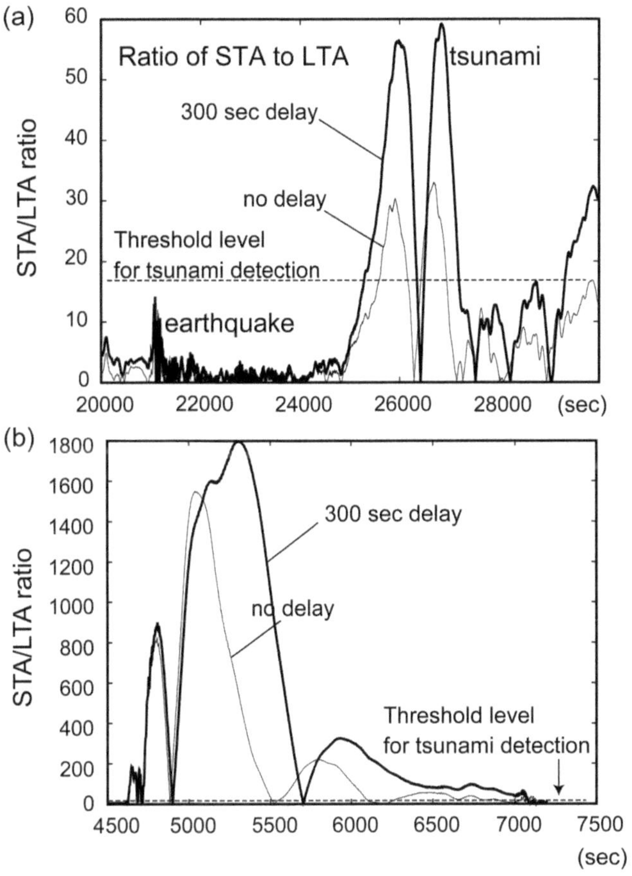

Fig. 2 Detection test of tsunami arrivals activated by 2011 off Tohoku earthquake. *Thin* and *thick lines* are ratio of STA and LTA using no del ay and 300 s delay, respectively. (**a**) Example using DONET C9 data. (**b**) Example using site GJT3 on off Tohoku region (Kido et al. 2011)

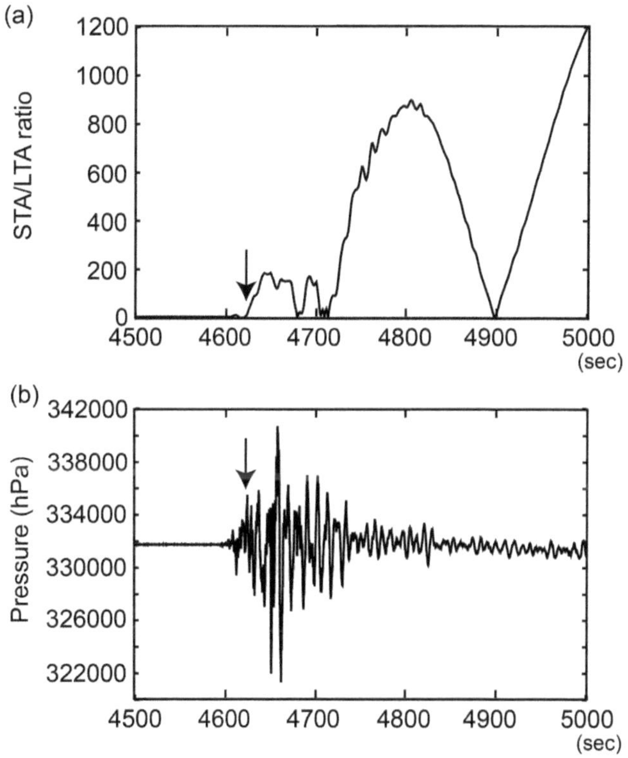

Fig. 3 Comparison using tsunami signals recorded by GJT3 between STA/LTA ratio profile and the original waveform around triggered tsunami arrivals. *Arrows* indicate the triggered tsunami according to Fig. 2. (**a**) Zoomed STA/LTA profile. (**b**) Zoomed original waveform

strong motion without time delay of the tsunami phase and long period signals like tsunami, respectively. In addition, we introduced a time delay to the LTA to detect the first arrival of signals from the tsunami. To confirm tsunami detection, we used tsunami signals after the 2011 off-Tohoku earthquake that were recorded by the pressure sensor of the DONET KMC09 station, which was at a water depth similar to that of our PSU. By trial and error we obtained the following parameters: 300 s for the LTA, 50 s for the STA, and 300 s delay of the LTA. In this case, the time difference between the first arrival and the peak of the amplitude was approximately 7 min. However, introduction of long time delay brought us larger tide effects. To save electrical batteries of the PSU, we aimed simple calculation for tsunami detection without tidal correction. Finally, we selected 300 s as the time delay. We set a tentative threshold level of 17 (Fig. 2a). We also confirmed that these parameters were applicable during the spring tides. The introduction of a time delay is effective when the distance between the hypocenter and the recording station is great enough to make it relatively easy to separate the tsunami signal from the strong-motion signal.

However, above example using DONET C9 data does not include strong motion because the DONET C9 locates at far from the source. Therefore, we have to confirm if above logic is reasonable for near-field tsunami. Therefore, we used ocean bottom-pressure data (tsunami data) of the 2011 off-Tohoku earthquake observed in the forearc rupture area, which were collected by Tohoku University in the forearc region near the Japan trench (Kido et al. 2011). Because this site is on the source fault of the 2011 earthquake, the tsunami signals and strong-motion signals were temporally close. The results of this near-field case confirmed that our logic of the tsunami mode can be used (Fig. 2b). In this case, the detected time calculated by using the time delay is almost identical to that with no delay. Because a tsunami is activated by the vertical component of seafloor motion, it is possible that the bottom-pressure data for the first actual signals of the tsunami are not recorded. We checked this possibility by using actual tsunami data from the 2011 off-Tohoku earthquake recorded by the GJT3 site. Figure 3 shows a comparison of these profiles, and the results indicate that the tsunami was detected 10–20 s after the first arrival of seismic signals. Thus we confirmed that our logic for tsunami detection using a example of the 2011 earthquake with relatively low frequency similar to that of tsunami

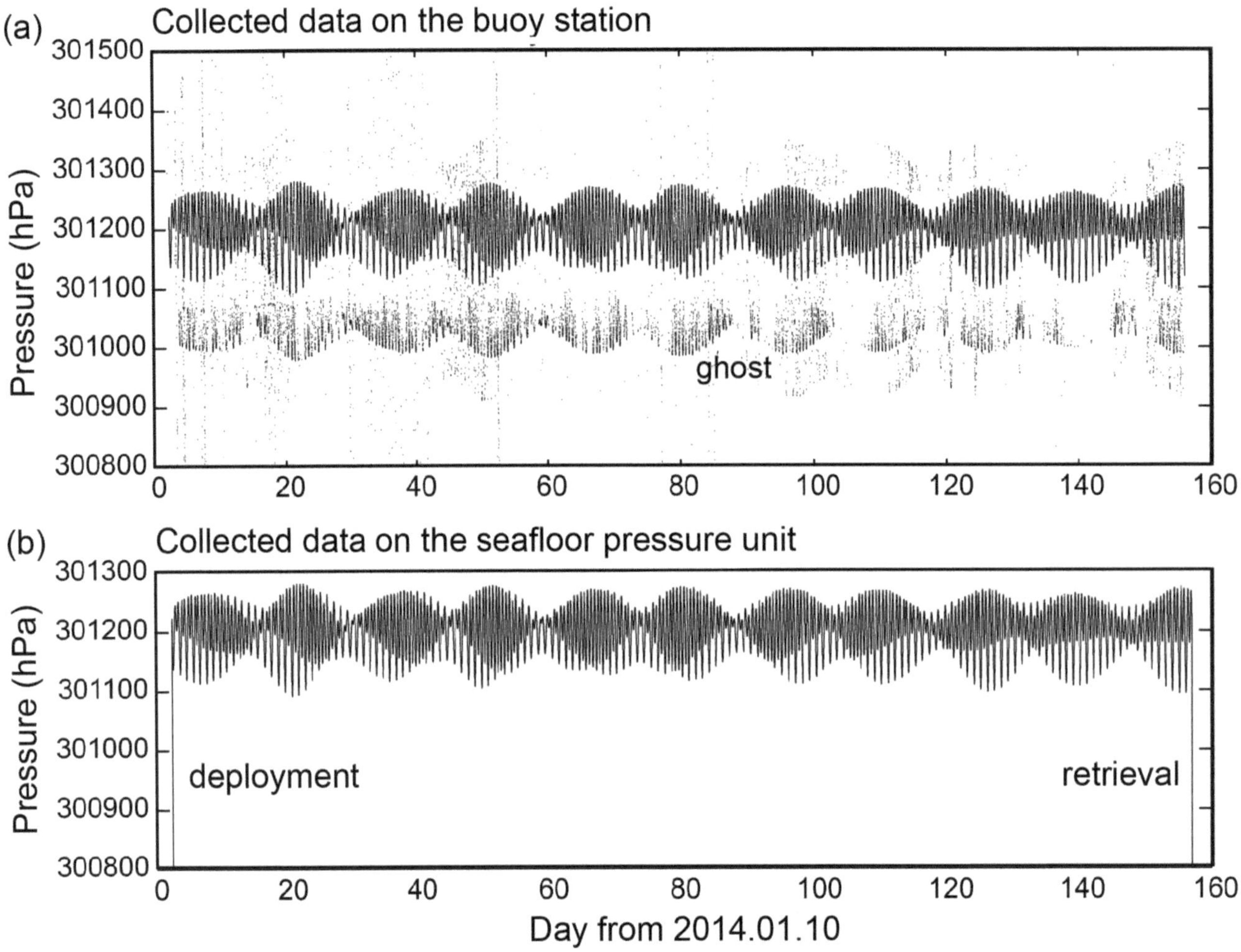

Fig. 4 Comparison of recorded data on the buoy station and the seafloor pressure unit. (**a**) Collected data on the buoy station after acoustic data transmission between the WES and PSU. (**b**) Collected data on the PSU before the acoustic data transmission

works correctly even if the waveform includes strong motion signals.

4 Result of Sea Trial

We carried out a 5-month sea trial from 12 January to 15 June 2014 to examine potential vulnerabilities within our system. Comparison of data recorded on the seafloor pressure unit with those collected at the WES show that approximately 85% of double pulses were correctly received at the buoy station (Fig. 4), but the rate of data collected at the buoy station decrease until approximately 54% due to damages on the wire rope. From the trial, we realized there were two important issues to be resolved.

The first of these is that the wire line between the WES and buoy station can be damaged by activities such as fishing. We found some holes tore in the wire rope by fishery hooks. In addition, possible reason to be able to give damages is rotation of the buoy. If much rotation of the buoy occurred, the torsion may accumulate on the wire rope. We prepared two ways of maintaining redundancy: by online transmission using an electrical cable, or by inductive transmission using the modem installed on the WES. If the wire line is physically damaged, data transmission is switched from online to inductive transmission. There remains the issue of the reliability of the inductive data transmission.

The second issue is difficulty with acoustic transmission between the WES and the PSU. Reasons of the acoustic transmission error are 1 ms difference from observed data by the PSU, wrong detection of reflection phases from the sea surface and sea bottom, spike noises due to ship track and so on. The total error rates for all transmitted data from the PSU is approximately 15%. The reason of the largest error is the difference of 1 ms, the error rate is approximately 7%. The difference of 1 ms is observed as ghost signals in Fig. 4. We will revise the assigned map and consider raising sensitivity of the acoustic signal detection. Second is the wrong detection of the reflection phases from the sea surface and the sea bottom. The WES sometimes detected the reflected phases as first arrivals. We calculated these reflected arrivals using navigation data. Because the buoy moved in a circle of 3,900 m diameter in response to surface currents, the maximum emission angle at the PSU was greater than 50°. We consider that the direct signals were sometimes weaker than the reflected signals, which were stronger than the first arrivals, causing the observed time differences for water depth to be reversed and ghost signals to be recorded. To resolve this, we are considering how to improve installation of the transponders for the PSU and WES, and we are evaluating the effect of multiple signals reflected from the sea surface. We are also studying how to reduce the slack ratio. This would improve the accuracy of observations of crustal deformation. Another potential cause of missing first arrivals is the noisy environment caused by the passage of a large ship.

5 Summary

We constructed a prototype of a buoy system for observations of tsunamis and crustal deformation. In a sea trial, we succeeded in receiving pressure data at a land station and transferring measurement data between the buoy and GPS/A seafloor transponders, and the PSU. We addressed the following issues: setting and implementation of tsunami mode, acoustic data transmission, data collection by the logger on the buoy, and data transmission to land via satellite communication. Although the basic technical elements of the system performed satisfactorily, there are reliability and stability issues still to be resolved. In particular, we will revise the wire rope with measures against the fisheries damages and the rotation of the buoy, and take measures for the acoustic data transmission issues. We will plan further improvements and tuning of the acoustic properties of the PSU and WES. We are preparing to add a precise point positioning system to separate tsunami signals from those of the vertical component of crustal deformation. In addition, we are seeking a way to reduce the slack ratio in order to improve the stability of acoustic data transmission and the accuracy of observations of crustal deformation. These issues will be resolved over the next 5 years through several 1-year sea trials.

Acknowledgements We greatly appreciate the contributions of the captain and crew of JAMSTEC's R/V Kaiyo and technicians at Marine Work Japan Ltd.

References

Aasen SE, Mustapha Z, Schjiberg P, Elliott T (2007) A deepwater tsunami survelliance system for Malaysia. In: Underwater technology and workshop on scientific use of submarine cables and related technologies 2007, pp 691–697. doi:10.1109/UT.2007.370822

Bernard E, Meinig C (2011) History and future of deep-ocean tsunami measurements. In: Proceedings of Oceans'11 MTS/IEEE, Kona, No. 6106894. IEEE, Piscataway, 19–22 September 2011, 7 p

Bernard E, Tang L, Wei Y, Titov V (2014) Impact of near-field, deep-ocean tsunami observations on forecasting the 7 December 2012 Japanese tsunami. Pure Appl Geophys 171(12):3483–3491. doi: 10.1007/s00024-013-0720-8

Hyodo M, Hori T (2010) Modeling of Nankai earthquake cycles: influence of 3D geometry of the Philippine Sea plate on seismic cycles. JAMSTEC Rep Res Dev 11:1–15

Imano M, Kido M, Ohta Y, Fukuda T, Ochi H, Takahashi N (2015) GPS/acoustic measurement using a multi-purpose moored buoy system (in this issue)

Kaneda Y (2010) The advanced ocean floor real time monitoring system for mega thrust earthquakes and tsunamis-application of DONET and DONET2 data to seismological research and disaster mitigation, MTS/IEEE OCEANS, doi:10.1109/OCEANS.2010.5664309

Kawaguchi K, Kaneda Y, and E. Araki (2008) The DONET: A real-time seafloor research infrastructure for the precise earthquake and tsunami monitoring, MTS/IEEE OCEANS, doi:10.1109/OCEANSKOBE.2008.4530918

Kido Y et al (2007) Detecting horizontal gradient of sound speed in ocean. Earth Planets Space 59:e33–e36

Kido M, Osada Y, Fujimoto H, Hino R, Ito Y (2011) Trench-normal variation in observed seafloor displacement associated with the 2011 Tohoku-Oki earthquake. Geophys Res Lett 38:L24303. doi:10.1029/2011GL050057

Pandoe WW, Herunadi B, Asvaliantina V (2012) The operational Indonesia tsunami buoy for the Indonesia tsunami early warning system (InaTEWS). In: Abstracts of tsunami workshop, Sentinel Asia

Takahashi N, Ishihara Y, Ochi H, Fukuda T, Tahara J, Maeeda Y, Kido M, Ohta Y, Mutoh K, Hashimoto G, Kogure S, Kaneda Y (2014) New buoy observation system for tsunami and crustal deformation. Mar Geophys Res 35:243–253

Improvement in the Accuracy of Real-Time GPS/Acoustic Measurements Using a Multi-Purpose Moored Buoy System by Removal of Acoustic Multipath

Misae Imano, Motoyuki Kido, Yusaku Ohta, Tatsuya Fukuda, Hiroshi Ochi, Narumi Takahashi, and Ryota Hino

Abstract

A component of the multi-purpose moored buoy has been improved for the instantaneous detection of seafloor displacement and possible tsunami generation upon the occurrence of large earthquakes. Here, improvements to the acoustic ranging component of this buoy, which is a key element of on-demand GPS/acoustic (GPS/A) measurements, are demonstrated. A 1-m positioning accuracy is required in GPS/A measurements using the buoy system for the detection of the horizontal seafloor crustal deformation associated with large earthquakes. Owing to the limitation of collecting data on only a limited range of sub-surface depth from a single point located far from the optimal location, resulting from the slack mooring system, obtaining the positioning accuracy is challenging. To overcome this challenge, we developed an automatic travel-time algorithm that reliably excludes the multipath from acoustic waves. Applying the algorithm to the data from 4 months of sea trial, we revealed that the short-period repeatability of the positioning improved from 4 to 0.5 m, while the long-term repeatability improved from 8 to 4 m, which is still beyond the required accuracy of 1 m. Because acoustic ranging under sub-optimal conditions will propagate any error in the data at intermediate steps into the uncertainty of the final positioning, approaches to reduce the errors at each step must be undertaken, such as determination of the pre-defined geometry of seafloor transponders more precisely.

Keywords

Automated algorithm • GPS/Acoustic measurements • Moored buoy • Positioning accuracy • Seafloor crustal displacement

1 Introduction

GPS/Acoustic (GPS/A) geodetic measurements have revealed the unexpectedly large extent of the 2011 Tohoku-Oki earthquake rupture near the Japan Trench (Kido et al. 2011; Sato et al. 2011). To reveal the earthquake mechanisms and occurrence, as well as co- and post-seismic behaviors, the observation of seafloor crustal deformation near the trench axis is very important. To date, horizontal seafloor geodetic measurements using GPS/A have mainly been carried out during campaigns held a few times a year, and the positioning accuracy of these has been on the order of a few centimetres (e.g., Spiess et al. 1998; Kido et al. 2006; Tadokoro et al. 2006). On the other hand, continuous observations of seafloor pressure and seismic waves are obtained by seafloor cable networks; e.g., Dense Ocean Floor Network system for Earthquakes and Tsunamis (DONET: Kaneda et al. 2007) and Seafloor Observation Network for Earthquakes and Tsunamis

M. Imano (✉) • Y. Ohta
Graduate School of Science, Tohoku University, Sendai, Japan
e-mail: misae.imano.s1@dc.tohoku.ac.jp

M. Kido • R. Hino
IRIDeS, Tohoku University, Sendai, Japan

T. Fukuda • H. Ochi • N. Takahashi
Japan Agency for Marine-Earth Science and Technology, Yokosuka, Japan

M. Hashimoto (ed.), *International Symposium on Geodesy for Earthquake and Natural Hazards (GENAH)*, International Association of Geodesy Symposia 145, DOI 10.1007/1345_2015_192

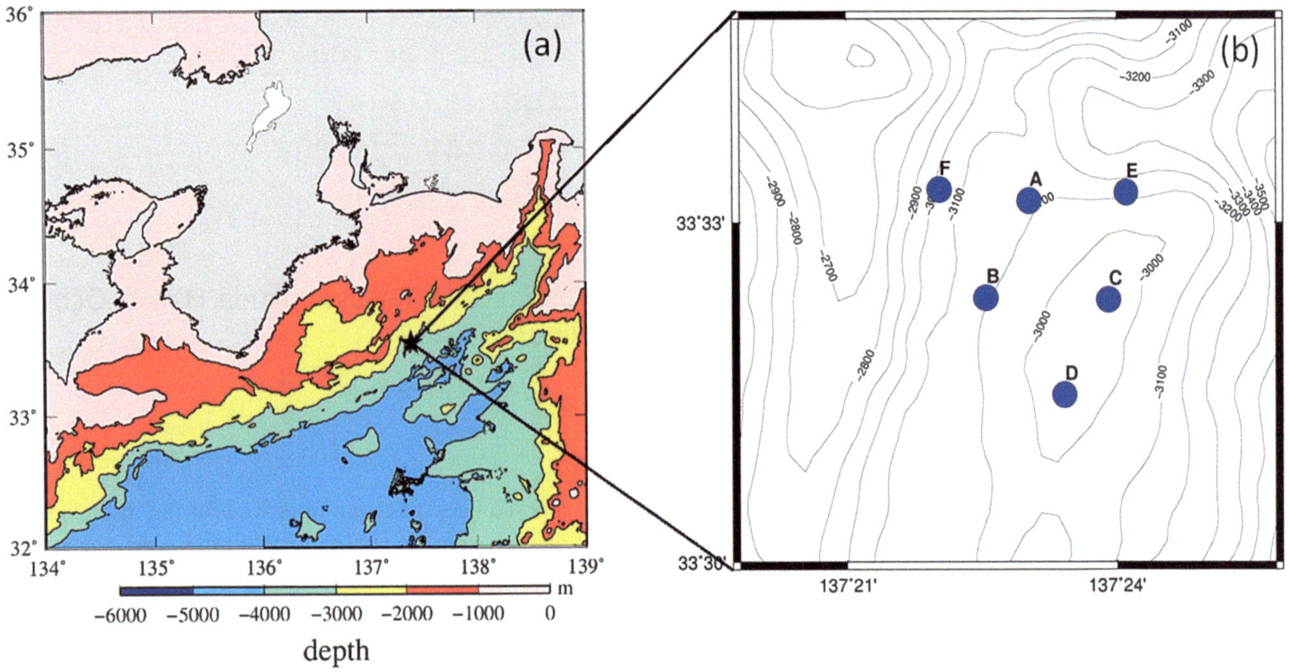

Fig. 1 (**a**) Location of the moored system in southwest Japan (*black star*) near the Nankai trough, and (**b**) Enlarged view of the *black star* in (a). *Blue points* indicate the six PXPs used in this study

along the Japan Trench (S-net: Fujimoto 2014). Real-time monitoring of the horizontal seafloor crustal deformation is becoming increasingly important for the on-demand, instantaneous detection of co- and post-seismic seafloor crustal deformation. Therefore, The Japan Aerospace Exploration Agency (JAXA), Japan Agency for Marine-Earth Science and Technology (JAMSTEC), and Tohoku University have co-developed a continuous horizontal and vertical crustal deformation and tsunami observation system, and have carried out two sea trials near the Nankai Trough (Fig. 1). The targeted positioning accuracy of this system is on the order of 1 m, since the buoy is moored by a slack cable, and the acoustic ranging is, therefore, carried out at an oblique angle relative to the center of the seafloor array. In the first trial, carried out over 4 months in 2013, hardware and software problems arose (e.g., raw GPS data were not properly recorded). In the second sea trial, following the improvements in the buoy system, no major problems were found (Takahashi et al. 2014). Some software issues, however, still remained to be solved so as to achieve efficient real-time acoustic ranging, such as detection and removal of the erroneous influence of the acoustic multipath. Here, we present the results from the second sea trial with specific focus on the development of an algorithm to detect the direct wave travel time. We have detailed the algorithm developed and have quantified the improvements reached in comparison with the previous method.

2 System Overview

The system consists of a GPS system using Precise Point Positioning (PPP; Zumberge et al. 1997) technique, a seafloor pressure sensor, and a GPS/A system (Fig. 2). Sea level and position of the buoy are monitored by the GPS system using PPP. A seafloor pressure sensor monitors tsunamis and detects vertical crustal deformation associated with large earthquakes by combining the GPS system as proposed by Ballu et al. (2010). Seafloor horizontal crustal deformation is monitored using the GPS/A technique. Seafloor pressure data are acoustically transmitted to the wire-end station, 1,000 m beneath the buoy, and then from the wire-end station to the surface buoy by an inductive modem (Takahashi et al. 2014). To estimate the offshore buoy position in real-time without onshore reference station data, PPP analysis is carried out. The GPS/A system measures the horizontal displacement of the seafloor benchmark, which consists of six seafloor transponders (PXPs) in our target area, by combining the GPS measurements that estimate the acoustic transducer

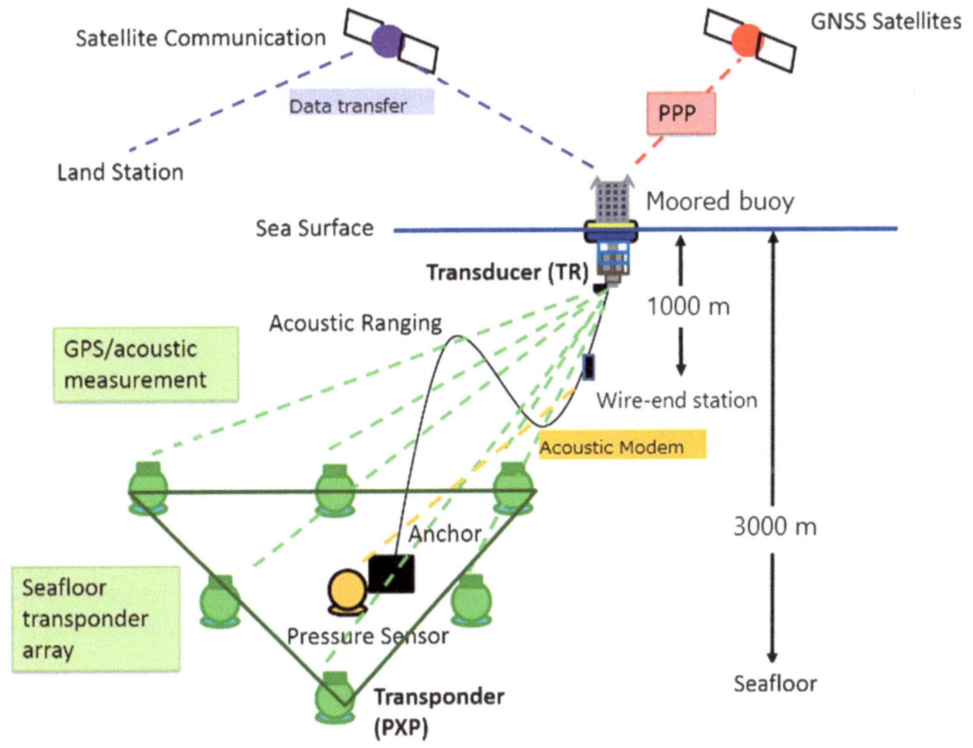

Fig. 2 Schematic illustration of the buoy system. Seafloor pressure data are transmitted by an acoustic modem from the logger in the seafloor pressure unit to the wire-end station at a depth of 1,000 m. Next, the data are transmitted from the wire-end station to the buoy station by the inductive modem (Takahashi et al. 2014). Acoustic signals in acoustic ranging are sent to the logger in the buoy and processed to the minimum but sufficient travel time data. The data (i.e., the processed travel time data, buoy position and attitude, and seafloor pressure), are transmitted to the land station through satellite communication

position on the buoy and the acoustic ranging between the buoy and PXPs. The obtained data are transmitted to the land station via satellite communication after being pre-processed and automatically compressed within the buoy for real-time monitoring. A set of acoustic measurement consists of 11 range pings over a 65-s interval. This cycle was repeated once a week, which was constrained by power limitation of the acoustic ranging. Note that all the six PXPs respond all at once to every single ping. With regard to the acoustic ranging signal, we employed a step sweep of 22 ms duration, in which the frequency increases stepwise by 0.5 kHz from 8.5 to 12.5 kHz. We confirmed that the undesired sidelobes adjacent to the correlogram peak can be more efficiently reduced than with the widely used pseudo-random coding, such as M-sequence.

3 Processing of the Acoustic Signal

Due to the low bit-rate of the satellite communication, it is not realistic to telemeter the full acoustic waveform to the land station. Therefore, data must be processed on the buoy and only the processed result was transmitted to the onshore station. In order to process the acoustic signal on the buoy with a low-powered device, it is necessary (1) to extract the recorded waveform over as small a window as possible, and (2) to send the processed data containing the minimum but sufficient information. To achieve (1) in the current system, the acoustic waveform recorded is extracted to a ± 20 ms window centered on the synthetic travel-time calculated from the provisional position of the buoy and that of each seafloor transponder. A cross-correlation between the transmitted signal and the received signal is then calculated to obtain an accurate travel-time. For (2), after the processing described in (1), the correlogram is used to generate a 1 ms window centered on the maximum correlation peak, which is sufficiently wide to include the sidelobes and to identify the primary peak. Since the correlation coefficient is represented in 8 bits and the sampling rate in 100 kHz, the resulting information to be transmitted can be reduced to only 101 bytes per single-ranging and individual PXP (Fig. 3). Acoustic ranging is carried out 11 times a week for the six PXPs as described previously. Therefore, roughly 7 kilobytes of data are transmitted to the land station once a week.

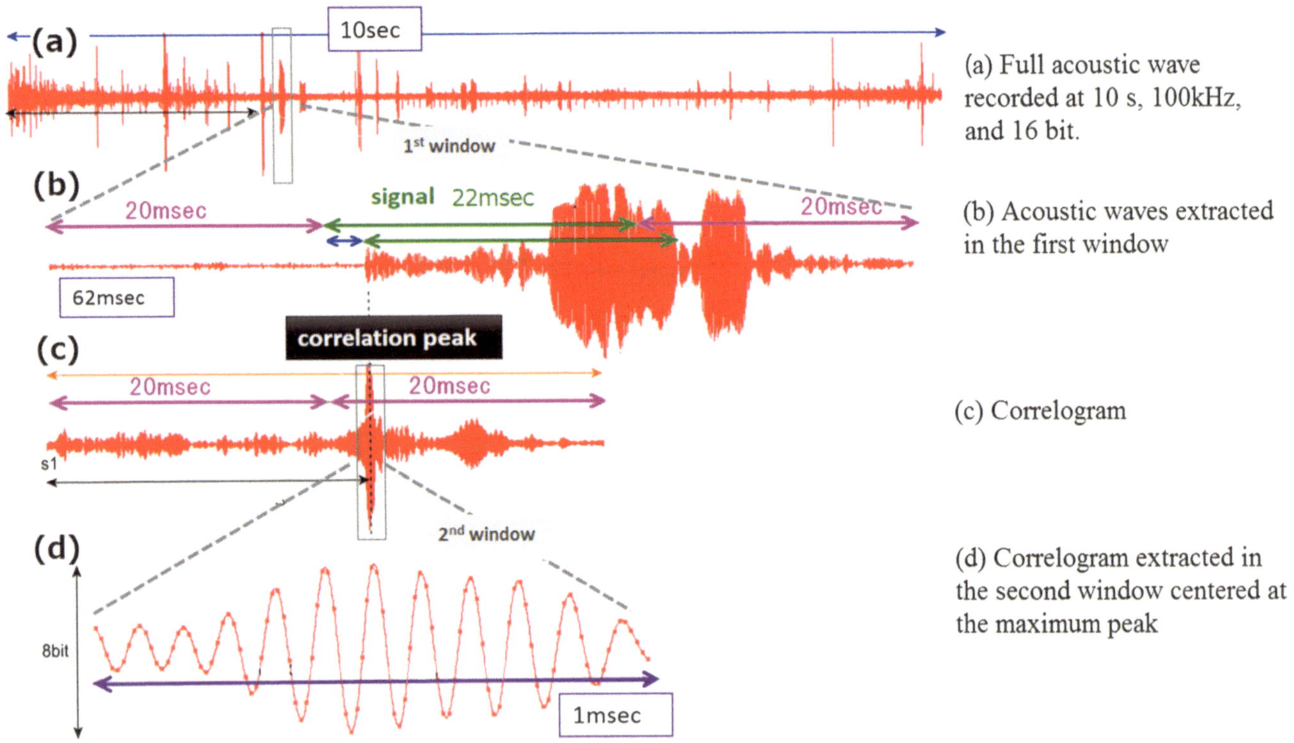

Fig. 3 Processing diagram of an acoustic wave on the buoy. (**a**) A full acoustic wave (sampling rate: 100 kHz, signal length: 10 s) of a single acoustic ranging. Response signals from the six seafloor PXPs can be seen. The letter on the top of the six peaks represents a reply from an individual PXP. (**b**) Acoustic wave extracted in the first window. (**c**) Correlogram between the transmitted and the extracted waves. (**d**) Correlogram extracted in second window (signal length: 1 ms) to be sent to the land station

However, in roughly 25% of the obtained correlogram and the major part of this paper, another significant peak is observed near the maximum peak following the recovery of the buoy and investigation of the full waveforms. This can be interpreted as a multipath at the sea surface. In some cases, a peak associated with a multipath is larger than that associated with a direct path (Fig. 4). In this case, it is not possible to obtain the travel-time of the direct wave automatically.

By investigating the correlograms, the time lag between twin peaks is found to be approximately about 3 ms (Fig. 4b). Then, we attempted to verify the cause of the twin peaks by assuming that the latter peak of twin peaks is caused by the multipath of the acoustic signal at the sea surface.

4 Verification of the Cause of Twin Peaks

To send the correct peak in the correlogram to the land station in real-time, we assessed whether the latter peak of the twin peaks was caused by the reflection of the acoustic signal at the sea surface. We demonstrated this by observing the time lag and the synthetic path difference between the direct wave and multipath as follows. First, given that direct and multipath signals are parallel to each other because the source is sufficiently far (Fig. 5), we can simply calculate the synthetic path difference (ΔL) from transducer depth from the sea surface (d_{tr}: 2.2 (m)) and incident angle to the vertical (θ) as follows:

$$\Delta L = d_{tr} \cos\theta \quad (1)$$

Second, we read the time lag (∼ a few microseconds) between the two peaks and convert it to distance using a sound speed of 1,500 m/s (Fig. 4). Then, we compared the observed direct path difference with the synthetic path difference (Fig. 6). Because it is larger than the synthetic path difference, we assumed that the buoy goes under water more deeply than the depth previously assumed (2.2 m) according to the tension of the mooring cable because of the strong Kuroshio Current. We considered that the scatter was caused by the transducer depth change due to the buoy tilt. Although the data are largely scattered, they correlate with the synthetic data as a function of the incident angles

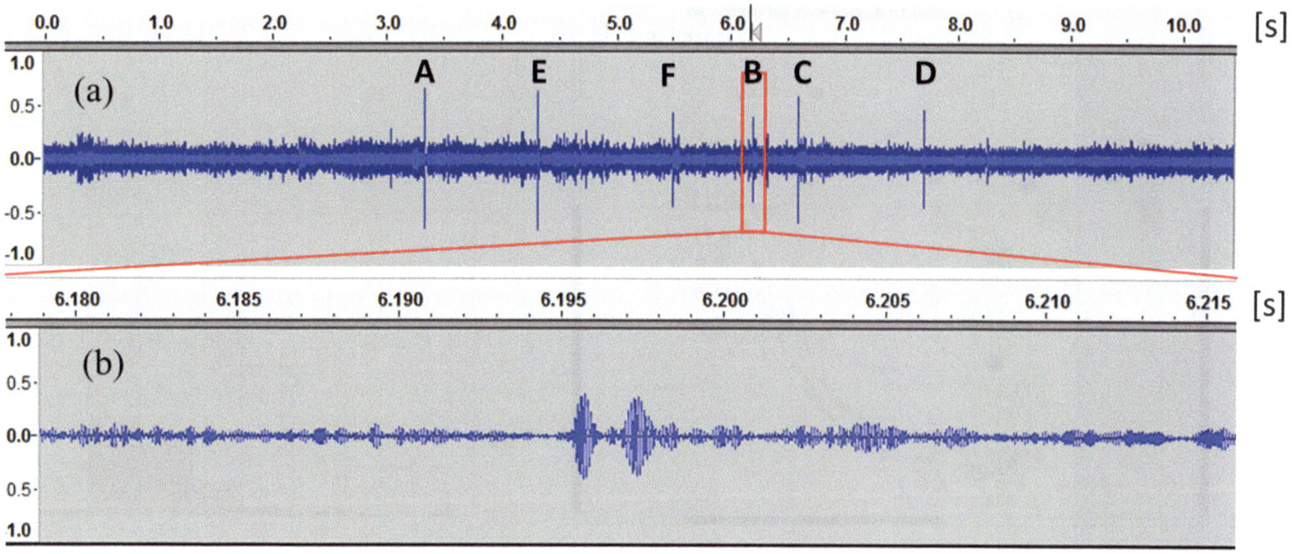

Fig. 4 (**a**) A correlogram between the transmitted signal and the received signal from the individual PXPs. In the correlogram, the travel times for the six PXPs are within in 10 s. (**b**) Twin peaks in one received signal. The first peak is the direct path and the second peak is multipath. The *vertical axis* represents the correlation coefficient. The *horizontal axis* indicates travel time in seconds. The time lag between the peaks is approximately 2 ms, which corresponds to a path difference of approximately 3 m

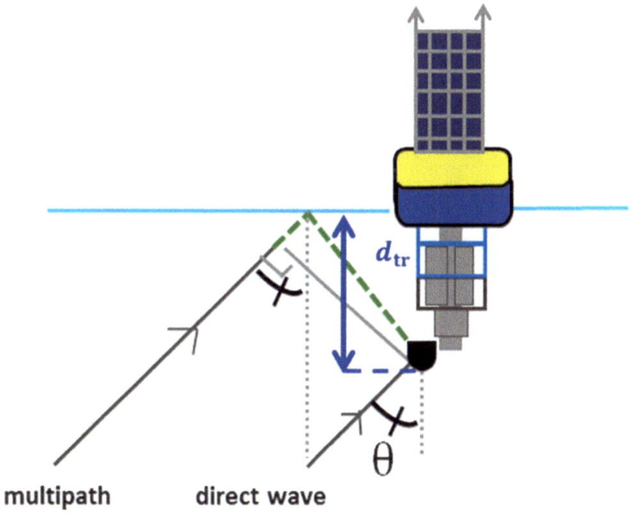

Fig. 5 Schematic of the path difference between the direct wave and the multipath. The depth of the transducer is assumed to be 2.2 m (*blue arrow*). The emergence angle and the incident angle are assumed to be equal (θ). The synthetic path difference between the direct wave and multipath is indicated by the *dashed green line*

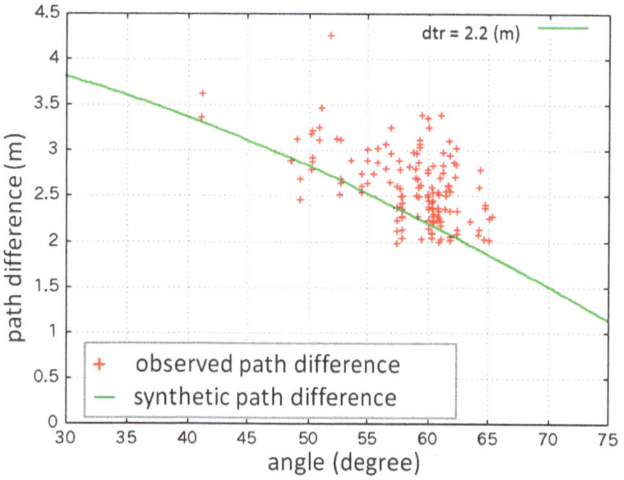

Fig. 6 Comparison between the synthetic path difference (*green line*; $d_{tr} = 2.2$ m) and the observed path difference (*red points*). The *horizontal axis* indicates the incident angle in degrees

(i.e., relative position between the buoy and PXPs). Figure 7 shows the positions of the buoy relative to the PXPs when twin peaks appear. It is obvious that twin peaks appear only when the buoy maintains a large distance from the target PXP. This is clear if we plot the histogram of the twin peak occurrence as a function of the incident angle. The twin peaks were found to occur only when the incident angle was larger than 50°.

This can be explained by the inherent directivity of the transducer as shown in Fig. 8. When the acoustic signal is emitted at small incident angles (Fig. 8a: green solid vector), the incident angle of multipath is also small (Fig. 8a: green dashed vector). In case of multipath at a small incident angle (Fig. 8b: green dashed vector), the sensitivity of the transducer was lower by 10 dB than that of the direct path

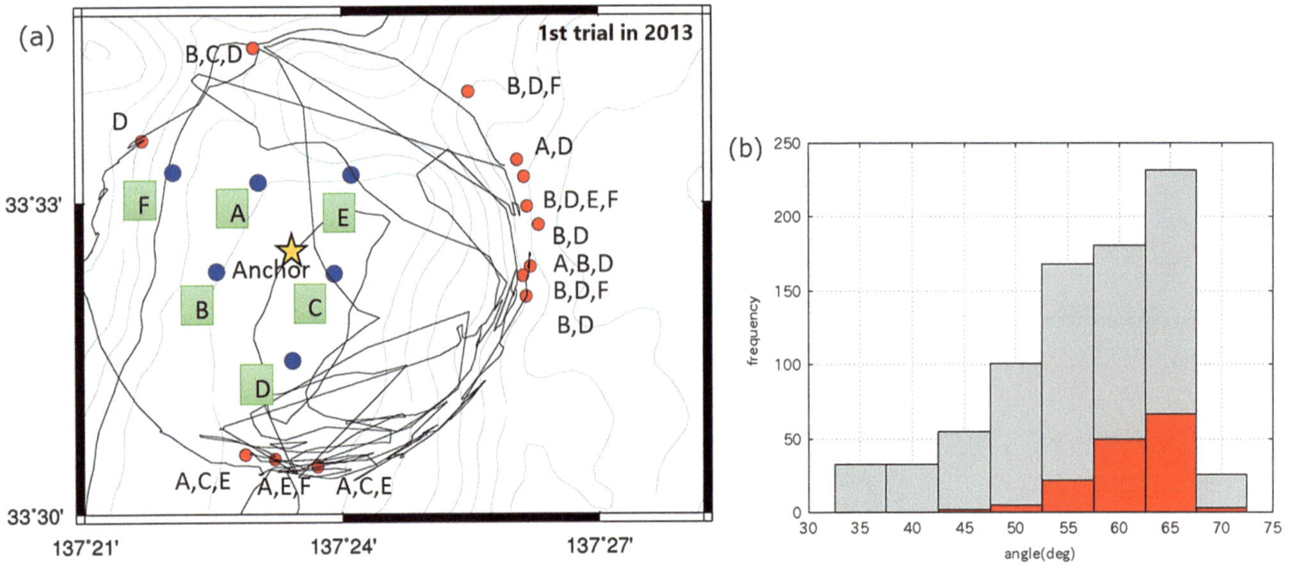

Fig. 7 (**a**) Spatial relationship between the buoy and the PXPs at the time of ranging during the first trial in 2013. Anchor (*yellow star*), PXPs (*blue circle*) with their IDs (*green rectangle*), buoy position during acoustic ranging (*red circle*), and buoy trace (*black line*) are shown. Letters next to the *red circles* are the IDs of individual PXPs for which twin peaks are observed. (**b**) Histogram of the frequency as a function of incident angle for each ping in an acoustic ranging in 2013. The *grey bars* indicate the frequencies of all incident angles. The *red bars* represent the frequencies of only those incident angles for which twin peaks appear

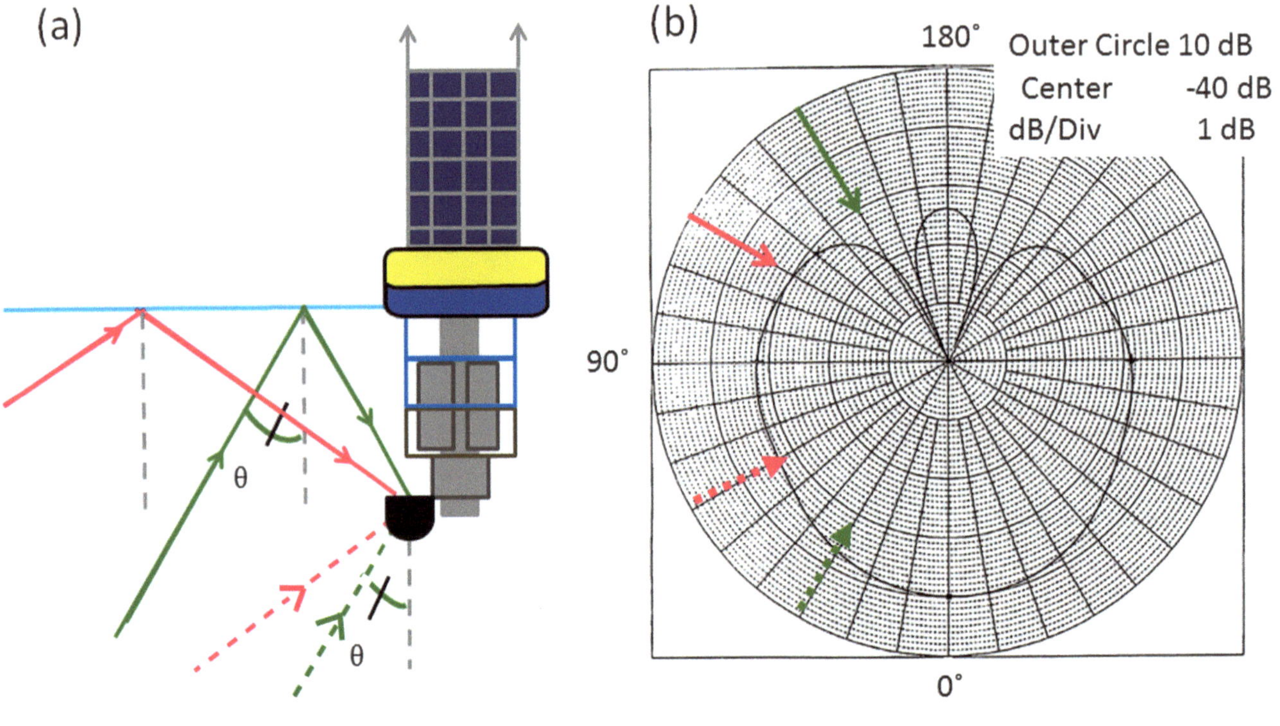

Fig. 8 (**a**) Schematic illustration of differences in the reflection angle by the incident angle. The vector from the seafloor to the transducer represents the direct path (*dashed line*) and the vector reflected at sea surface represents multipath (*solid line*). *Pink vectors* are the paths of large (60°) incident angles and *green vectors* are the paths of small (30°) incident angles. (**b**) Sensitivity of the transducer as a function of incident angle (θ). *Pink* and *green arrows* represent the same as in (**a**)

Table 1 Standard deviation of the estimated array positions over the entire analysis period (13 weeks) and over a set of an acoustic ranging during approximately 10 min using (A) the travel time data processed using the new method described in the text, and (B) the data obtained using the conventional method

Standard deviation	(A) in the entire analysis period	(B) in the entire analysis period	(A) in a set of acoustic ranging	(B) in a set of acoustic ranging
E–W (m)	3.7	5.2	0.45	3.7
N–S (m)	2.2	3.9	0.34	2.6

(Fig. 8b: green solid vector). This resulted in a peak in the correlogram due to multipath being small or almost diminished. Therefore, the multipath detection rate at small incident angles was lower than that at large incident angles, as shown in Fig. 7b.

We then reprocessed the travel time data as follows. After cross-correlation processing, (Fig. 3c), we checked the existence of another peak over the correlation coefficient value 0.3 (the noise level), 3 ms before the maximum peak. If the other peak was found, the earlier peak would be transmitted to the land station. In this manner, we could detect the travel time of the only direct waves.

5 Estimation of the PXP Array Position

The array position of the six PXPs was estimated following the methodology presented in Kido et al. (2006) using (A) the travel time data processed using the new method described above, and (B) the data obtained using the conventional method for comparing their accuracy. The buoy position was estimated by the kinematic PPP analysis using GIPSY/OASIS II software (Stephen et al. 1996). The PPP estimation accuracy was ~2 to 3 cm horizontally and ~3 to 4 cm vertically (Ohta et al. 2006). Because we considered the positioning error by acoustic ranging away from the array center to be larger than that of the positioning error of the buoy position, we believe that the positioning accuracy of the buoy has little effect on the estimated array position.

We estimated 53 array positions in total using data for 13 weeks. The standard deviation of the estimated array positions over the entire analysis period (13 weeks) and over a set of the acoustic ranging during approximately 10 min are shown in Table. 1, and the time series and map view of the estimated array position for 1 ping are plotted in Fig. 9a–c. From these results, the new method was found to effectively enhance the accuracy of the remote acoustic ranging. The positioning accuracy within a set of acoustic ranging (10 min) was improved from 4 to 0.5 m, while the positioning accuracy during the entire analysis period (13 weeks) was improved from 8 to 4 m. The current accuracy of the positioning, however, was insufficient compared with initial target value (~1 m in the horizontal component). The effect of the array geometry error appears depending on the buoy position during acoustic ranging. By comparing Fig. 9c (the estimated array positioning with its date) and Fig. 9d (the buoy position), the emergence of the array positioning error depending on the buoy position was verified. The estimated array position was assumed to vary widely because acoustic ranging was carried out far from the array center. As a result, the error by individual PXP position uncertainty was increased. Therefore, we should re-determine individual PXP position precisely based on the method proposed by Kido et al. (2006) using the data obtained by a moving survey (Spiess et al. 1998) in a campaign observation to reduce the error and improve the positioning accuracy in GPS/A measurement using the buoy system.

6 Conclusion

We have demonstrated that the second twin peak in the correlogram originated from sea surface multipath. Therefore, we have proposed the improvement of an algorithm to detect automatically the true correlation peak (the first peak).

Applying the algorithm to the data obtained in the second sea trial, we revealed that the positioning accuracy of a set of acoustic ranges was improved from 4 to 0.5 m while the positioning accuracy over 13 weeks was improved from 8 to 4 m.

This algorithm will be adopted in the next trial as real-time processing. The scatter of the estimated array position, however, remains on the order of few meters over the entire observation period compared to the desired 1 m accuracy. Therefore, it is necessary to determine individual PXP position (accuracy = ~10 cm) using a campaign observation data obtained in a moving survey through a research vessel in order to improve array positioning accuracy in GPS/A measurements using the buoy system for the detection of horizontal seafloor crustal deformation, achieving ~1 m accuracy.

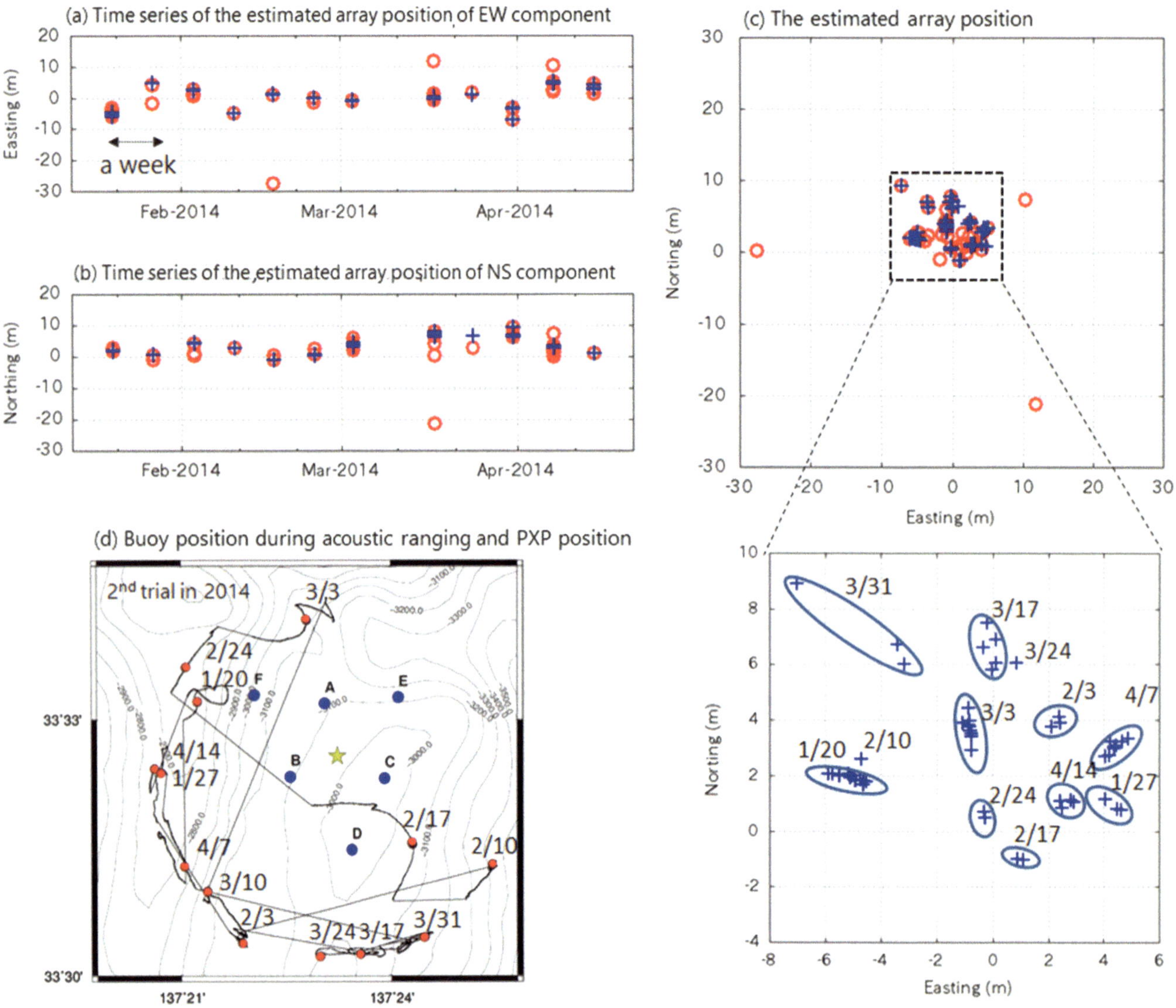

Fig. 9 Time series plot of the estimated array position. The *red open circle* represents the array position estimated using the current processing method. The *blue cross* represents the estimated array position applying the improved processing: (**a**) EW component. (**b**) NS component. (**c**) Projection of the estimated array position of EW and NS components for map view. *Dashed rectangle* shows the enlarged array position estimated using the improved processing. The array position estimated using the data during the same acoustic ranging set is surrounded by the *blue ellipse*. The letter at the right top of the *blue ellipse* shows the date (month/day) when an acoustic ranging is carried out. (**d**) Buoy position during acoustic ranging (*red closed circle*) and individual PXP position (*blue closed circle*) and the array center (*yellow star*)

References

Ballu V, Bouin MN, Calmant S, Folcher E, Bore JM, Ammann J, Pot O, Diament M, Pelletier B (2010) Absolute seafloor vertical positioning using combined pressure gauge and kinematic GPS data. J Geod 84(1):65–67. doi:10.1007/s00190-009-0345-y

Fujimoto H (2014) Seafloor geodetic approaches to subduction thrust earthquakes. Monographs on environment. Earth Planet 2(2):23–63. doi:10.5047/meep.2014.00202/0023

Kaneda Y, Kawaguchi K, Arai E, Matsumoto H, Nakamura T, Kamiya S, and Hori T (2007) Dense ocean floor network system for earthquakes and tsunamis – towards the understanding of mega thrust earthquakes. OCEAN2007, Vancouver

Kido M, Fujimoto H, Miura S, Osada Y, Tsuka K, Tabei T (2006) Seafloor displacement at Kumano-nada caused by the 2004 off Kii Peninsula earthquakes, detected through repeated GPS/Acoustic surveys. Earth Planets Space 58:911–915

Kido M, Osada Y, Fujimoto H, Hino R, Ito Y (2011) Trench-normal variation in observed seafloor displacements associated with the 2011 Tohoku-Oki earthquake. Geophys Res Lett 38:L24303. doi:10.1029/2011GL50057

Ohta Y, Sagiya T, Kimata F (2006) Assessment of the long-term stability of the PPP kinematic GPS. J Geod Soc Jpn 52(4):309–318

Sato M, Ishikawa T, Ujihara N, Yoshida H, Fujita M, Mochizuki M, Asada A (2011) Displacement above the hypocenter of the 2011 Tohoku-Oki earthquake. Science. doi:10.1126/science.1207401

Spiess FN, Chadwell CD, Hildebrand JA, Young LE, Purcell GH Jr, Dragert H (1998) Precise GPS/Acoustic positioning of seafloor reference points for tectonic studies. Science. doi:10.1016/s0031-9201(98)00089-2

Stephen ML, Sever Y, Bertiger WL, Heflin M, Hurst K, Muellerschoen RJ, Wu SC, Yunk T, Zumberge J (1996) GIPSY-OASIS II: a high precision GPS data processing system and general satellite orbit analysis tool. Jet Propulsion Laboratory, California Institute of Technology, Pasadena, CA

Tadokoro K, Ando M, Ikuta R, Okuda T, Besana GM, Sugimoto S, Kuno M (2006) Observation of coseismic seafloor crustal deformation due to *M7* class offshore earthquakes. Geophys Res Lett. doi:10.1029/2006GL26742

Takahashi N, Ishihara Y, Ochi H, Fukuda T, Tahara J, Maeda Y, Kido M, Ohta Y, Mutoh K, Hashimoto G, Kogure S, Kaneda Y (2014) New buoy observation system for tsunami and crustal deformation. Mar Geophys Res 35:243–253. doi:10.1007/s11001-014-9235-7

Zumberge JF, Heflin MB, Jefferson D, Watkins MM, Webb FH (1997) Precise point positioning for the efficient and robust analysis of GPS data large networks. J Geophys Res 102:5005–5017

Part IV

Interaction of Earthquakes and Volcanoes

Pressure Sources of the Miyakejima Volcano Estimated from Crustal Deformation Studies During 2011–2013

Miyo Fukui, Takeshi Matsushima, Natsumi Yumitori, Jun Oikawa, Atsushi Watanabe, Takashi Okuda, Taku Ozawa, Yuhki Kohno, and Yousuke Miyagi

Abstract

Following the eruption of the Miyakejima Volcano in the Izu Islands, Japan, in the year 2000, a continuous GPS network observed the ongoing contracting crustal deformation. Subsequently, a slight inflation of the island was detected from around 2006, and we initiated a campaign of dense GPS observations around the volcano from 2011. Precise crustal deformation studies indicated inflation in the southern part of the island and deflation around the center of the crater. Using these observations, we estimated that three magma sources (a shallow deflation sill under the crater, a southern inflation dyke, and a deep inflation spherical source) were activated during 2011–2013. In particular, the presence of an inflation dyke at an intermediate depth had not been inferred by previous studies. Accordingly, we posit that the supply of magma from a deep spherical source to the new dyke source has been initiated only recently.

Keywords

Crustal deformation • GPS • Miyakejima • Pressure source • Volcano

1 Introduction

Miyakejima is a volcanic island approximately 8 km in diameter and located 180 km south of Tokyo, Japan (Fig. 1). Over the past 1,000 years, at least 15 eruptions have been recorded (Tsukui and Suzuki 1998), with eruptions almost every 20 years over the past 100 years (in 1940, 1962, 1983, and 2000). At the center of the island, a caldera of 500 m depth and 1,600 m diameter was formed by the eruption in the year 2000 (Nakada et al. 2001). The eruption type during this period was markedly different from that of the past 100 years. The most recent eruption resulted in the formation of a caldera (for the first time in 2,500 years) and the emission of large quantities of volcanic gas forced the inhabitants to evacuate the island for a period of four and half years.

Just after the 2000 eruption, many GPS were introduced in the island and several models for magma-plumbing systems of the eruption were estimated using GPS and gravity surveys (e.g., Furuya et al. 2003; Irwan et al. 2006; Ueda et al.

M. Fukui • T. Matsushima (✉) • N. Yumitori
Institute of Seismology and Volcanology, Faculty of Sciences, Kyushu University, 2-5643-29 Shin'yama, Shimabara 855-0843, Japan
e-mail: miyo.f@sevo.kyushu-u.ac.jp; mat@sevo.kyushu-u.ac.jp

J. Oikawa • A. Watanabe
Earthquake Research Institute, The University of Tokyo, 1-1-1 Yayoi, Bunkyo-ku, Tokyo 113-0032, Japan
e-mail: oikawa@eri.u-tokyo.ac.jp; atsushi@eri.u-tokyo.ac.jp

T. Okuda
Earthquake and Volcano Research Center, Graduate School of Environmental Studies, Nagoya University, D2-2 (510), Furo-cho, Chikusa-ku, Nagoya 464-8601, Japan
e-mail: okuda@seis.nagoya-u.ac.jp

T. Ozawa • Y. Kohno • Y. Miyagi
National Research Institute for Earth Science and Disaster Prevention, 3-1 Tennodai, Tsukuba, Ibaraki 305-0006, Japan
e-mail: taku@bosai.go.jp; yuhki-k@bosai.go.jp; m_yousuke@bosai.go.jp

M. Hashimoto (ed.), *International Symposium on Geodesy for Earthquake and Natural Hazards (GENAH)*, International Association of Geodesy Symposia 145, DOI 10.1007/1345_2015_150

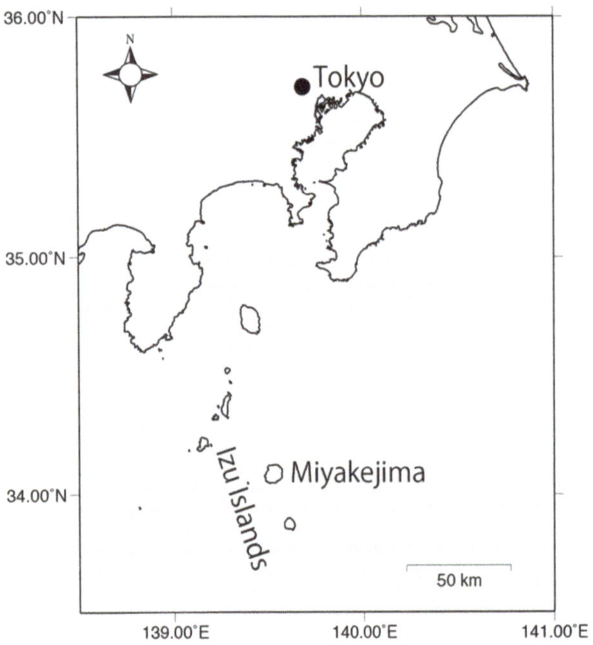

Fig. 1 Map of Miyakejima Island in the Izu Island arc

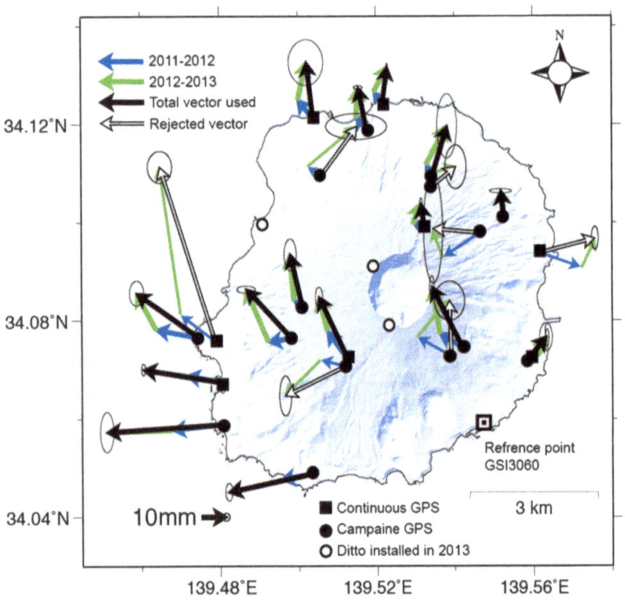

Fig. 2 GPS observation points and observed displacement vectors for 2011–2013. *Squares* show continuous GPS sites observed by other organizations. Some of displacement vectors contain a large error, we did not use these vectors for the following model analysis

2005). Since the associated volcanic activity had almost subsided, crustal deformation has been observed independently by the Japan Meteorological Agency (JMA, 4 sites), Geospatial Information Authority of Japan (GSI, 4 sites), Japan Coast Guard (JCG, 1 sites), and National Research Institute for Earth Science and Disaster Prevention (NIED, 5 sites) (Fig. 2). However, each institution analyzed their own data independently; the existing observation networks were considered insufficient to produce a detailed magma accumulation model. To address this, our dense GPS observation campaign was initiated in 2011, in association with Kyushu University, the University of Tokyo, Nagoya University, and NIED. Here, based on the observations of this campaign, we present the precise crustal deformation and estimated pressure source models for the Miyakejima Volcano during 2011–2013.

2 Data

We conducted static measurements during our GPS observation campaign in September 2011, 2012, and 2013. The sampling interval of the GPS measurements was 5 s, elevation mask was 10°, and we used L1 and L2 frequency types of receiver and the survey time was approximately 3 or 4 days at each observation point. In 2013, we studied 21 points in total, including 2 new points near the summit crater, where no observations had been obtained since the eruption in the year 2000. Integrating the continuous observation points and our dense campaign points, it has enabled an accurate assessment of the island-wide ongoing crustal deformation.

The area around Onohara Island, about 10 km west–southwest of Miyakejima, was seismically active during April 17–18, 2013. An earthquake (Mjma 6.2) occurred at 17:57 JST on April 17, producing slight crustal deformation and generating a tsunami that hit the coasts of Miyakejima. We obtained GPS observations at Onohara Island two months after this earthquake and used the resulting data to produce two fault models in our previous study (Matsushima et al. 2013) (Fig. 3). In the present study, we estimated the amount of displacement at each GPS observation point using these two fault models in order to remove the influence of the earthquake that was included in our displacement data.

In particular, we used the data obtained from nine continuous GPS installations monitored by GSI, JMA, and JCG, along with the 15 sites mentioned before. These GPS data covered the observation periods September 6–9, 2011 and September 2–6, 2013. The data were analyzed using the RTKLIB ver. 2.4.1 software (Takasu et al. 2007). Because

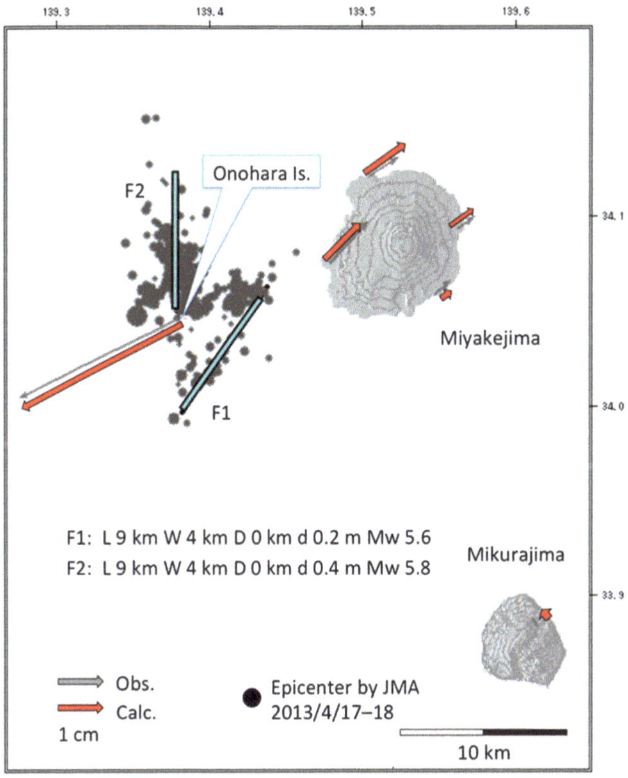

Fig. 3 Two earthquake fault models (*blue lines*) and calculated (*red*) deformation vectors near Miyakejima Volcano on April 17, 2013 (Matsushima et al. 2013)

the distance between each site is less than 10 km, we ignored the ionospheric and tropospheric delays during analysis. IGS precise ephemeris was also used during processing. We calculated the baseline vector of each site using the GSI 3060 site as reference. In order to obtain the displacement vector which has the high signal-to-noise ratio, we used the displacement vector sum for 2011 to 2012 and 2012 to 2013. Since there is a clear difference between these 2-year vectors, shown as open arrows in Fig. 2, we did not use these vectors for our modeling.

Figure 4 illustrates the observed deformation vector and areal strain of the Miyakejima Volcano that was analyzed using the Magnetic and Geodetic data Computer Analysis Program for Volcano (MaGCAP-V) software (Meteorological Research Institute 2008). It indicates that the southern part of the island underwent considerable expansion and a narrow area of around the crater underwent crustal shortening.

3 Modeling

We estimated the position and volume change of the underground pressure source (magma chamber) of the Miyakejima Volcano using an elevation-modified Mogi model (Fukui et al. 2003) and two open crack models (Okada 1992) that took into account the elevation of the observation points from the horizontal crustal deformation data for 2011 to 2013 (Fig. 5). We used MaGCAP-V in this analysis. The default value of this experiment was from the model parameter of GSI, and the source parameters were estimated by trial and error method, i.e., the comparison of these estimations with surface crustal deformation calculated by forward modeling (Table 1). Presence of a dyke has been inferred to explain the crustal movement in the southwestern part of the island. However, the displacement around the crater has been attributed to the presence of a sill. Based on the results, the estimated magma sources include a shallow deflation sill under the crater (0.3 km BSL), a southern inflation dyke (4–7 km BSL), and a deep inflation spherical Mogi (13 km BSL).

4 Discussion

Figure 6 shows the displacement vectors calculated for each source model. While the shallow sill contributes to the displacement around the crater, the dyke of intermediate depth greatly contributes to the crustal movement of the southern part of the island. The deep Mogi source contributes to the displacement of the northern parts of the island. Figure 7 depicts residual sum of squares for the dyke source parameters during the grid search inversion. While the depth and dip angle of the dyke source are sensitive for model fitting, the width of the dyke cannot be determined if the width is 4 km or more because of the limited GPS network coverage within the island.

Based on InSAR time series analysis, Ozawa and Ueda (2011) demonstrated that during 2006–2011, subsidence along with crustal shortening occurred around the caldera. Our model indicates a sill source under the caldera near the sea level, consistent with the result presented by Ozawa and Ueda. We consider the observed shortening to be a result of volume shrinkage due to the ongoing release of large amounts of volcanic gases (e.g., Kazahaya et al. 2004).

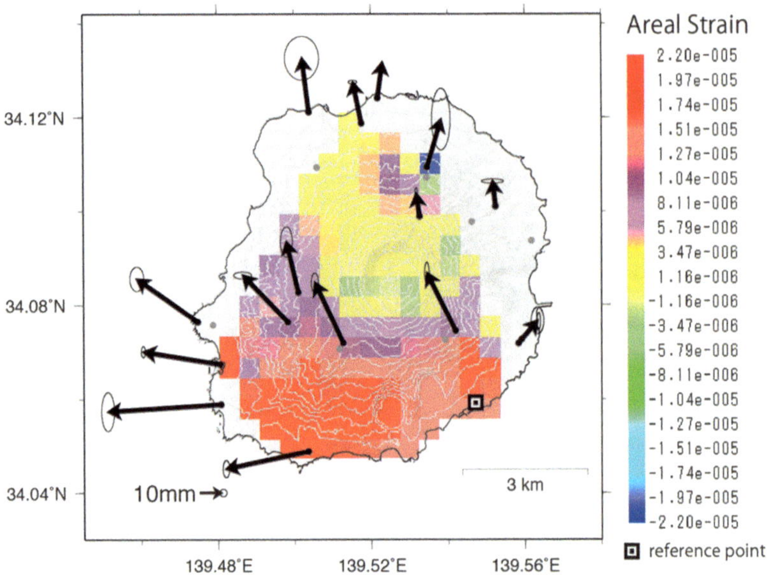

Fig. 4 Observed deformation vectors and areal strain for Miyakejima Volcano. An area of high expansion is visible in the southern part of the island

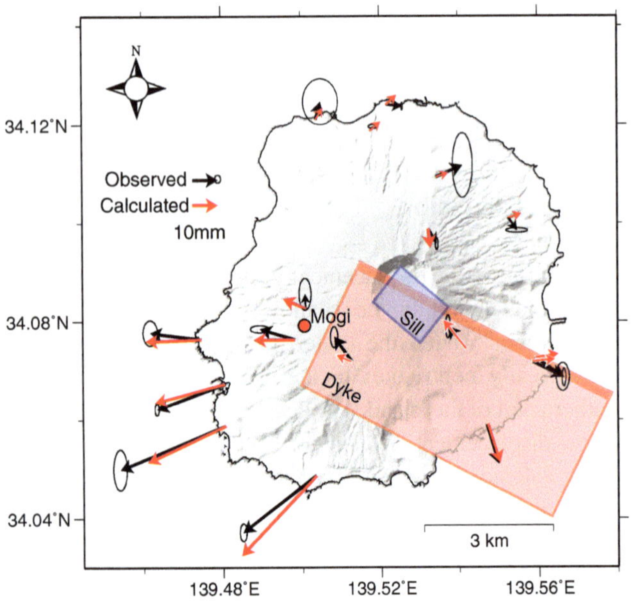

Fig. 5 Estimated three-magma-source model derived in the present study. The calculated (*red*) and observed (*black*) deformation vectors with 1σ error are also shown. The observation vectors are horizontally shifted to be a same vector as the model-calculated vectors at GSI3060 site in order to easy to compare observation and calculated vector. The estimated pressure sources are a shallow sill (*blue rectangle*), a dyke at intermediate depth (*red rectangle*), and a deeper Mogi source (*red filled circle*)

Volcanic earthquakes in Miyakejima, distributed only in the vicinity of the crater with focal depth of 0.5 to 2 km BSL (National Research Institute for Earth Science and Disaster Prevention (NIED) 2013), causes the stress change around the conduit under the sill. From seismicity data we could not detect the movement of the dyke and Mogi sources.

In a more recent study, Geospatial Information Authority of Japan (GSI) (2011) developed pressure source models using the continuous GPS observation data of GSI and JMA for the period 2009–2010. The developed models consist of a spherical expansion source in the deep underground area and a shallow contraction source area under the crater. However, deformation data for 2011–2013 cannot be explained using only the deep spherical source estimated by GSI. Rather, we inferred the presence of an expanded dyke at an intermediate depth to explain the expansion vectors in the southern part of the island.

Kimata (2001) pointed out using GPS that displacement during 1998–1999 was caused by an inflating dyke at the depth range of 3–5 km beneath the southwestern part of the island. And Murase et al. (2009) pointed out using GPS and leveling survey that displacement during 1983–2000 was caused by an inflating dyke at the depth range of 6–7 km also beneath the southwestern part of the island. Although strike

Table 1 Model parameters of three Magma sources beneath the Miyakejima volcano estimated in this study

< Sill source>Deflation	
Latitude (center)	34.083 N
Longitude (center)	139.528 E
Length (EW)	1.5 km
Width (NS)	1.0 km
Depth	0.3 km
Dip	0°
Open	−0.4 m
Volume (2 yr)	-5.6×10^5 m
<Dyke source>Inflation	
Latitude (center)	34.066 N
Longitude (center)	139.540 E
Length (top)	6.7 km
Width	5.0 km
Depth (top)	4.0 km
Strike (N to E)	120°
Dip	50°
Open	0.4 m
Volume (2 yr)	$+1.3 \times 10^7$ m
<Mogi source>Inflation	
Latitude	34.079 N
Longitude	139.500 E
Depth	13.0 km
Volume (2 yr)	$+1.4 \times 10^7$ m

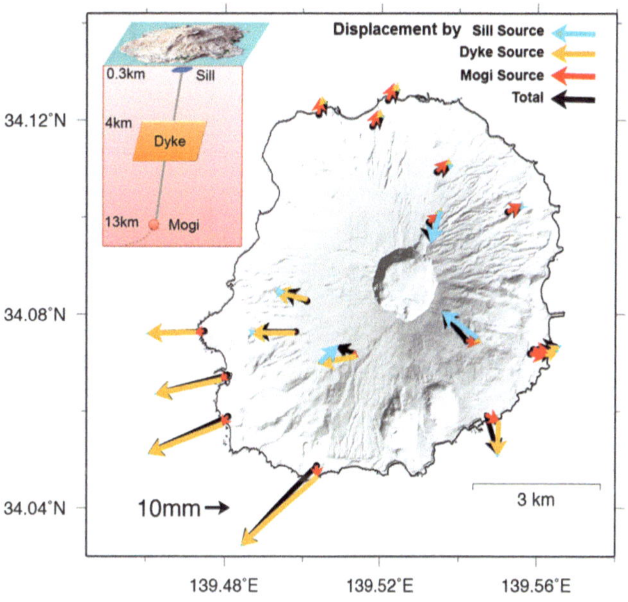

Fig. 6 Displacement vectors calculated from each source model

and dip of the dyke is different from the estimation of Kimata and Murase et al. before the eruption in the year 2000, it is shown that the magma under the Miyakejima Volcano is accumulated in an inflating dyke of several km depth.

Based on the modeling results presented here, we suggest that the magma supply may have been re-activated to the magma chamber associated with the eruption in the year 2000, although the cumulative volume of magma currently present is likely much smaller than the magma discharge during the 2000 eruption. Therefore, it is important to continue the GPS observations to monitor the volcanic activity at Miyakejima and obtain additional information.

5 Conclusion

A campaign to obtain dense GPS observations around the Miyakejima Volcano has been underway since 2011. From the precise crustal deformation data obtained, we were able to detect the inflation in the southern region of the island and deflation around the center of the crater. Using the deformation data, we estimated that the three magma sources (a shallow deflation sill under the crater, a southern inflation dyke, and a deep inflation spherical source) were activated during 2011–2013. In particular, the inflation dyke at the intermediate depth had not been confirmed in the previous studies, and we suggest that the supply of magma to the new dyke source may have been initiated from the deep spherical source. To ensure the monitoring of the inflation sources in future, it will be necessary to enhance the southern observation network and obtain more detailed geodetic data.

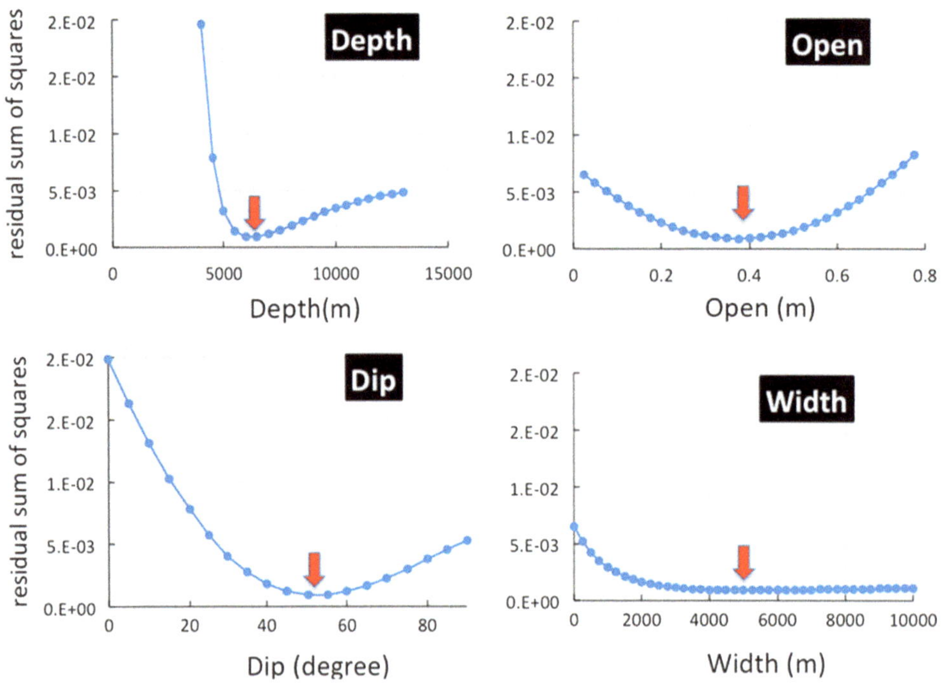

Fig. 7 Example of residual sum of squares for the dyke source parameters during the grid search inversion

Acknowledgments This work was assisted by the University of Tokyo Earthquake Research Institute cooperative research program. We used the GPS data provided by the GSI and JMA. Some of figures were prepared using Generic Mapping Tools (Wessel and Smith 1998). We thank members of Seismology and Volcanology Research Department, Meteorological Research Institute, JMA for technical support. We would like to thank Editage (www.editage.jp) for English language editing.

References

Fukui K, Sakai T, Yamamoto T, Fujiwara K, Takagi A, Churei M (2003) Availability and limitations of an elevation-modified Mogi model to estimate the pressure source from volcanic crustal deformation. In: Volcanological Society of Japan 2003 Fall Meeting 35. Meeting is held in Fukuoka City on October 11–13, 2003, p 35

Furuya M, Okubo S, Sun W, Tanaka Y, Oikawa J, Watanabe H, Maekawa T (2003) Spatiotemporal gravity changes at Miyake-jima Volcano, Japan: Caldera collapse, explosive eruptions and magma movement. J Geophys Res 108(B4):2219. doi:10.1029/2002JB001989

Geospatial Information Authority of Japan (GSI) (2011) Crustal deformations around Miyake Volcano. Report of coordinating Committee for Prediction of Volcanic Eruption, vol 106. pp 87–97

Irwan M, Kimata F, Fujii N (2006) Time dependent modeling of magma intrusion during the early stage of the 2000 Miyakejima activity. J Volcanol Geother Res 150(1):202–212. doi:10.1016/j.jvolgeores.2005.07.014

Kazahaya K, Shinohara H, Uto K, Odai M, Nakahori Y, Mori H, Iino H, Miyashita M, Hirabayashi J (2004) Gigantic SO_2 emission from Miyakejima volcano, Japan, caused by caldera collapse. Geology 32:425–428. doi:10.1130/G20399.1

Kimata F (2001) Magma intrusion system estimated from the ground deformation at Miyake Volcano in and after the 1983 eruption. Abstract 2001 Japan Earth Planetary Science Joint Meeting V0-004

Matsushima T, Fukui M, Oikawa J, Watanabe A, Ohminato T, Ozawa T, Miyagi Y, Kohno Y, Okuda T (2013) GPS Observation of 2013 Miyakejima-kinkai earthquake (Mj6.2). In: Volcanological Society of Japan 2013 Fall Meeting 166. Meeting is held in Inawashiro Town on September 29–October 1, 2013, p 166

Meteorological Research Institute (2008) Development of Software for analyzing crustal activity in volcanoes. Technical reports of the Meteorological Research Institute, vol 53. pp 123–140 (in Japanese with English abstract)

Murase M, Takano K, Kimata F (2009) Episodic magma supply changing eruption style and eruption frequency estimated from long-term geodetic measurements. In: Lewis N, Moretti A (eds) Volcanoes: formation, eruptions and modelling. Nova Science, New York, USA, 379 pp. ISBN: 978-1-60692-916-2

Nakada S, Nagai M, Yasuda A, Shimano T, Geshi N, Ohno M, Akimasa T, Kaneko T, Fujii T (2001) Chronology of the Miyakejima 2000 eruption: characteristics of summit collapsed crater and eruption products. J Geogr (Chigaku Zasshi) 110(2):168–180. doi:10.5026/jgeography.110.2_168 (in Japanese with English abstract)

National Research Institute for Earth Science and Disaster Prevention (NIED) (2013) Seismic activity and crustal deformation at Miyake-jima volcano. Report of coordinating Committee for Prediction of Volcanic Eruption, vol 111. pp 63–68

Okada Y (1992) Internal deformation due to shear and tensile faults in a half-space. Bull Seismological Soc Am 82:1018–1040

Ozawa T, Ueda H (2011) Advanced interferometric synthetic aperture radar (InSAR) time series analysis using interferograms of multiple orbit tracks: a case study on Miyake-jima. J Geophys Res 116, B12407. doi:10.1029/2011JB008489

Takasu T, Kubo N, Yasuda A (2007) Development, evaluation and application of RTKLIB: a program library for RTK-GPS. In: GPS/GNSS symposium 2007. pp 213–218 (in Japanese)
Tsukui M, Suzuki Y (1998) Eruptive history of Miyakejima volcano during the last 7000 years. Bull Volcanol Soc Jpn 43:149–166
Ueda H, Fujita E, Ukawa M, Yamamoto E, Irwan M, Kimata F (2005) Magma intrusion and discharge process at the initial stage of the 2000 activity of Miyakejima, Central Japan, inferred from tilt and GPS data. Geophys J Int 161(3):891–906
Wessel P, Smith WH (1998) New, improved version of generic mapping tools released. Eos Trans Am Geophys Union 79(47):579

Part V

Natural Hazards

Application of A10 Absolute Gravimeter for Monitoring Land Subsidence in Jakarta, Indonesia

Yoichi Fukuda, Jun Nishijima, Yayan Sofyan, Makoto Taniguchi, Mahmud Yusuf, and Hasanuddin Z. Abidin

Abstract

To investigate the cause of land subsidence occurring in Jakarta, Indonesia, we have conducted absolute gravity measurements near or at the existing GPS points in 2008, 2009, 2010, 2011 and 2013, by using a portable absolute gravimeter Micro-G LaCoste A10 serial number 017 (MGL A10-017). The A10 measurements, particularly those conducted before 2010, were adversely affected by the influence of the high-temperature and high-humidity climate, which caused problems with the ion vacuum pump, laser controls and unstable behavior of the data processing personal computer. Moreover, some of the gravity points were lost mainly due to construction work. For these reasons, the number of reliable gravity data was restricted. Nevertheless, the observed gravity changes along the northern coastal area of Jakarta showed an increase in gravity, which was consistent with the GPS results. The analysis of the height changes versus the gravity changes suggested that the subsidence was caused by soil compaction and/or groundwater level changes, which are likely connected with excessive groundwater extraction.

Keywords

A10 absolute gravimeter • Absolute gravity measurements • Groundwater extraction • Land subsidence

1 Introduction

Because land subsidence is an urgent problem in many urbanized cities in Indonesia, many studies have recently been conducted in various cities by using various techniques such as leveling, GPS, and InSAR (e.g., Abidin et al. 2004, 2012, 2013; Ng et al. 2012). Although land subsidence in Jakarta was first reported in 1926, it became widely recognized by the public when buildings and a fly-over bridge were cracked in 1978 as a result (Djaja et al. 2004). Subsequently, repeated leveling measurements were performed in the northern part of Jakarta, and the Local Mines Agency reported that a cumulative subsidence of 20 to 200 cm occurred during 1982–1997. GPS surveys began in the 1990s (Abidin et al. 2001) and reported the rate of subsidence to be more than 10 cm/year near the coastal area (Abidin et al. 2011).

As reported by Djaja et al. (2004), the impact of land subsidence in Jakarta is serious, resulting in cracked buildings and other permanent structures. Moreover, subsidence has expanded the areas of flooding, seawater intrusion, and tidal flooding, particularly along the northern coastal areas. These

Y. Fukuda (✉)
Graduate School of Science, Kyoto University, Kyoto, Japan
e-mail: fukuda@kugi.kyoto-u.ac.jp

J. Nishijima • Y. Sofyan
Department of Earth Resources Engineering, Kyushu University, Fukuoka, Japan

M. Taniguchi
Research Institute for Humanity and Nature, Kyoto, Japan

M. Yusuf
Meteorology Climatology and Geophysical Agency, Jakarta, Indonesia

H.Z. Abidin
Institute of Technology Bandung, Bandung, Indonesia

M. Hashimoto (ed.), *International Symposium on Geodesy for Earthquake and Natural Hazards (GENAH)*,
International Association of Geodesy Symposia 145, DOI 10.1007/1345_2016_221

consequences have caused harmful effects on human life in addition to large economic losses.

To mitigate the detrimental effects of subsidence, it is very important to understand the causes. Murdohardono and Sudarsono (1998) suggested that four different causes could explain subsidence in Jakarta, including groundwater level changes, construction load, natural consolidation of the alluvial soil layer, and tectonic crustal deformation. Among these, groundwater level change due to excess pumping is the most likely cause. Jakarta is the capital city of Indonesia, and the metropolitan area has developed very rapidly. Consequently, water requirements have increased drastically, resulting in excess pumping of the groundwater (e.g., Kagabu et al. 2013). Abidin et al. (2011) suggested a strong linkage between land subsidence in Jakarta and the urban development process including groundwater extraction. They also showed that the elevation changes are strongly correlated with the groundwater level changes

Conversely, the second and the third causes should also be carefully investigated because the Jakarta area is relatively flat and is widely covered by alluvial soil. Compaction of the soil layer, particularly the soft sediment layer near the coastal areas, due to natural consolidation or to the load construction could be a possible cause. Although the fourth cause, tectonic movement, is less probable, it cannot be ignored completely.

Thus far, subsidence in Jakarta has been investigated mainly by spirit leveling, GPS, and InSAR techniques. These geometric methods can measure the amount of land movements with high accuracy. However, geometric methods alone are not sufficient for revealing the cause of subsidence; therefore, other methods or data should be combined. To investigate the cause of land subsidence, precise gravity measurement is a promising tool because it can reveal mass movements or mass changes associated with the subsidence.

Assuming that land subsidence is caused by groundwater level changes, compaction of the soil layer or other phenomena associated with mass changes, observed gravity changes can be written as follows:

$$\delta g\,(\mu\mathrm{Gal}) = (-3.086 + 2\pi G\rho)\,\delta h\,(\mathrm{cm}) \qquad (1)$$

where δg is the gravity change in μGal (10^{-8} m/s^2), G is Newton's gravitational constant, ρ is the density of the material causing the height change, and δh is the height change in cm. Equation (1) shows that ρ can be determined by the rate of gravity change versus height change ($\delta g/\delta h$). It should be noted that δh does not directly correspond to groundwater level change even if it causes height change. The height change due to the groundwater change can be interpreted as the compaction of the porous part of the soil after the drainage of water. If we observe both drainage and compaction processes, the observed ρ should be the density of water (1 g/cm^3). On the contrary, the later compaction process does not cause mass change. In a different case, tectonic movements may give the density of the crust (typically 2.67 g/cm^3). In this way, the observed ρ or $\delta g/\delta h$, more specifically, the combination of GPS measurements and gravity measurements can be used to identify the cause of subsidence and is therefore the main motivation of this study.

An additional purpose of this study is to confirm the applicability and benefits of the A10 portable absolute gravimeter for this type of field survey. Although the A10 has recently been employed for a rather wide range of studies (e.g., Fukuda et al. 2010; Falk et al. 2012; Sousa and Santos 2010; Dykowski et al. 2014), its application to land subsidence studies remains limited, particularly in a noisy urban city with a high-temperature and high-humidity climate.

In this study, we first describe the A10 absolute gravimeter and several technical issues related to its practical use in high-temperature urban conditions. Next, we report the results of the measurements and discuss the cause of the subsidence.

2 Measurements

2.1 A10 Absolute Gravity Measurements

Similar to the well-known FG-5, the A10 absolute gravimeter is a ballistic absolute gravimeter that measures the gravity value as the vertical acceleration of a dropping corner cube (test mass). As the corner cube freely falls in a vacuum chamber, its dropping distances and times are measured with a laser interferometer and a rubidium atomic clock, respectively (MGL 2008; Fukuda 2011). Although the nominal accuracy of A10 (10 μgal) is not as good as that of FG-5, the former has several desirable characteristics as a field absolute gravimeter such that it can be set up quickly and operated easily with a 12 V DC battery. Figure 1 shows an example of the A10-017 measurements in Jakarta using a roadside vehicle.

By using the A10-017 portable absolute gravimeter, we have thus far conducted gravity measurements in Jakarta in 2008, 2009, 2010, 2011 and 2013. However, the A10 measurements were not satisfactory, particularly in the earlier years, mainly due to instrumentation problem. For example, a problem occurred with the ion vacuum pump in 2008. Originally, the A10-017 was equipped with a single ion vacuum pump to keep the dropping chamber in a high vacuum. However, the efficiency of the ion pump decreased under high temperature and failed to keep the dropping chamber in the high vacuum. To resolve this problem, a second ion pump was installed when the A10 was returned to the manufacturer for overhaul after the 2008 measurements. High temperature also caused other instrumental problems such as unstable laser controls and thermal runaway of the personal computer

Fig. 1 A photograph of road-side absolute gravity measurements by using A10-017

(PC). These problems were symptomatically treated individually by using a sunshade and a PC cooler. Nearly all of these problems have been resolved following the 2010 measurements.

An additional problem was detected during the 2011 measurements in which smaller gravity values were systematically observed. The cause of this problem was shifting of the atomic clock frequency and the laser frequency, perhaps due to exposure to helium gas during the measurements in the superconducting gravimeter observation site in Cibinong. The problem was confirmed when we compared the absolute gravity values at a gravity reference station in Japan. We thus conducted calibrations of the rubidium clock and the laser frequency, which revealed that the clock and the laser caused shifts of −23 μGal and −20 μGal, respectively. Moreover comparisons of the gravity values at the same gravity points before, during, and after the 2011 survey suggested that the shift in the gravity values was time dependent. Therefore, we finally assumed a simple linear function for correcting the 2011 measurements to minimize the gravity differences and to explain the result of the calibrations. Thus, although gravity values of the 2011 measurements are less reliable, we included those values in this study because they showed consistency with other measurements, as is discussed subsequently.

2.2 Measurements Points in Jakarta

Jakarta is the capital of Indonesia and the country's largest city with an area of approximately 650 km^2. Its population is approximately 10 million in the city and more than 20 million including surrounding areas. Jakarta is located in the lowland area of the northern coast of West Java Province. The area is relatively flat, with topographical slopes ranging between 0° and 2° in the northern and central parts, and between 0° and 5° in the southern part, the altitude of which is approximately 50 m above sea level. Subsidence has been reported mainly in the northern coastal parts to the central parts; few changes have been reported in the southern areas. Therefore the GPS points set by Institute Teknologi Bandung (ITB) group were selectively distributed in these subsidence areas (Abidin et al. 2008).

We primarily planned to select the gravity points from the existing GPS points, and some were indeed selected accordingly. However many of the GPS points were not suitable for A10 measurements because some were difficult to access by a vehicle, and some were too noisy for obtaining gravity measurements. For selection of gravity points, we therefore adopted the following criteria: The site must have a flat and solid base with sufficient space for A10 measurements, with as little ground vibration as possible. In addition the site must be expected to remain in the same condition in the future, and it should be within several tens of meters from the existing GPS points.

According to these criteria, we first selected approximately 10 absolute gravity points. However, some were lost mainly due to construction work, even though we paid careful attention to the third criterion described above. Although we added new points for supplement, we have not obtained gravity change data at those points thus far. Consequently, gravity changes obtained at only six gravity points were employed in this study. Figure 2 shows the location map of the selected gravity points. As shown in the map, four points are located along the northern coastal area, and other two are located in the central area. In addition to these points, we conducted measurements in Cibinong at the absolute gravity point established by Badan Informasi Geospasial (BIG), the national geospatial information agency of Indonesia (in English); this site is approximately 50 km south of Jakarta. Because no significant gravity change was expected in Cibinong, these values were employed to verify the reliability of the gravity measurements.

In addition to the A10 measurements, we also conducted relative gravity measurements by using a Scintrex gravimeter (CG-5); additional gravity points were selected for the relative measurements. However, due to a high amount of urban noise such as ground vibration in addition to large drifts, the data of relative measurements often showed discrepancies on the order of several tens of micro-Gals. Because these data were not as accurate as the A10 data, we employed only the results of absolute gravity measurements in this study.

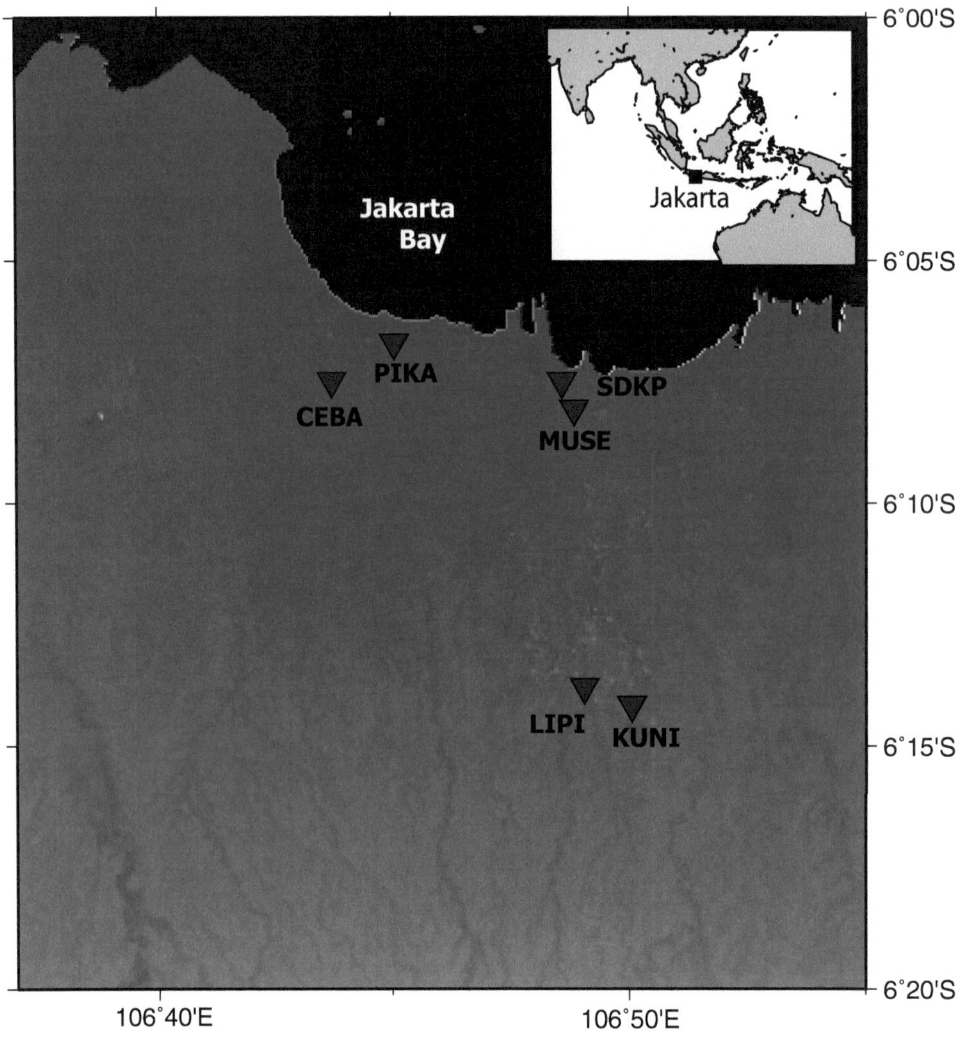

Fig. 2 Location map of the gravity points in Jakarta, Indonesia

3 Results and Discussion

3.1 Gravity Changes

A10 automatically lifts up and drops the test mass to obtain a single drop gravity value. We combined 100 drops to get a so-called set value, and conducted 10 sets of measurements to obtain an absolute gravity value as we ordinarily conducted (Fukuda et al. 2011). Table 1 summarized the absolute gravity values obtained at the gravity points. It should be noted that the errors in Table 1 are not the formal accuracy but are so called set scatters that show the deviations of the set values from the mean value. Thus these are good indicators of the noise level and the stability of the points. Table 1 clearly shows that the noise level in Jakarta was larger than at BIG in Cibinong.

As previously described, instrumental problems resulted in a lack of data in 2008, and only two points of data (CEBA and LIPI) were included in 2009. Although logistical issues resulted in a lack data of these points in 2013, we added a few more absolute points during that year for future studies. In order to verify and to increase the reliability of the obtained values, we attempted to obtain measurements more than twice within a few days at the same point in each year for as long as possible. In Table 1, the differences of the gravity values at the same point in the same year show the overall accuracy of the measurements. Although some data showed larger differences, the majority were within ± 10 μGal which is the nominal accuracy announced by the manufacturer. Thus, we employed all of the data in Table 1 to calculate the gravity change rates.

Figure 3 depicts the gravity changes at each gravity point. The rates of the changes with their estimation errors

Table 1 Summary of the absolute gravity measurements. The absolute gravity values at the specific point and in the specific year are given as the sum of the value in "Grav. mgal" plus the values of μGal in the same column. All of the gravity values were calculated at 100 cm above ground level with the dg/dz value of 3.086 μGal/cm

	CEBA	PIKA	SDKP	MUSE	LIPI	KUNI	BIG
Lat(deg)	−6.1247	−6.1117	−6.1250	−6.1343	−6.2298	−6.2360	−6.4903
Lon(deg)	106.7285	106.7510	106.8099	106.8141	106.8178	106.8343	106.8498
Grav. (mgal)	+978148 mgal	+978149 mgal	+978119 mgal	+978149 mgal	+978139 mgal	+978140 mgal	+978114 mgal
	μgal	μgal	μgal	μgal	μgal	μgal	μgal
2009.7	139.4 ± 10.8				873.5 ± 10.7	075.0 ± 6.2	
	155.6 ± 12.1				857.2 ± 10.7		
2010.7	147.4 ± 11.0	334.0 ± 11.3	647.8 ± 3.0	670.5 ± 11.6	856.7 ± 11.1	068.1 ± 7.8	528.4 ± 3.4
	178.5 ± 11.2		649.2 ± 6.2		861.9 ± 11.1	080.6 ± 4.3	531.1 ± 2.2
	153.9 ± 10.9		641.0 ± 7.8		859.5 ± 10.8		
					849.8 ± 10.7		
					859.2 ± 10.7		
2011.7	154.3 ± 13.9	336.7 ± 12.7	632.1 ± 5.9	663.1 ± 11.5	857.7 ± 10.7	082.0 ± 5.9	528.6 ± 2.3
	180.0 ± 12.6						534.7 ± 3.8
							531.7 ± 1.8
							532.8 ± 2.8
							526.3 ± 7.6
2013.7		433.7 ± 11.4	663.2 ± 10.1	700.1 ± 11.0		103.6 ± 10.1	526.3 ± 7.6

are summarized in the third column of Table 2. With the exception of LIPI, gravity increased at all points. The largest rate of 35.5 ± 11.4 μGal/year was observed at PIKA which is located in the northern coastal area. The gravity changes at the reference point of BIG were nearly neutral, as was expected.

3.2 Height Changes

As previously stated, the ITB group has been conducting GPS surveys in the Jakarta area since the 1990s. Moreover, recent results of the land subsidence rates were reported in Abidin et al. (2011). Their results (Fig. 6 of Abidin et al. 2011) showed very large subsidence rates along the coastal area, particularly near PIKA (Fig. 2), and markedly large rates in the central area as well. By using height change data at the nearest GPS points, we estimated the height change rates at the absolute gravity points; the results are summarized in the second column of Table 2. Although it was difficult to accurately evaluate the errors of the height change rates, we considered that they were about 1 cm/year or better. The spatial pattern of the subsidence was nearly the same as that reported previously, and with the exception of LIPI, rather large subsidence rates were found. Similar to that in observed in the gravity changes, PIKA showed the largest subsidence rate of more than 10 cm/year, which is consistent with the gravity rate.

3.3 Gravity Versus Height Changes

Figure 4 depicts the relationship of the height changes versus gravity changes summarized in Table 2. In the figure, the vertical axis shows the rate of gravity changes and the horizontal axis shows that of height changes. Solid lines labeled F, W, and C corresponded to the gravity gradients of free-air, the Bouguer reduction for water density (1.00 g/cm^3), and mean crust density (2.67 g/cm^3), respectively. As shown in the figure, most points are near the F and W lines. The points on the W line indicate that the height changes were caused by groundwater level changes. By using the preliminary results of the absolute gravity data obtained before 2010, Fukuda et al. (2011) argued that these groundwater level changes are the main cause of the subsidence in Jakarta. In this study, however, the gradients at PIKA, CEBA and MUSE were more likely associated with free-air gradients. The free-air gradient means that there is no mass change associated with the height changes. Thus, the height changes can be attributed to compaction of the soil layer or the crushing of porous parts. As shown in Table 1, the errors at PIKA, CEBA and MUSE were larger than those at SDKP and KUNI. The larger errors indicate that those three points could be located on the soft soil layer near the coast line. Therefore, natural compaction would be a possible cause of the subsidence at these points. In addition, excess groundwater pumping may have accelerated the subsidence. Rapid groundwater pumping may have withdrawn water from the porous medium,

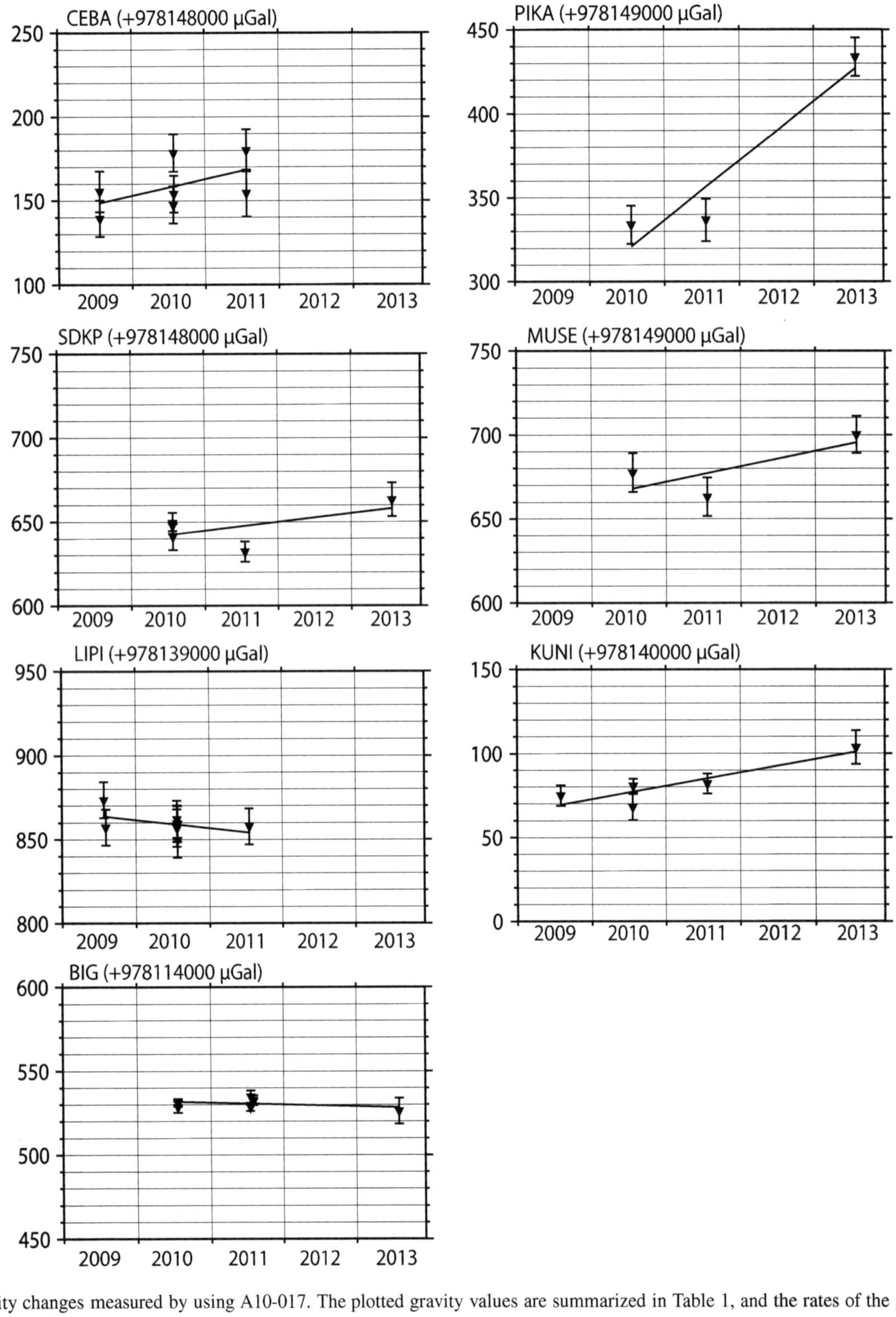

Fig. 3 Gravity changes measured by using A10-017. The plotted gravity values are summarized in Table 1, and the rates of the gravity changes are shown in Table 2

which may have been followed by compaction. Thus, the gravity changes may show the free-air gradient because the water may have previously been extracted. Although the large error bars in Fig. 4 are too large to lead to a definite conclusion, the cause of the subsidence in Jakarta could be a mixture of such mechanisms.

4 Conclusions

To investigate the cause of the land subsidence in Jakarta, we have performed absolute gravity measurements by using A10-017. We were able to overcome technical issues such as

Table 2 Rates of height changes and gravity changes

	Height (cm/year)	Gravity (μgal/year)
CEBA	−3.2	9.9 ± 7.1
PIKA	−11.4	35.5 ± 11.4
SDKP	−2.2	5.2 ± 4.1
MUSE	−2.6	9.1 ± 8.2
LIPI	1.1	−4.8 ± 3.8
KUNI	−3.2	7.9 ± 2.3
BIG	–	−1.1 ± 1.2

All of the gravity values were calculated at 100 cm above ground level with the dg/dz value of 3.086 μGal/cm

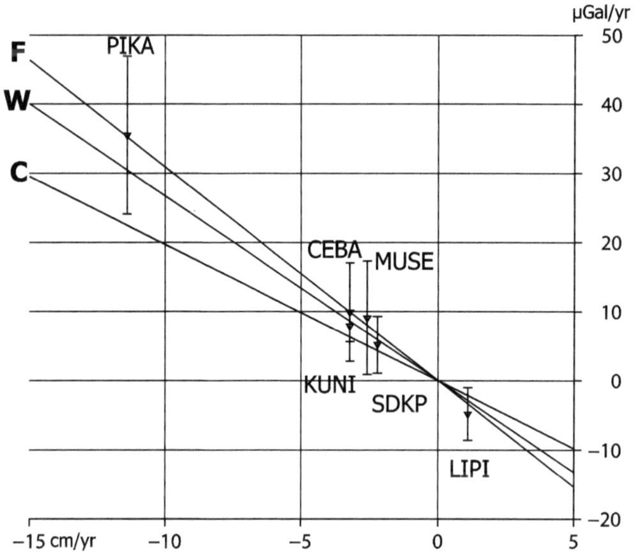

Fig. 4 Relationship of height change versus gravity change. The *solid lines* labeled F, W, and C correspond to gravity gradients of free-air, the Bouguer reduction for water density (1.00 g/cm^3), and mean crust density (2.67 g/cm^3), respectively. The plotted values are summarized in Table 2

high temperature, high humidity and noisy urban conditions to successfully detect gravity changes due to the subsidence. The obtained gravity change rates were compared with the height change rates observed by GPS measurements. The results suggest that soil compaction and the groundwater level changes, which could be connected to excess pumping of groundwater, most likely caused the subsidence.

This result demonstrates the most important advantage of the gravity method, which detects mass changes or mass movements to obtain information independent of that gained by geometric methods such as leveling, GPS, and InSAR. In addition, we have proved that absolute gravity measurement in urban conditions is feasible by using A10. Similar to that with other gravity methods, A10 can be applied for various studies in urban conditions.

Moreover, we have determined that the data obtained by A10 is more reliable than that obtained by using a relative gravimeter because the absolute measurement was free from instrumental drift and gave an independent gravity value for each measurement. As demonstrated in this study, A10 is a promising tool for the studies of land subsidence, groundwater level changes, geothermal fluid, volcanic activities, and other natural and human-made hazards.

Acknowledgements This research was partially supported by JSPS KAKENHI Grant No. 20403007. The A10 absolute gravity measurements were supported by many colleagues and students of Kyoto University, Kyushu University, ITB, and other institutions. We thank all of those involved in this research.

References

Abidin HZ, Djaja R, Darmawan D, Hadi S, Akbar A, Rajiyowiryono H, Sudibyo Y, Meilano I, Kusuma MA, Kahar J, Subarya C (2001) Land subsidence of Jakarta (Indonesia) and its geodetic based monitoring system. Nat Haz 23(2–3):365–387

Abidin HZ, Djaja R, Andreas H, Gamal M, Hirose K, Maruyama Y (2004) Capabilities and constraints of geodetic techniques for monitoring land subsidence in the urban areas of Indonesia. Geomat Res Aust 81:45–58

Abidin HZ, Andreas H, Djaja R, Darmawan D, Gamal M (2008) Land subsidence characteristics of Jakarta between 1997 and 2005, as estimated using GPS surveys. GPS Solut 2(1):23–32

Abidin HZ, Andreas H, Gumilar I, Fukuda Y, Pohan YE, Deguchi T (2011) Land subsidence of Jakarta (Indonesia) and its relation with urban development. Nat Hazards. doi:10.1007/s11069-011-9866-9

Abidin HZ, Andreas H, Gumilar I, Sidiq TP, Fukuda Y (2012) Land subsidence in coastal city of Semarang (Indonesia): characteristics, impacts and causes. Geomatics Nat Hazards Risk. doi:10.1080/19475705.2012.692336

Abidin HZ, Gumilar I, Andreas H, Murdohardono D, Fukuda Y (2013) On causes and impacts of land subsidence in Bandung Basin, Indonesia. Environ Earth Sci 60:1545–1553. doi:10.1007/s12665-012-1848-z

Djaja R, Rais J, Abidin HZ, Wedyanto K (2004) Land subsidence of Jakarta Metropolitan Area. In: 3rd FIG regional conference, Jakarta, Indonesia, 3–7 October 2004. p 7

Dykowski P, Krynski J, Sekowski M (2014) Testing the suitability of the A10-020 absolute gravimeter for the establishment of new gravity control in Poland. In: IAG symposia, vol 141, pp 11–17. doi:10.1007/978-3-319-10837-7_2

Falk R, Müller J, Lux N, Wilmes H, Wzionek H (2012) Precise gravimetric surveys with the field absolute gravimeter A-10. IAG Symp 136:273–279. doi:10.1007/978-3-642-20338-1_33

Fukuda Y, Nishijima J, Hasegawa T, Sofyan Y, Taniguchi M (2010) Monitoring groundwater variations using A10 absolute gravimeter. In: Proceedings of the IAG symposium on terrestrial gravimetry: static and mobile measurements (TG-SMM2010) 22–25 June 2010, Saint Petersburg

Fukuda Y (2011) Monitoring groundwater variations using precise gravimetry on land and from space. In: Taniguchi M (ed) Groundwater and subsurface – human impacts in Asian Coastal Cities. Springer, pp 85–112

Fukuda Y, Nishijima J, Taniguchi M (2011) Applications of a field absolute gravimeter for monitoring temporal gravity changes. In: Proceedings of the 10th SEGJ international symposium, Nov., 20–23, Kyoto, Japan

Kagabu M, Shimada J, Delinom R, Nakamura T, Taniguchi M (2013) Groundwater age rejuvenation caused by excessive urban pumping in Jakarta area, Indonesia. Hydrol Process 27:2591–2604

MGL (2008) A-10 portable gravimeter user's manual. http://www.microglacoste.com/pdf/A-10Manual.pdf. 58 pp

Murdohardono D, Sudarsono U (1998) Land subsidence monitoring system in Jakarta. In: Proceedings of the symposium on Japan–Indonesia IDNDR project: volcanology, tectonics, flood and sediment hazards, Bandung, 21–23 September, pp 243–256

Ng AH, Gea L, Li X, Abidin HZ, Andreas H, Zhang K (2012) Mapping land subsidence in Jakarta, Indonesia using persistent scatterer interferometry (PSI) technique with ALOS PALSAR. Int J Appl Earth Obs Geoinf 18:232–242. doi:10.1016/j.jag.2012.01.018

Sousa M, Santos A (2010) Absolute Gravimetry on the Agulhas Negras Calibration Line. Revista Brasileira de Geofisica 28(2):165–174

Introduction to the Gravity Database (GALILEO) Compiled by the Geological Survey of Japan, AIST

A. Miyakawa, K. Nawa, Y. Murata, S. Ito, S. Okuma, and Y. Yamaya

Abstract

The Geological Survey of Japan (GSJ), which is a section of the National Institute of Advanced Industrial Science and Technology (AIST), has conducted gravity surveys throughout Japan. The GSJ launched the online gravity database known as "GALILEO" on the GSJ website. The GALILEO source data are included in the Gravity CD-ROM/DVD of Japan from the GSJ. GALILEO supports three main functions: (1) browsing and downloading of raster data (with or without accompanying descriptions) or vector data (e.g., KMZ files), (2) comparisons of gravity maps with geological or topographic maps of the same area (using a JAVA applet), and (3) on-demand mapping using Generic Mapping Tools, which enables GALILEO users to visualize Bouguer anomaly maps on different assumed densities of the surface rocks. These quick views of the gravity anomaly maps through GALILEO are useful for users accessing the database. Gravity data compiled by the GSJ have contributed towards a better understanding of various aspects of the geology of Japan, especially as related to disaster prediction and response.

Keywords

Bouguer anomaly • Gravity database • Japan • Web database

1 Introduction

The gravity method has been used for a wide variety of purposes over the past century (Hinze et al. 2013). The Geological Survey of Japan (GSJ), a section of the National Institute of Advanced Industrial Science and Technology (AIST), has conducted gravity surveys to map gravity anomalies throughout Japan. The GSJ has published detailed Bouguer anomaly maps at a scale of 1:200,000 (Gravity Map Series), and digital gravity databases (on CD-ROM and DVD). The gravity data compiled by the GSJ have contributed towards a better understanding of various aspects of the geology of the Japanese Islands, including resource geology (Feebrey et al. 1998), volcanology (Komazawa 1995; Komazawa and Mishina 2002; Yamawaki et al. 2004), subsurface structures (e.g., Komazawa 2004; Li 2011), fault systems (Panayotopoulos et al. 2014), and terrain density distributions (Murata 1993; Nawa et al. 1997). Furthermore, the gravity database of the GSJ has been applied to disaster research, including the probabilistic estimation of long-term volcanic hazards (e.g., on Kyushu Island, Japan; Jaquet et al. 2012). The gravity database of the GSJ provides promising avenues and possibilities for research in earth sciences, and for disaster research applications.

To promote the use of the GSJ gravity database, the GSJ recently posted the online gravity database on the GSJ website, hereafter referred to as GALILEO (Gravity AnomaLIes Learning Online). GALILEO contains the following: (1) published Bouguer anomaly maps at a scale of 1:200,000, with explanatory notes; (2) local Bouguer

A. Miyakawa (✉) • K. Nawa • Y. Murata • S. Ito • S. Okuma • Y. Yamaya
Geological Survey of Japan, AIST, Higashi-1-1-1, Tsukuba, Ibaraki Prefecture, Japan
e-mail: miyakawa-a@aist.go.jp

M. Hashimoto (ed.), *International Symposium on Geodesy for Earthquake and Natural Hazards (GENAH)*,
International Association of Geodesy Symposia 145, DOI 10.1007/1345_2015_112

anomaly maps; and (3) on-demand plots of the gravity anomaly maps. The GALILEO gravity database, which is accessed via the GSJ database site (https://gbank.gsj.jp/gravdb/), is available to all users, based on a Creative Commons License "CC BY 2.1 JP", except for some "CC BY-ND 2.1 JP" content that contains Geospatial Information Authority of Japan (GSI) publications. In the remainder of this paper, we introduce and describe GALILEO, so as to promote and enhance its availability for use in earth science and disaster research. In addition, we outline future plans for the database, as GALILEO is currently in a startup phase.

2 Gravity Database in Japan

Several organizations in Japan have also conducted terrestrial gravity surveys and have made their work open to the public. The Geospatial Information Authority of Japan (GSI) established the Japan Gravity Standardization Net 1996 (abbreviated to JGSN96) and have published the gravity data on their web page (http://www.gsi.go.jp/ENGLISH/page_e30132.html). The Gravity Research Group in Southwest Japan has published the "Gravity Database of Southwest Japan" (Shichi and Yamamoto 2001). The Kanazawa University Gravity Research Group has published the "Gravity Database of Kanazawa University" (Honda et al. 2012).

The GSJ has published detailed Bouguer anomaly maps at a scale of 1:200,000 (Gravity Map Series) for 31 districts and published 3 local gravity maps (https://www.gsj.jp/Map/EN/geophysics.html). The GSJ previously published the digital gravity database on CD-ROM/DVD, which includes gridded gravity data, a gravity measurements data file, Bouguer anomaly maps, and free-air gravity anomaly maps (Geological Survey of Japan 2004, 2013). The CD-ROM/DVD contains parts of previously described datasets of other organizations. The images comprise a Bouguer anomaly map, a free-air anomaly map, and a geological map based on the seamless digital geological map of Japan at a scale of 1:200,000 (Wakita et al. 2009). The DVD, which is an updated version of the CD-ROM, includes an enlarged dataset of gravity measurements and map and image files.

3 Gravity Database on the GSJ Website (GALILEO)

3.1 Data Sources for GALILEO

The data sources for GALILEO, which are used for constructing gravity anomaly maps and/or images, cite the gravity dataset of Japan recorded on CD-ROM/DVD (published by the GSJ). The gravity data are based on 1,039,745 and 1,708,881 gravity measurement points included on the CD-ROM and DVD, respectively. A total of 657,538 and 1,037,330 points on the CD-ROM and DVD, respectively, were obtained by the GSJ, and the remainder were obtained and provided by other organizations for mapping by the Geological Survey of Japan (2004; 2013). The gravity data from other organizations was checked by mapping the datasets, and anomalous data or outliers on the map were eliminated by the GSJ; however, the quality of the original data in acquisition phase could not be completely controlled by the GSJ. Because of restricted access to gravity data from some external organizations (private companies, research centers, etc.), approximately half of the gravity measurement data are restricted-access data; because of this, users cannot download the actual data points at this time.

The corrections applied to the gravity measurement point data follow the "Standard Procedure for Evaluating the Correction of Gravity 1988" (SPECG1988); SPECG1988 was developed by the GSJ specifically for compiling the gravity map data (Komazawa 1988; GSJ Gravity Survey Group 1989). The procedure involves a calculation scheme and formulations for normal gravity, free-air reduction, atmospheric correction, lithospheric correction, and Bouguer and terrain correction. An initial goal for the accuracy of the approximation of each correction was a value of 0.1 mGal. All measured gravity data were referred to the Japan Gravity Standardization Net 1971 (JGSN75), which is based on the International Gravity Standardization Net 1971 (IGSN71), and normal gravities were estimated according to the Gravity Formula 1980. The WGS-84 datum, based on the GRS80 ellipsoid, was used for the calculations.

To obtain the grid values for mapping, the point gravity data were interpolated using two-dimensional quadratic polynomials with distance-dependent weighting, which was defined so as to fit the local surface. The grid spacings in the Gravity CD-ROM and DVD data are 1 km and 500 m, respectively. A polyconic projection was used to construct the grid values. The WGS-84 datum based on the GRS80 ellipsoid was used as the geodetic reference system. The origin of the projection is 36°N latitude and 136°E longitude. At present, we have used only the 1-km grid values (CD-ROM version) for the on-demand mapping because of server processing limitations. The 500-m grid values (DVD version) will be available in the near future.

3.2 Major Functions of GALILEO

3.2.1 Browse and Download

Users can browse and download images and maps in the GALILEO gravity database through the internet, including

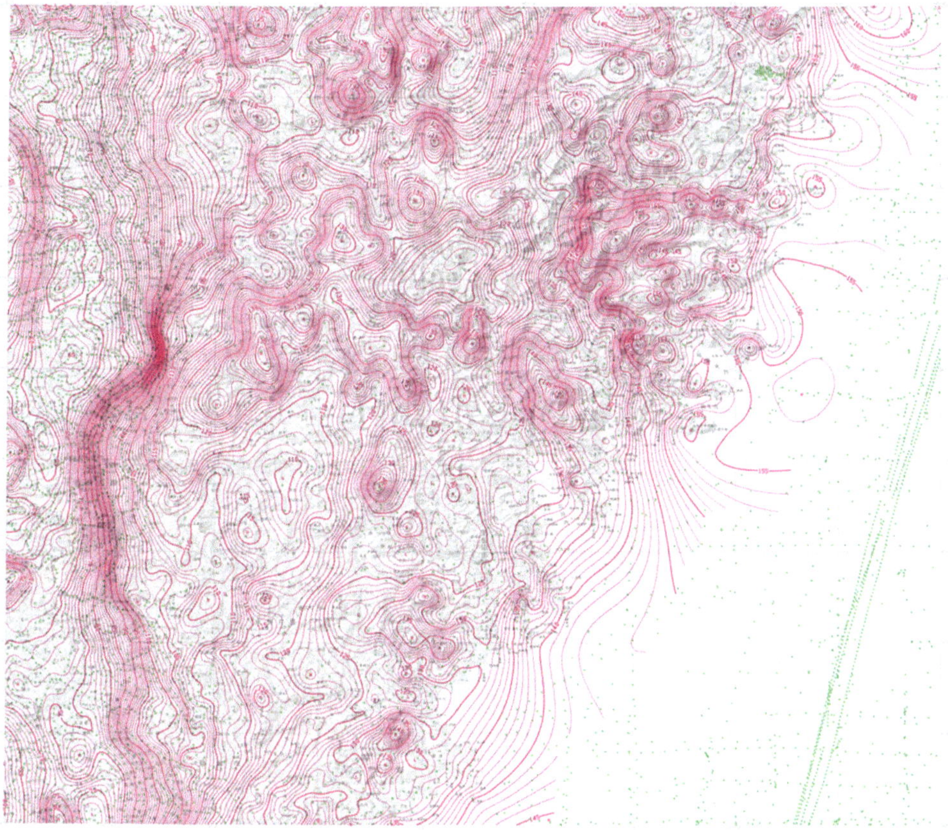

Fig. 1 Gravity map showing Bouguer gravity anomalies, accessed via the function for browsing and downloading images and maps

the contents of the Gravity Map Series (several maps have associated descriptions), which is the same as that published by the GSJ. The image/map database, with assumed densities of the surface rocks of 2.00, 2.30, and 2.67 g/cm^3, and a seamless regional geological map at a scale of 1:200,000 are available (the Japanese Islands are divided into 73 areas by the GSI). Users can select a standard Gravity Map (Fig. 1), a regional map area, or an area based on coordinates entered by the user. The maps and images available in the Gravity Map Series are then displayed on the screen. Users can select the target image/map to download and the format of data files (either JPG or KMZ). Downloaded KMZ files can be visualized using GIS software by georeferenced coordinates.

3.2.2 Comparisons of Gravity Maps and Geological/Topographic Maps

Bouguer anomaly and geological maps of the same area can be compared side-by-side using a Java applet, as shown in Fig. 2. For a given area in the "Gravity Map List", users can select (from the available options) which map will appear in the left and right panels. This function requires users to activate the Java runtime environment (JRE) in advance. Users can zoom in or out and can scroll around the image/map. The left button of the mouse is for zooming in and the right button is for zooming out when the user has selected the "zoom mode". The cursor cross points to the same point on both the left and right panels, so that users can compare exactly the same point on the gravity/geological/topographic maps. Scrolling is possible by dragging on the images/maps when the user has selected the "scroll mode". Users can scroll the image/map on either the left or right panel, and the same scrolling operations will occur in the other panel.

3.2.3 On-Demand Mapping

On-demand mapping is also available in order to browse any specified range of area, although the images/maps are prepared in a pre-fixed range of areas (as explained in Sects. 3.2.1 and 3.2.2). This function is enabled with the aid of Generic Mapping Tools (GMT), version 4 (Wessel and Smith 1998). A user selects "on-demand mapping" from the home page, and the user then selects the area to view (Fig. 3). Users

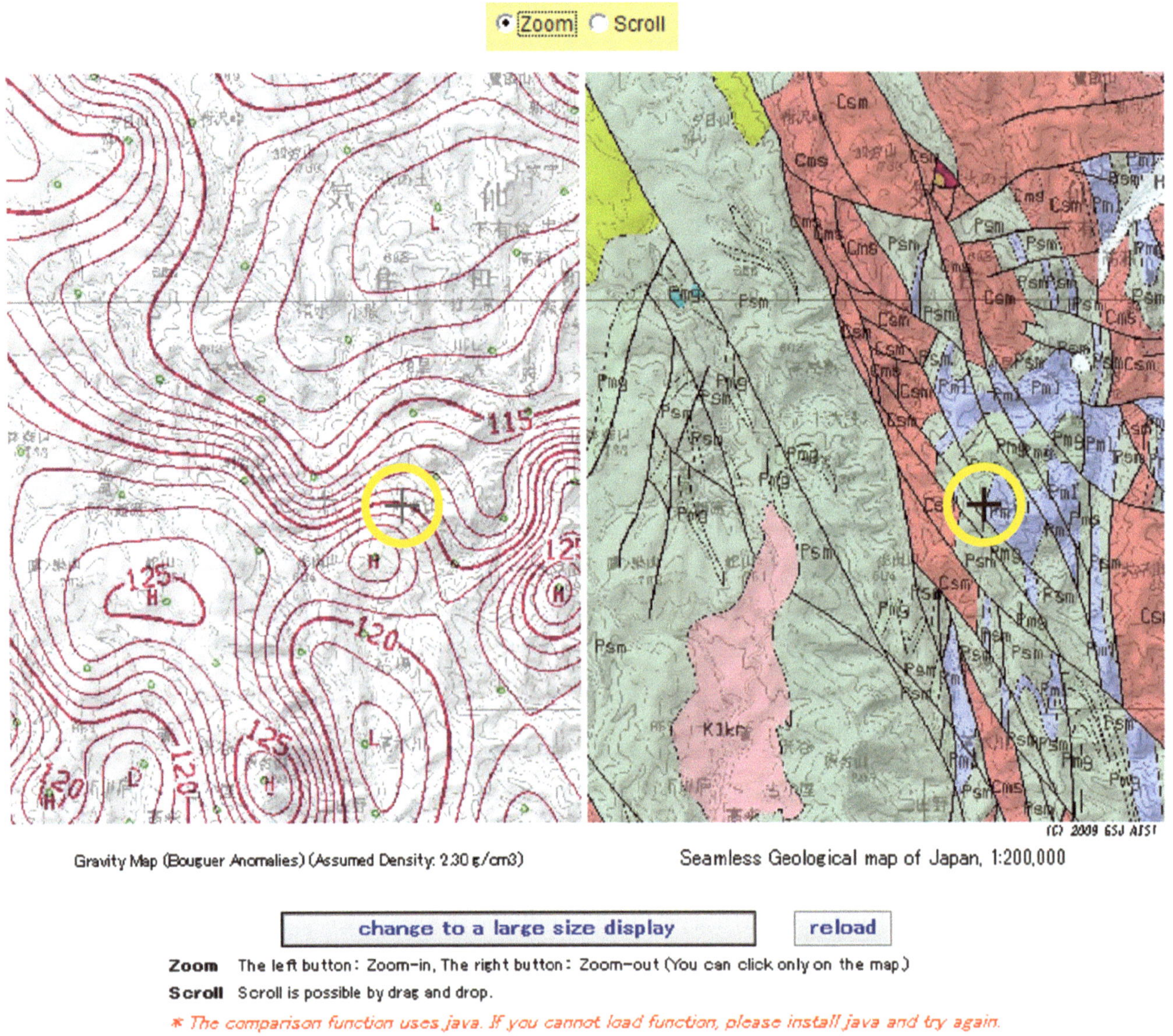

Fig. 2 Screen capture illustrating the function for comparing gravity, geological, and topographic maps. The Bouguer anomalies are displayed at assumed densities of 2.30 g/cm^3. (*Right*) Geological map. Both maps shows the same area. *Black crosses* (in *yellow circles*) are in the same position on the both maps

can select from any of the following types of map via a drop-down list: "point map", "Bouguer map", "upward map", "residual map", or "gradient map". On-demand mapping in GALILEO enables users to see the Bouguer anomaly maps at any assumed density of the surface rocks (although the density is fixed in the print medium). We used two types of gravity data for on-demand mapping depending on the proprietary of the data: (1) original point data gravity measurements obtained by the GSJ and some open data from other organizations, and (2) grid values constructed from point-data gravity measurements that are both open and restricted in their availability. The gravity point data can be continuously merged at intervals of 0.025°, as the GSJ attempts to provide at least one measurement point per 1′ interval. We preferentially used the open-access point data for the maps, and, if no open-access point data were available within 0.025°, we used the grid values. The region using point data or grid values can be seen in the map (Figs. 3 and 4). To calculate merged grid values on the Bouguer anomaly map, merged point data from original measurements and grid values were smoothed to a 90-s grid using a "blockmedian" command to avoid aliasing short wavelengths (see the GMT manual); then, the data are gridded on 20-s grids with the "surface" command of the GMT.

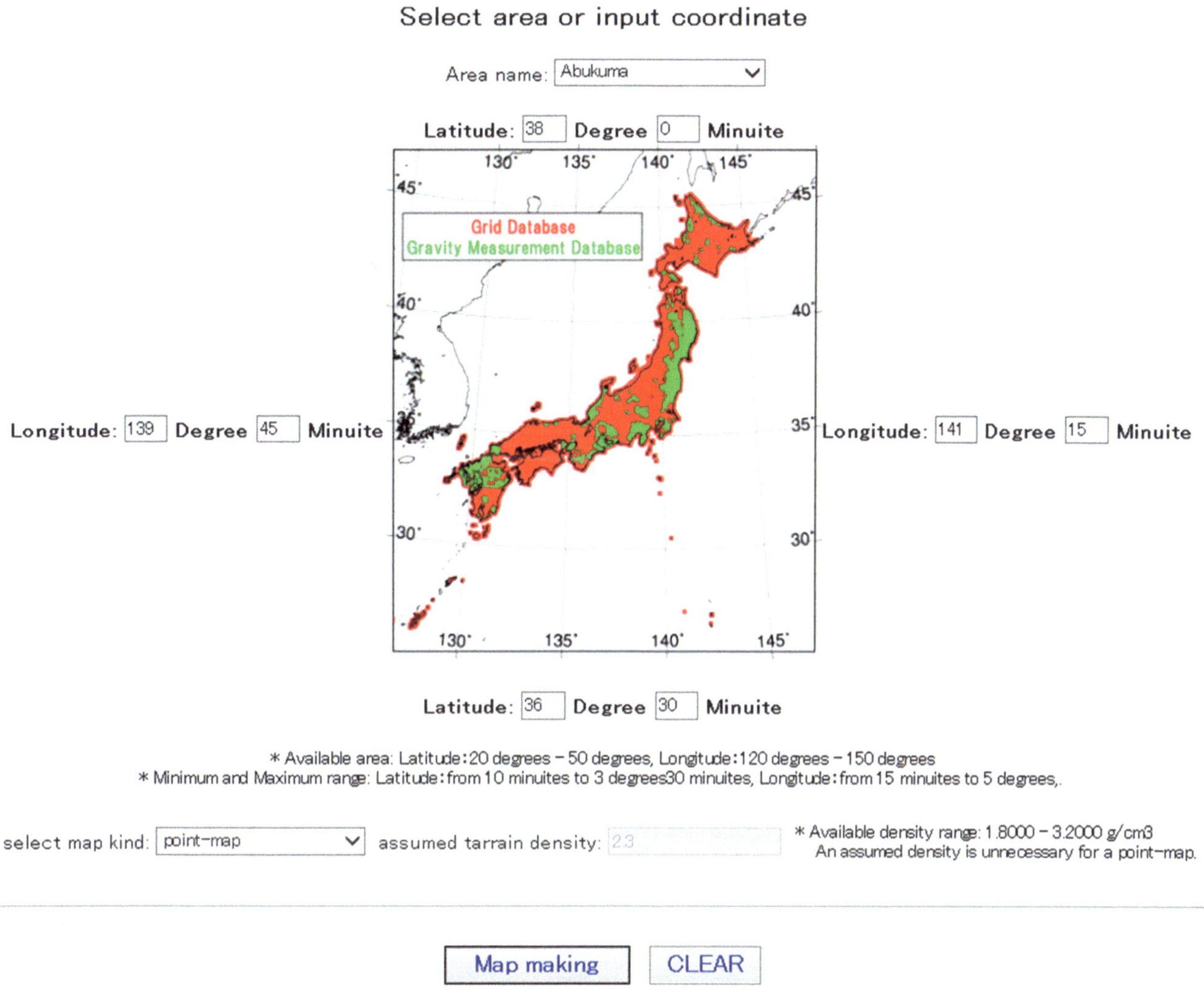

Fig. 3 Screen capture illustrating the on-demand mapping function in GALILEO. Coordinates of the map area and the assumed density the surface rocks are entered by the user. The region using measurement point data and the region using grid values used for generating the gravity anomaly maps are shown in *green* and *red*, respectively

The "point map" is a map that displays gravity measurement points and the grid used to image the gravity anomaly map. The "Bouguer map" shows a map of Bouguer anomalies in the selected area. The "upward map" is a map of Bouguer anomalies continued upwards to 2 km above the ground. The "residual map" displays the residual Bouguer anomalies obtained by subtracting the value calculated at 2 km above ground from the original Bouguer anomaly values. The "gradient map" is a map obtained by calculating horizontal gradients of the Bouguer anomaly values.

Pre-processed gravity maps on active fault areas are available as examples of the on-demand mapping function (Fig. 5). These pre-processed gravity maps cover 66 areas, including segments of active faults in Japan, which are listed on the right-hand side of the web page. When an active fault area is selected, the user is taken to a new page with a list of available gravity maps. When an individual map is selected, the map opens. There is a hyperlink to the Visualization System for Subsurface Structures, which is one of the GSJ database systems (https://gbank.gsj.jp/subsurface/english/index.html), at the bottom of the page that display the map.

3.3 Applications to Earthquake and Natural Hazard Research

The information on gravity anomalies provided through GALILEO is useful for studies and research on earthquakes

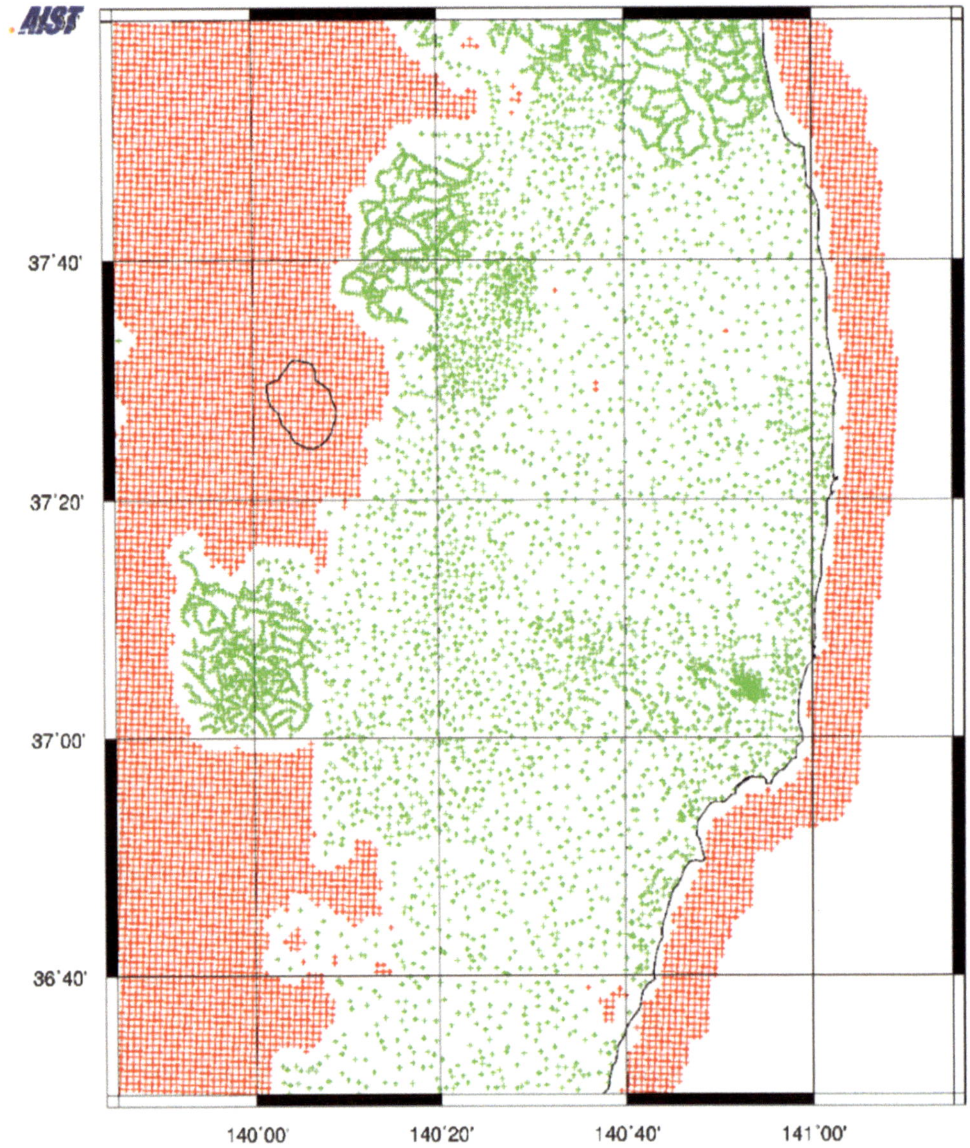

Fig. 4 Distribution of gravity measurement point data and grid values in the Abukuma area, Japan. The *green crosses* show the locations of measurement point data and the *red crosses* show the locations of grid values used for generating the gravity anomaly maps

and natural hazards. For example, the gravity database compiled by the GSJ contributes to long-term evaluations of the Umi Fault, which is one of the active faults in the Fukuoka Prefecture, Japan (Headquarters for Earthquake Research Promotion 2013). Determining the lengths of faults is important in seismic hazard evaluation, as the length corresponds to the size of the largest likely earthquake on the fault. The estimated length of the Umi Fault is constrained by the structure of gravity anomalies, as represented in the gravity database.

Quick views of gravity anomalies using GALILEO are available not only to researchers undertaking emergency surveys but also to the public, who may wish to view the geological conditions in a given area. For example, a Mw 6.8 earthquake occurred along the Kamishiro Fault (Headquarters for Earthquake Research Promotion 2014) in Nagno Prefecture, Japan, on 22 November 2014. GALILEO had already been used to prepare pre-processed gravity maps around the Kamishiro Fault before the earthquake occurred, as mentioned above (Fig. 6). High Bouguer anomaly gradients were observed from the south to the center of the map. The epicenter of the earthquake was located at the northern tip of the high Bouguer-anomaly gradient structure. The high-gradient structure corresponds

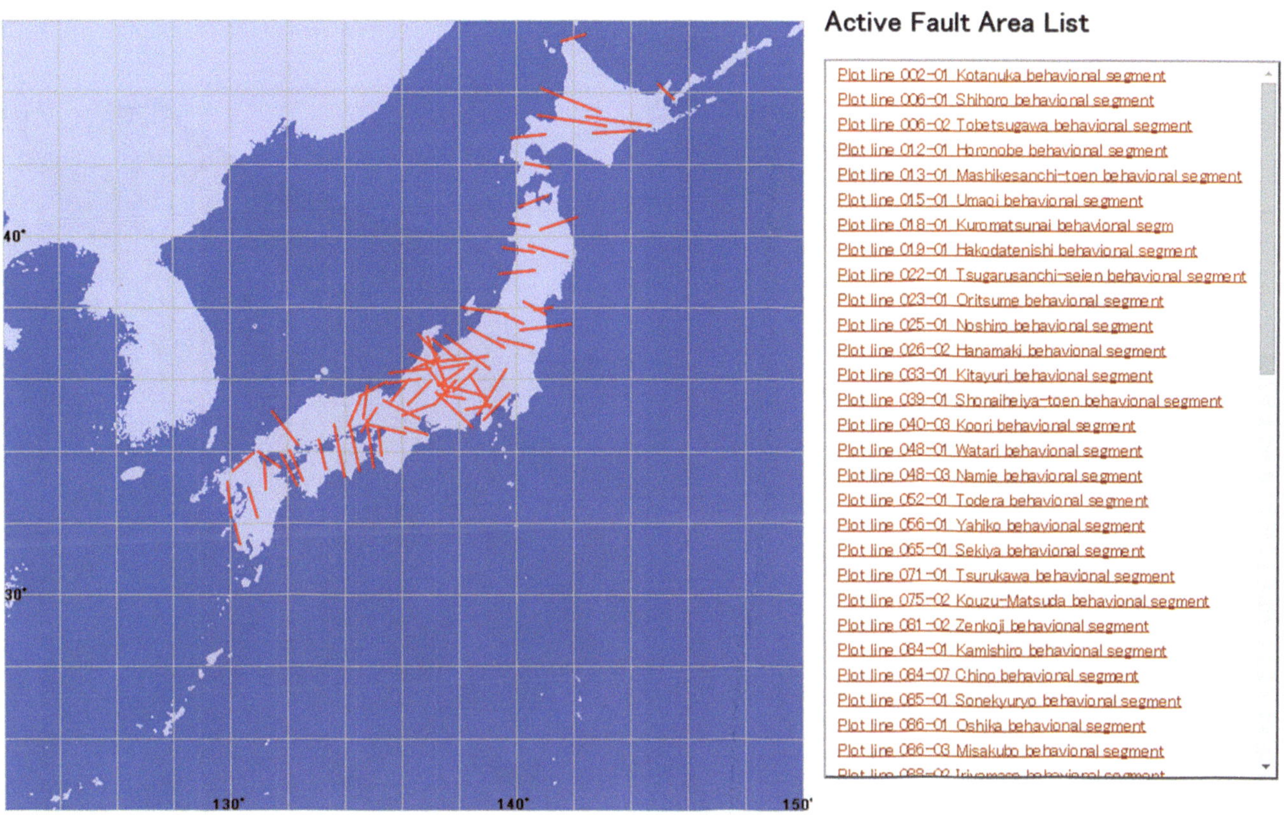

Fig. 5 Screen capture showing pre-processed gravity maps for active fault areas. Pre-processed gravity maps are available by selecting active fault segments, either from the map or from the menu on the right-hand side of the page. The *red lines* show the profile lines crossing active faults. The subsurface structure images are plotted along lines beforehand in the Visualization System for Subsurface Structures

to the Itoigawa–Shizuoka Tectonic Line (ISTL), which is one of the largest faults in Japan. On the other hand, the high-gradient structure is vague to the north of the epicenter. Consequently, the earthquake may have occurred along a structural boundary within the ISTL, although further analysis is required to determine the details of the structures involved. However, the important point here is that users can employ GALILEO to see a gravity anomaly map around the Kamishiro Fault immediately after the earthquake.

4 Summary and Future Work

GALILEO has been able to provide gravity information to online users, especially images and maps of gravity anomalies, although improvements to the software are on-going; e.g., the development of a more user-friendly interface and providing a geo-referenced image database (JPG data). We are preparing to add other interactive visualization functions to the system, such as the ability to extract Bouguer anomaly along profile lines on anomaly maps. Some digital data of the GSJ, including the point dataset of gravity measurements, have been distributed through digital media (CD-ROM/DVD) and the Geological Information Index of Southeast Asia (G-INDEX); however, the G-INDEX will soon be unavailable. Our next goal, to be implemented soon, is to add a function for data distribution (e.g., downloading of gravity measurement point data that can be opened and processed as grid values). When the point or grid gravity dataset is accessed by an online user, it is necessary for us to show the reliability of individual data points, for example using statistical methods (Shoberg and Stoddard 2013). Other websites, such as that of the North American Gravity and Magnetic Database (e.g., Hinze et al. 2005), provide excellent examples of data distribution options that we will refer to during improvements of the GSJ gravity database. Using the data distribution function, users can access up-to-date gravity data immediately as the dataset is updated by the GSJ. The accessed data will be useful to a variety of researchers studying earthquakes and natural hazards.

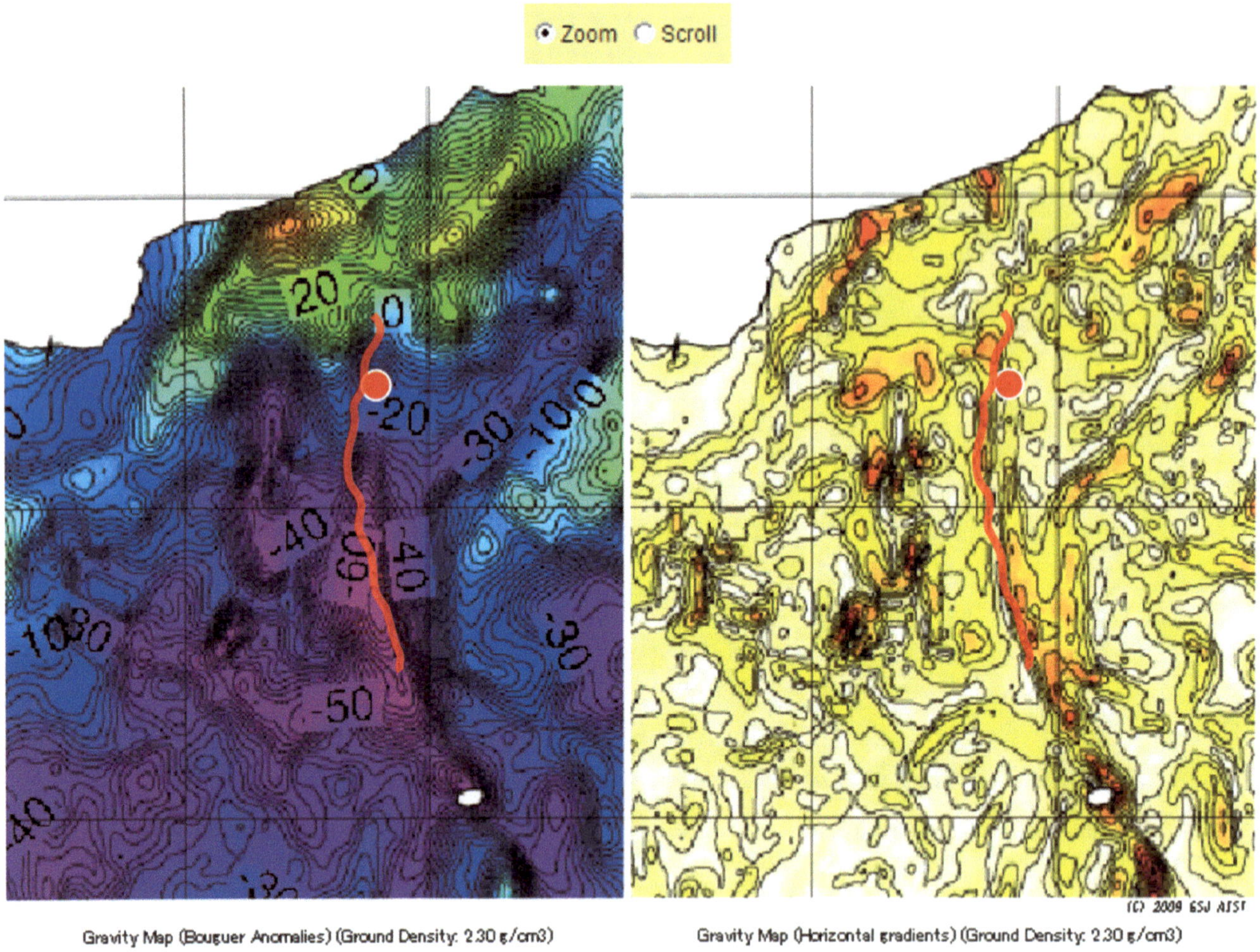

Fig. 6 Pre-processed gravity maps in the vicinity of the Kamishiro Fault. (*Left*) Bouguer anomalies displayed at an assumed density of 2.30 g/cm^3. (*Right*) Horizontal Bouguer anomaly gradients at an assumed density of 2.30 g/cm^3. Both maps show the same area around the Kamishiro Fault. *Red lines* on both maps display the location of the Kamishiro Fault. *Red dots* on both maps display the epicenter of the Mw 6.8 earthquake that occurred in Nagano Prefecture, Japan, on 22 November 2014

Acknowledgment We thank the Associate Editor of the journal of the International Association of Geodesy Symposia, and three anonymous reviewers for providing constructive reviews that greatly improved the manuscript.

References

Feebrey CA, Hishida H, Yoshioka K, Nakayama K (1998) Geophysical expression of low sulphidation epithermal Au–Ag deposits and exploration implications–examples from the Hokusatsu region of SW Kyushu, Japan. Resour Geol 48:75–86

Geological Survey of Japan (ed) (2004) Gravity CD-ROM of Japan Ver. 2, Digital Geosci. Map P-2 (CD-ROM). Geol Surv Jap

Geological Survey of Japan (ed) (2013) Gravity database of Japan DVD edition, Digital Geosci. Map P-2 (DVD-ROM). Geol Sur Jap

GSJ Gravity Survey Group (1989) On the standard procedure SPECG1988 for evaluating the correction of gravity at the Geological Survey of Japan. Bull Geol Surv Jap 40:601–611

Headquarters for Earthquake Research Promotion (2013) Report for the long-term evaluation on Umi fault. http://www.jishin.go.jp/main/yosokuchizu/katsudanso/reg_kyushu_04_umi.htm (in Japanese). Accessed 6 March 2015

Headquarters for Earthquake Research Promotion (2014) Report for the earthquake occurred in northern part of Nagano prefecture on November 22 2014. http://www.jishin.go.jp/main/chousa/14dec_nagano/index.htm (in Japanese). Accessed 13 January 2015

Hinze WJ, Aiken C, Brozena J, Coakley B, Dater D, Flanagan G, Forsberg R, Hildenbrand T, Keller GR, Kellogg J, Kucks R, Lee X, Mainville A, Morin R, Pilkington M, Plouff D, Ravat D, Roman D, Urrutia-Fucugauchi J, Veronneau M, Webring M, Winester D (2005) New standards for reducing gravity data: the North American gravity database. Geophysics 70:J25–J32

Hinze WJ, Von Frese RRB, Saad AH (2013) Gravity and magnetic exploration: principles, practices, and applications. Cambridge University Press, Cambridge, p 20

Honda R, Sawada A, Furuse N, Kudo T, Tanaka T, Hiramatsu Y (2012) Release of gravity database of the Kanazawa University. J Geod Soc Jap 58(4):153–160 (in Japanese with English abstract)

Jaquet O, Lantuéjoul C, Goto J (2012) Probabilistic estimation of long-term volcanic hazard with assimilation of geophysics and tectonic data. J Volcanol Geotherm Res 235:29–36

Komazawa M (1988) A gravimetric terrain correction method by assuming an annular prism model. J Geod Soc Jap 34:11–23

Komazawa M (1995) Gravimetric analysis of Aso volcano and its interpretation. J Geod Soc Jap 41(1):17–45 (Japanese journal; written in English)

Komazawa M (2004) Density structure inferred from gravity anomalies in the eastern area of the Itoigawa-Shizuoka Tectonic Line, central Japan. Earth Planets Space 56:1309–1314

Komazawa M, Mishina M (2002) Caldera structure inferred from gravity anomalies west of Nagamachi-Rifu Fault, Northeast Japan. Earth Planets Space 54:1049–1053

Li CF (2011) An integrated geodynamic model of the Nankai subduction zone and neighboring regions from geophysical inversion and modeling. J Geodyn 51(1):64–80. doi:10.1016/j.jog.2010.08.003

Murata Y (1993) Estimation of optimum average surficial density from gravity data: an objective Bayesian approach. J Geophys Res Solid Earth 98(B7):12097–12109. doi:10.1029/93JB00192

Nawa K, Fukao Y, Shichi R, Murata Y (1997) Inversion of gravity data to determine the terrain density distribution in southwest Japan. J Geophys Res Solid Earth 102(B12):27703–27719. doi:10.1029/97JB02543

Panayotopoulos Y, Hirata N, Sato H, Kato A, Imanishi K, Kuwahara Y, Cho I, Takeda T, Asano Y (2014) Investigating the role of the Itoigawa-Shizuoka tectonic line towards the evolution of the Northern Fossa Magna rift basin. Tectonophysics 615–616:12–26. doi:10.1016/j.tecto.2013.12.014

Shichi R, Yamamoto A (2001) Gravity database of southwest Japan (CD-ROM). Bull Nagoya Univ Spec Rept 9

Shoberg T, Stoddard PR (2013) Integrating stations from the North America gravity database into a local GPS-based land gravity survey. J Appl Geophys 89:76–83

Wakita K, Igawa T, Takarada S (eds) (2009) Seamless geological map of Japan at a scale of 1: 200,000 (DVD edition). Digital Geoscience Map G-16, Geological Survey of Japan, National Institute of Advanced Industrial Science and Technology

Wessel P, Smith WH (1998) New, improved version of generic mapping tools released. EOS Trans Am Geophys Union 79(47):579. doi:10.1029/98EO00426

Yamawaki T, Tanaka S, Ueki S, Hamaguchi H, Nakamichi H, Nishimura T, Oikawa J, Tsutsui T, Nishi K, Shimizu H, Yamaguchi S, Miyamachi H, Yamasato H, Hayashi Y (2004) Three-dimensional *P*-wave velocity structure of Bandai volcano in northeastern Japan inferred from active seismic survey. J Volcanol Geotherm Res 138:267–282. doi:10.1016/j.jvolgeores.2004.07.010

Recent Developments of GPS Tsunami Meter for a Far Offshore Observations

Yukihiro Terada, Teruyuki Kato, Toshihiko Nagai, Shunichi Koshimura, Naruyuki Imada, Hiromu Sakaue, and Keiichi Tadokoro

Abstract

A new tsunami observation system using GPS buoys has been developed, which employs the RTK-GPS technique to detect and monitor tsunamis in real-time before they reach the coast. A series of experimental GPS buoys succeeded in detecting three tsunamis with amplitudes of about 10 cm. Following this success, since 2007, the Japanese government has established GPS buoy systems for monitoring sea waves at 19 sites around the Japanese coast. These systems succeeded to detect 11th March 2011 Tohoku-Oki earthquake tsunami. Through these experiences, we recognized two problems that need to be solved in order to deploy buoys at farther distances from the coast: one is that positioning accuracy decreases as the distance increases and the data transmission by radio becomes difficult for a long distance. In order to overcome these difficulties, first, a new algorithm of PPP-AR for real-time application was employed. The test analysis showed that the positioning accuracy may attain a few centimeters even if the reference GPS network that generates precise orbits and clocks is farther than 1,000 km. Then, a satellite communication system was experimentally used to send data in both directions between the land base and buoy. The data that was obtained on the buoy was transmitted to the land base and was shown on a webpage in real-time, successfully. This kind of system may have further applications of earth science; for example, we are trying to implement GPS/acoustic apparatus for continuous monitoring of ocean floor crustal deformations.

Keywords

GPS • GPS buoy • Precise point positioning • Satellite communication • Tsunami

1 Introduction

If a buoy with a sensor for detecting sea surface height change is set offshore distant enough from the coast and data are transmitted in real-time, detection of a tsunami can be effectively used for disaster mitigation. We have developed a GPS buoy system for this purpose, equipped

Y. Terada (✉)
Kochi National College of Technology, Kochi, Japan
e-mail: terada@ce.kochi-ct.ac.jp

T. Kato
The University of Tokyo, Tokyo, Japan
e-mail: teru@eri.u-tokyo.ac.jp

T. Nagai
Echo Corporation, Tokyo, Japan
e-mail: t-nagai@ecoh.co.jp

S. Koshimura
Tohoku University, Sendai, Japan
e-mail: koshimura@irides.tohoku.ac.jp

N. Imada
Hitachi Zosen Corporation, Osaka, Japan
e-mail: imada_n@hitachizosen.co.jp

H. Sakaue
Non, Tokyo, Japan
e-mail: say22hero@gmail.com

K. Tadokoro
Nagoya University, Nagoya, Japan
e-mail: tad@seis.nagoya-u.ac.jp

M. Hashimoto (ed.), *International Symposium on Geodesy for Earthquake and Natural Hazards (GENAH)*,
International Association of Geodesy Symposia 145, DOI 10.1007/1345_2015_151

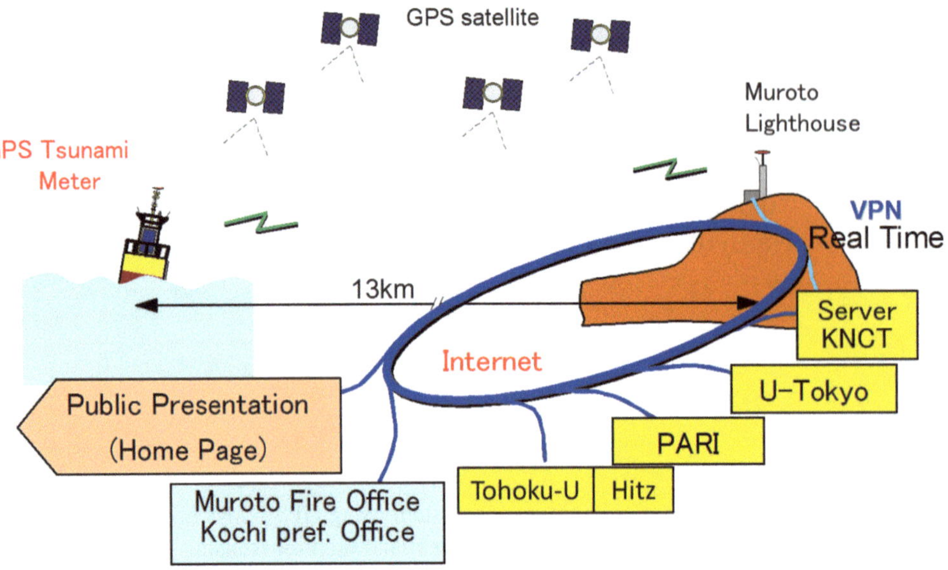

Fig. 1 Layout of the experimental GPS tsunami observation system

with a GPS antenna at the top of the buoy floating at the sea surface (Kato et al. 2000). Data taken on the buoy was transmitted to the land base where another GPS set was placed and the baseline mode real-time kinematic (RTK) GPS technology was used for data analysis. The estimated position of the sea surface relative to a stable location on land were monitored at the local governmental offices and the data were disseminated through internet in real-time as well. Figure 1 shows the GPS tsunami observation system we have used.

This system was first deployed off Ofunato city, northeast Japan, for the period between 2001 and 2003, and then moved to off Cape Muroto, southwest Japan, for the period between 2004 and 2007 and between 2008 and 2012. These buoys succeeded in detecting three tsunamis whose amplitudes were nearly 10 cm. They are the 23rd June 2001 Peru earthquake, the 26th September 2003 Tokachi earthquake and the 5th September 2004 Kii earthquake (Nagai et al. 2003; Kato et al. 2005). These results showed that the GPS buoys are useful for early detection of tsunamis. See Kato et al. (2000, 2005, 2011), and Terada et al. (2011) for other details of developments.

Following the success of our GPS buoy experiments, the Ministry of Land, Infrastructure, Transport and Tourism (MLIT) of Japan has established the GPS buoy system for monitoring sea waves at 19 sites around the Japanese coast since 2007 as a part of the nationwide wave monitoring system called Nationwide Ocean Wave information network for Ports and HArbourS (NOWPHAS) system. These systems also succeeded in detecting the 28th February 2010 Chile earthquake. Then, a giant earthquake of Mw9.0 occurred offshore Tohoku district, northeast Japan, on 11th March 2011 and the generated tsunami devastated the whole area along the Pacific coast of the district, which caused about 18,000 people of deaths and missing in the area. Figure 2 shows the real-time tsunami records by this earthquake, observed at the Off South Iwate (Kamaishi) station of NOWPHAS system (see Fig. 3 for the location). The tsunami measured over 6 m in height and was observed about 15 min after the earthquake. The Japan Meteorological Agency used the data for updating the level of the tsunami warning to the greatest value. After the highest wave was observed, real-time monitoring of the tsunami data stopped suddenly. Since the data of all GPS buoys installed in this area stopped at once, it was thought that the cause was interception of the communication network by electric power loss. However, as the backup power supplies worked effectively at the each land base stations, recordings of the tsunami were continued and the data were retrieved later. Figure 4 shows thus obtained examples of full wave form recordings of the Tohoku-Oki earthquake including the later part of retrieved records. Almost all of the observation sites of NOWPHAS successfully recorded the full wave form of the tsunamis.

2 Problems Found in Case of 2011 Tohoku-Oki Earthquake Tsunami

Through these experiences of GPS buoys for detecting tsunamis, we have recognized two problems that need to be solved for more effective use of GPS buoy for disaster mitigation.

The first problem is that the buoys established at less than 20 km from the coast is not far enough for disaster mitigation. In case of Tohoku-oki earthquake on 11th March

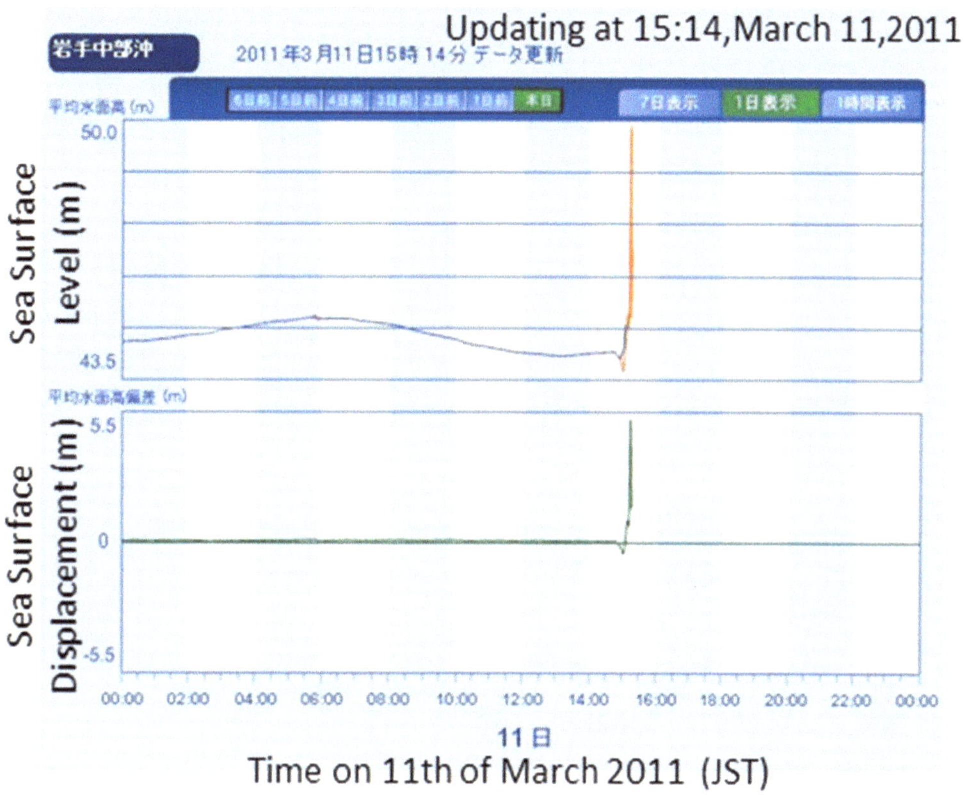

Fig. 2 The tsunami record due to the Tohoku-Oki earthquake of 11th March 2011. Figure is captured from real-time web display of NOWPHAS at Off South Iwate (Kamaishi) site. (http://www.milit.go.jp/kowan/nowphas/) *Horizontal axis* is shown as Japan Standard Time (UTC + 9 h) of 11th March 2011

2011, the update of tsunami alert was conducted probably only a few minutes before it reaches the coast. At that time, it was probable that electric power was already shut down and updated tsunami alert might have not reached to all of the relevant people. Though this latency could be improved by much closer, ideally automatic, monitoring of offshore buoys, farther placement of the buoy from the shore would provide a longer lead time for more effective evacuation.

The second problem is that the data stopped recording due to power and/or Internet failure soon after the earthquake. Though we secured recordings even after power failure by preparing an uninterruptible-power-supply (UPS) for a while, which enabled us to recover the record after the earthquake, the real-time dissemination of the record was not attained.

Thus, in this study, we tried to overcome these problems to deploy GPS buoys at, say, farther than 100 km, at least, from the coast, by introducing a new positioning algorism and a satellite communication system for more robust data transmission to farther distance to avoid power/internet interruptions.

3 Improvements of Positioning Strategy

The system has had a limit of distance up to about 20 km from the coast as it employs a conventional baseline mode kinematic algorithm called RTK GPS. The system used broadcasted satellite orbits and clocks for real-time applications, so that the positioning accuracy decreases rapidly as the distance exceeds 20 km. In order to improve the distance limit, in this study, we employed a new algorithm called PPP-AR (precise point positioning with ambiguity resolution) for real-time kinematic applications (Mervart et al. 2008).

Precise Point Positioning (PPP) technology was first introduced for static analysis (e.g., Zumberge et al. 1997) and the technology has been widely used for geodetic applications due to its superior characteristics of less computational burden and easier interpretations of crustal displacement as it does not require any prior information of the motion of referenced stations, as compared with the baseline mode analysis. The most precise estimation of position is obtained by resolving ambiguities inherent to the phase observation

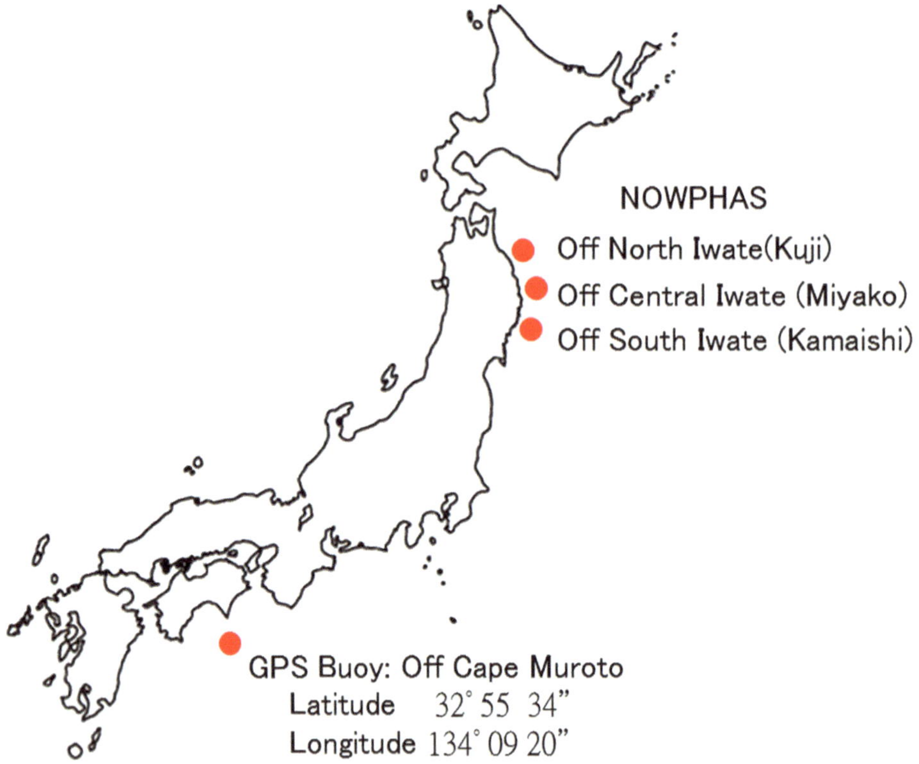

Fig. 3 GPS buoy locations used in this study

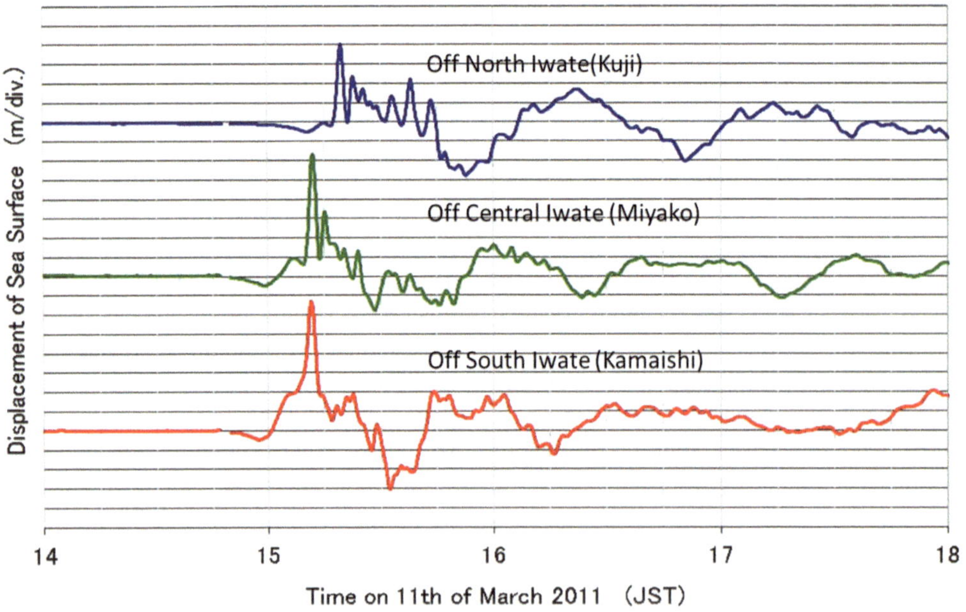

Fig. 4 Full wave form of Tohoku-Oki earthquake tsunami observed by NOWPHAS (Observed by MLIT and Analyzed by PARI: Kawai et al. 2011). Only the first onset was monitored in real-time as shown in Fig. 2. Later part of records were retrieved afterwards

of GPS. Resolving ambiguity for PPP can be achieved by obtaining precise orbits and clocks of GPS satellites from IGS after the observation. Then, the technology of PPP has been developed for the real-time kinematic applications. Mervart et al. (2008) for example used satellite single differenced data of phase and code observations to estimate corrections for RTK PPP. By employing precise orbits and clocks, it can then resolve ambiguities, which we call PPP-AR

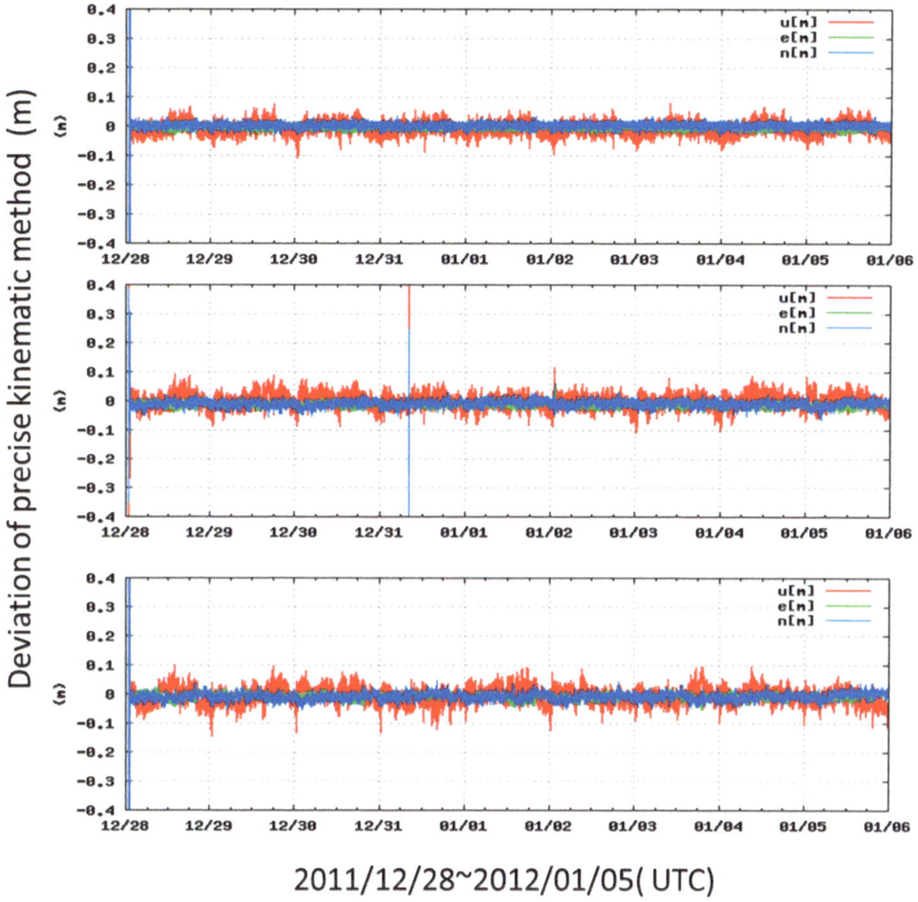

Distance of Rover and reference area (Hokkaido): 150km

Standard deviation (σ)	N-S(mm)	8.9
	E-W(mm)	8.1
	Up(mm)	18.3
Fix rate (%)		99.95

Distance of Rover and reference area (Kanto): 1000km

Standard deviation (σ)	N-S(mm)	11.0
	E-W(mm)	7.5
	Up(mm)	22.6
Fix rate (%)		99.80

Distance of Rover and reference area (kyusyu): 1500km

Standard deviation (σ)	N-S(mm)	13.2
	E-W(mm)	11.4
	Up(mm)	26.3
Fix rate (%)		99.81

Fig. 5 Results of PPP-AR at Shimokawa GEONET site for different distances of reference networks that generates precise orbits and clocks: (*upper*) 150 km, (*middle*) 1,000 km, and (*lower*) 1,500 km, respectively. Observation period is from 28 December 2011 to 5 January 2012 (Courtesy by Hitachizosen Corporation 2012)

RTK. This algorithm was implemented in the software called RTNet developed by GPS Solutions Inc., which we used in this study. The advantage of the software is that it can provide precise orbits and clocks from regional reference networks like GEONET (GNSS Earth Observation Network System) of Japan. In addition the RTNet tool allows a real-time analysis capability, which is favorable for our purpose.

First, we examined the accuracy of PPP-AR using RTNet software with post-processed high-accuracy baseline analysis. Figure 5 shows the estimated time series of a GEONET site at Shimokawa, with different distances from the reference network for generating precise orbits and clocks for corrections. The figure clearly shows that the variability is only a few centimeters, at most, for even more than 1,000 km distance from the regional reference network. These results are encouraging to us that GPS buoys could be put at farther offshore, say, more than 100 km or even more than 1,000 km from the coast.

We then implemented the RTNet system to our experimental buoy system off Cape Muroto (see Fig. 3 for the location). Figure 6 shows a comparison of coordinates obtained with conventional RTK GPS (upper inset) and PPP-AR (lower inset). In this case, the buoy is set about 35 km south off Cape Muroto. The sample data is taken for 1 h from 12:00 to 13:00 on 18th May in 2012. 1 Hz sampled data of vertical components of the buoy position is plotted. The upper record is obtained by the conventional RTK-GPS, while the lower is obtained by the PPP-AR algorithm. It is readily seen that the short term fluctuations in both records looks highly correlated, suggesting that both records could precisely recover the higher frequency wind waves of the sea-level heights. However, RTK-GPS records show significant fluctuations for the longer period of several minutes and even sudden offsets and faulty data, probably due to mis-fix of ambiguity, intermittently, which is not seen in the record of PPP-AR (lower plots of Fig. 6). Clearly the sea surface fluctuations during this time period is only several tens of centimeters which suggest calm weather, so that we cannot expect any sudden offsets or long-period of fluctuations with periods of several minutes or longer.

Since we do not see such bad record in RTK-GPS if the distance is less than 20 km, we judge that the reason of

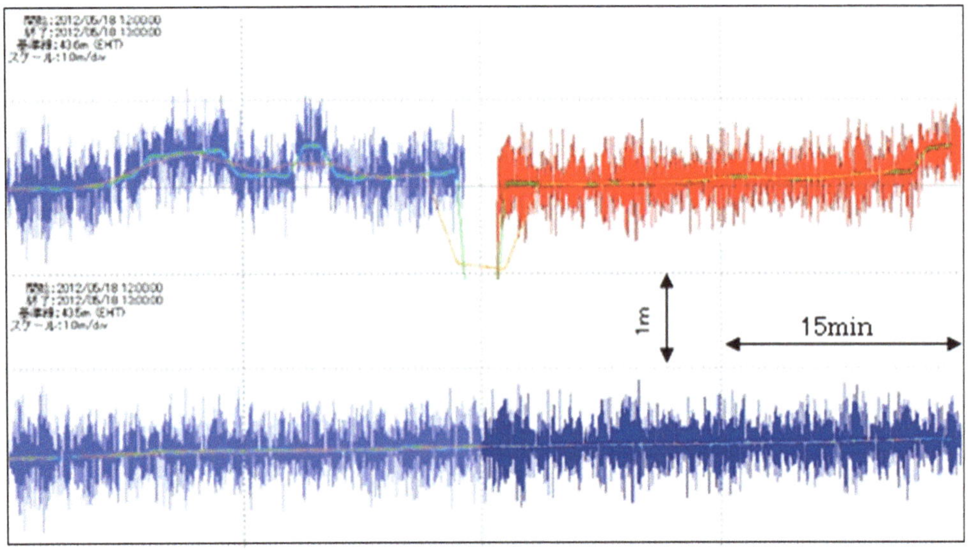

Fig. 6 Comparison between RTK-GPS (Upper) and PPP-AR (Lower) GPS data: Muroto experimental site (Fig. 3)

such result stems from long distance environments of RTK-GPS. In such circumstances, there occurs frequent mis-fix of ambiguities and cause jumps in the estimated positions. The stability of PPP-AR suggests that the ambiguity in phase is totally resolved to give correct estimation of position. These situations clearly show the effectiveness of PPP-AR positioning method relative to conventional RTK GPS, in particular, for the far offshore deployments of GPS buoys.

4 Utilization of Satellite Communication System

Given that above new system is feasible and is established far offshore from the coast, then another problem is how the data obtained at the buoy can be transmitted to land. The ground based radio transmission system that is currently used, cannot be used for such long distance of more than 100 km. Then, a satellite data transmission will be the solution. If satellite data transmission is possible, the data have not to be sent to the area nearest to the coast, which allows us that the data can be safely recorded without interruptions due to earthquakes and/or tsunamis. Thus, data transmission using satellites may solve the second problem of unexpected interruption of data at the critical time when tsunamis approach the coast.

Experiments for such satellite data transmission from the buoy to the ground data management center were conducted under another independent cooperative research project with Japan Aerospace Exploration Agency (JAXA), National Institute of Information and Communications Technology (NICT) and Hitachizosen Corporation (Yamamoto et al. 2014). The first preliminary experiment was conducted from 24th October to 6th November in 2012. The Japanese engineering test satellite (ETS VIII) was used for the purpose of data transmission. In this experiment unidirectional data transmission from the buoy to the land base was conducted. The result showed about 10^{-2} to 3×10^{-1} of Frame Error Rate (FER) depending on wave height. This result was thought satisfactory for continuous monitoring of sea surface height changes, though some more improvements could be made.

Then, we conducted another experiment for bi-directional data transmission for testing PPP-AR. Figure 7 shows the flow of data in this experiment. In this experiment, Japanese Quasi Zenith Satellite System (QZSS) was used to transmit correction information for PPP-AR from the ground to the buoy. The data transmission from buoy back to land was carried out using ETS VIII as is shown in Fig. 7. The experiment was conducted from 3rd to 5th of January and from 1st to 21st of June in 2014. Figure 8 shows the obtained data by this experimental system; the upper inset is non filtered sea surface height which shows wind waves and tide, while lower data is low pass filtered data which is mostly tidal component. First, the upper time series includes all of sea surface waves. During this time period, it is readily seen that the vertical motion of the buoy has amplitude (peak-to-peak) of about 0.6 m, quite calm environments. Then, the amplitude of short period waves was getting bigger to about 1.3 m toward noon in Japan time, which is a little rough water. If tsunami of a few centimeter in amplitude is overlapped on the wave, it may not be visible in the record. However, as the periods of tsunami are from several minutes to longer than one hour, the wave form of tsunami can be extracted if a low pass filter is applied as shown in the lower plot of Fig. 8. Tsunami, if superimposed on a smooth curve

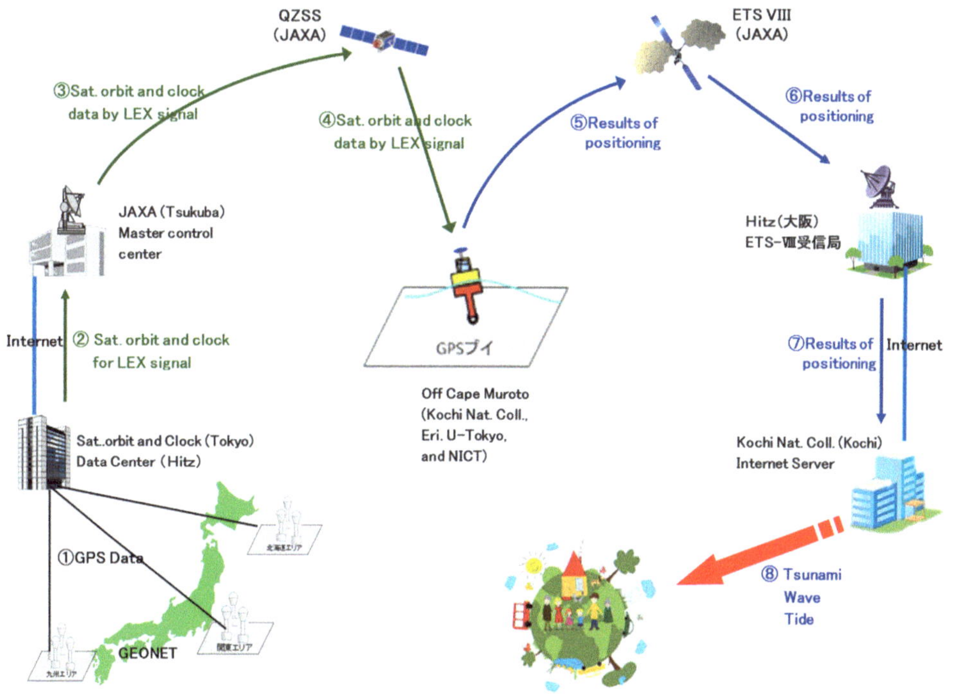

Fig. 7 Plan of data transmission test via satellite communication systems of QZSS and ETSVIII (Yamamoto et al. 2014)

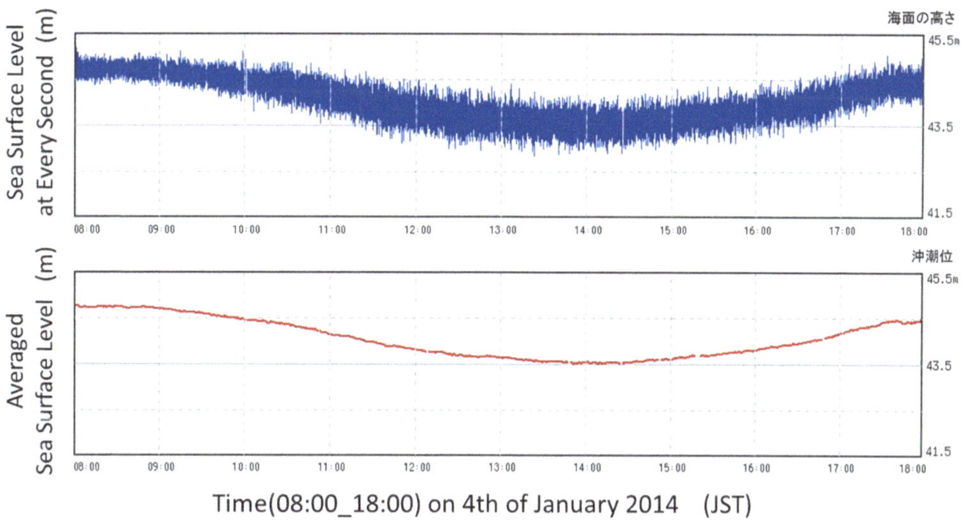

Fig. 8 Results of data transmission by satellite communication system of QZSS and ETSVIII (*Upper*: raw data, *Lower*: low-pass filtered data). Unit: (*horizontal*) hour in Japan Standard Time, (*vertical*) meter

of tidal change of the lower figure, may not be overlooked if its amplitude is larger than a few centimeters.

However, the data transmission capability is still a problem as it varies depending on the wave height. Figure 9 shows the relation of data transmission rates and significant wave heights in the case of the first experiment in 2014. The wave height was very low on the first day of experiment, less than 50 cm. That became higher and higher on the 2nd day, for which all the observed data are shown in Fig. 8. The significant wave height became about 2.3 m. The maximum wave height at this time was 4.6 m. On the contrary, data transmission rate is getting worse as the significant wave height was getting higher and it became better when the wave height became lower during the last day. This result is the same as that obtained at the preliminary experiment in 2012. Thus, it seems clear that there is a reverse correlation between significant wave height and data transmission rate. It seems that the rate of data transmission error changes due to

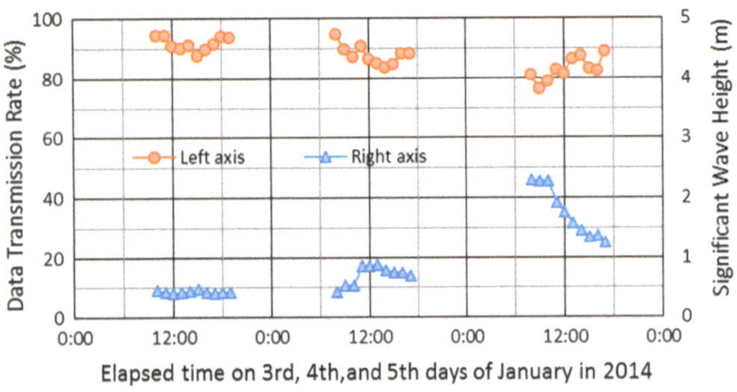

Fig. 9 Relation between data transmission rate (*circle* and *left hand axis*) and significant wave height (*triangle* and *right hand axis*)

the tilting of buoy due to sea surface variations. As we used a planer antenna for signal transmission and acquisition, it has a directional property depending on zenith angle. If the buoy inclines larger, the gain of signal gets lower and the frame error rate increases. This characteristics may be improved by better designing the antenna for data transmission and/or hardware settings. (see Yamamoto et al. (2014) for further details of experiments).

Given that a satellite communication link is feasible, the data loss that we experienced at the time of the 2011 Tohoku-Oki earthquake may be resolved by putting the reference network much farther from the epicentral area. In this study, we used three regional network areas for reference as is shown in Fig. 5. We could run the routine in parallel at any time and use the most effective reference network when any of them has a problem as it is hard to imagine that all of Japanese GEONET could suffer from a blackout.

5 Further Applications of GPS Buoy for Continuous Monitoring of Ocean Bottom Crustal Movements

In addition to the above developments, we are trying to apply our buoy system for continuous monitoring of ocean bottom crustal movements. Many previous researchers have developed the GPS-acoustic system for this purpose (e.g., Spiess et al. 1998; Kido et al. 2006; Fujita et al. 2006; Sato et al. 2011a, b; Tadokoro et al. 2012). The basic idea of a GPS-acoustic system combines GPS and acoustic ranging to estimate the position of the seafloor. Figure 10 illustrates a plan of GPS-acoustic system for monitoring ocean floor crustal movements. First, GPS is used to estimate the position of the floating body (such as vessel or buoy), and then, the lengths between the floating body and the ocean bottom positions are measured by acoustic ranging. Though the estimated lengths between the sea surface and the sea bottom changes greatly, the position of the center of the multiple ocean bottom sensors (more than three) does not change much. Thus we take its position as the position of the seafloor.

As the previous systems have used vessels to measure the position of the ocean surface, the measurements have been only intermittent. GPS buoys could be used for continuous monitoring of ocean bottom crustal movements, if the GPS buoy is equipped with an acoustic system. Based on this idea, we started experiments using our GPS buoy off Cape Muroto in 2013. A similar idea with slightly different system design was introduced by Takahashi (2014). Our approach, different from their approach, is the integration of GPS buoys for multiple use of GPS technology for better cost effectiveness and wider contributions to earth science, not only for tsunami and seafloor crustal deformation but also for atmospheric and ionospheric research. The experimental observation system in this study, shown in Fig. 10, continuously received acoustic data at the land base station from 3rd of August to 23rd of November. The data acquisition rate is 96% during the first 10 days. Moreover, the acoustic data obtained at the buoy showed clear signal suggesting high signal-to-noise ratio, probably because there was no disturbing engine noise in case of using a ship. Although the detailed examination of data is left for future studies, it was shown that a clear waveform can be observed if we use a buoy for a floating platform. We will try to continue experiments for tackling the problem.

6 Conclusion and Remarks

The tsunami monitoring system using a GPS buoy has been developed for more than eighteen years. This system clearly showed that tsunamis can be measured within the accuracy of several centimeters. After the 11th March 2011 Tohoku-Oki earthquake tsunami, the GPS buoy system has been improved to solve two problems to deploy the buoys at distances farther offshore:

1. Introduction of a new algorithm for GPS data analysis, precise point positioning with ambiguity resolution method (PPP-AR), greatly improved the distance limitation from the coast to GPS buoys to more than 1,000 km.
2. Satellite data transmission between buoys and ground data management centers using ETS VIII and QZSS was successfully implemented for tsunami observation in the outer ocean, which enables avoiding data loss due to power failures right after large earthquakes.
3. Integration of the system with a GPS/acoustic system for continuous monitoring of sea-floor crustal movements, together with wider applications for earth science, will make the system much more cost effective for ocean geophysical exploration.

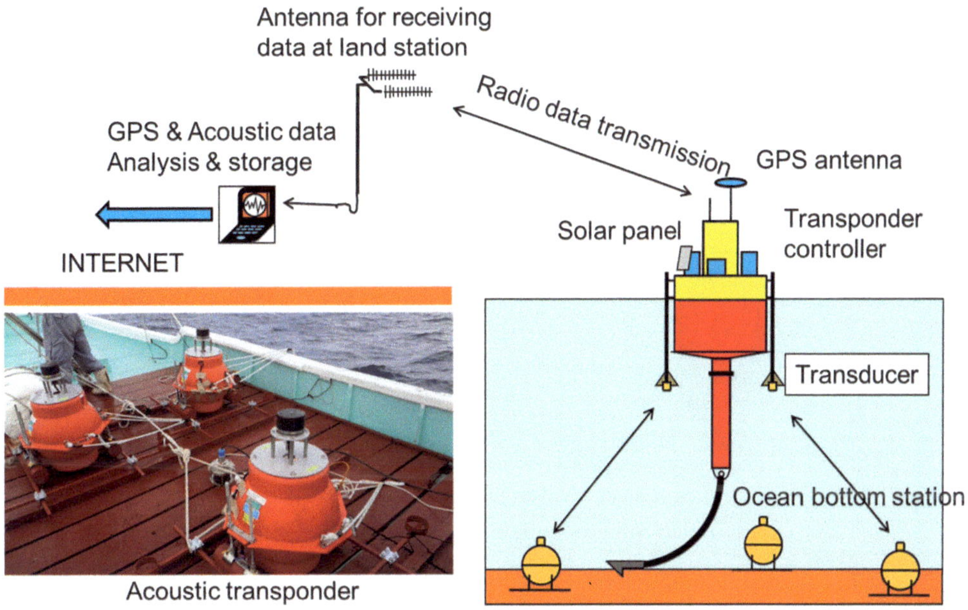

Fig. 10 Plan of continuous monitoring system of crustal movements at the ocean floor

Acknowledgments The authors would like to express deepest thanks to Mr. G. Hashimoto of and Mr. O. Motohashi of JAXA, Mr. S. Yamamoto of NICT, Mr. A. Wada of Hitachizosen Corporation and many other people for their support to this research. The authors also would like to express sincere thanks to three reviews whose comments significantly improved the manuscript. This work was supported by Grant-in-Aid for Scientific Research (S) 21221007 and the contracted research of Ministry of Education, Culture, Sports, Science and Technology of Japan.

References

Fujita M, Ishikawa T, Mochizuki M, Sato M, Toyama S, Katayama M, Kawai K, Matsumoto Y, Yabuki T, Asada A, Colombo OL (2006) GPS/Acoustic seafloor geodetic observation: method of data analysis and its application. Earth Planets Space 58:265–275

Kato T, Terada Y, Kinoshita M, Kakimoto H, Issiki H, Matsuishi M, Yokoyama A, Tanno T (2000) Real time observation of tsunami by RTKGPS. Earth Planets Space 52(10):841–845

Kato T, Terada Y, Ito K, Hattori R, Abe T, Miyake T, Koshimura S, Nagai T (2005) Tsunami due to the 2004 September 5th off the Kii peninsula earthquake, Japan, recorded by a new GPS buoy. Earth Planets Space 57:297–301

Kato T, Terada Y, Nishimura H, Nagai T, Koshimura S (2011) Tsunami records due to the 2010 Chile Earthquake observed by GPS buoys established along the Pacific coast of Japan. Earth Planets Space 63:e5–e8

Kawai H, Satoh M, Kawaguchi K, Seki K (2011) Characteristics of the 2011 off the Pacific Coast of Tohoku Earthquake Tsunami. Report of the Port and Airport Research Institute, vol 50. pp 3–64

Kido M, Fujimoto H, Miura S, Osada Y, Tsuka K, Tabei T (2006) Seafloor displacement at Kumano-nada caused by the 2004 off Kii Peninsula earthquakes, detected through repeated GPS/Acoustic surveys. Earth Planets Space 58:911–915

Mervart L, Lukes Z, Rocken C, Iwabuchi T (2008) Precise point positioning with ambiguity resolution in real-time. In: Proceedings of the 21st international technical meeting of the Satellite Division of The Institute of Navigation (ION GNSS 2008), Savannah, pp 397–405

Nagai T, Ogawa H, Nukada K, Kudaka M (2004) Characteristics of the observed 2003 Tokachi off earthquake tsunami profile. Coast Eng J 46(3):315–327

Sato M, Ishikawa T, Ujihara N, Yoshida S, Fujita M, Mochizuki M, Asada A (2011a) Displacement above the hypocenter of the 2011 Tohoku-oki earthquake. Science 332:1395

Sato M, Saito H, Ishikawa T, Matsumoto Y, Fujita M, Mochizuki M, Asada A (2011b) Restoration of interplate locking after the 2005 Off-Miyagi Prefecture earthquake, detected by GPS/acoustic seafloor geodetic observation. Geophys Res Lett 38, L01312. doi:10.1029/2010GL045689

Spiess FN, Chadwell CD, Hildebrand JA, Young LE, Purecell GH Jr, Dragart H (1998) Precise GPS/Acoustic positioning of seafloor reference points for tectonic studies. Phys Earth Planet Int 108:101–112

Tadokoro K, Ikuta R, Watanabe T, Ando M, Okuda T, Nagai S, Yasuda K, Sakata T (2012) Interseismic seafloor crustal deformation immediately above the source region of anticipated megathrust earthquake along the Nankai Trough, Japan. Geophys Res Lett 39, L10306. doi:10.1029/2012GL051696

Takahashi N, Ishihara Y, Ochi H, Fukuda T, Tahara J, Maeda Y, Kido M, Ohta Y, Mutoh K, Hashimoto G, Kaneda Y (2014) New buoy observation system for tsunami and crustal deformation. Mar Geophys Res 35(3):243–253. doi:10.1007/s11001-014-9235-7

Terada Y, Kato T, Nagai T, Koshimura S, Miyake T, Nishimura H, Kunihiro S (2011) Development of a tsunami monitoring system using a GPS buoy. In: Proceedings of the international global navigation satellite systems society (IGNSS) symposium, Peer Reviewed Paper, vol 14. pp 1–14

Yamamoto S, Kawasaki K, Terada Y, Kato T, Hashimoto G, Motohashi O, Saida Y, Matsuzawa R (2014) Data transmission experiment from the buoy using the Engineering Test Satellite VIII (ETS-VIII) – the aim of early detection of TSUNAMI. IEICE Technical Report, 2014-8, pp 5–10. (in Japanese with English abstract)

Zumberge JF, Heflin MB, Jefferson DC, Watkins MM, Webb FH (1997) Precise point positioning for the efficient and robust analysis of GPS data from large networks. J Geophys Res 102:5005–5017

Ground Deformation in the Kyoto and Osaka Area During Recent 19 Years Detected with InSAR

Manabu Hashimoto

Abstract

We previously found local uplift in the Kyoto basin and subsidence along the Arima-Takatsuki Tectonic Line (ATTL), an active fault in the northern Osaka plain by the analysis of ALOS/PALSAR images acquired during 2007–2010. In order to reveal the history of ground deformation in these regions before the operation of ALOS/PALSAR, ERS 1/2 and Envisat images acquired during 1992 to 2007 were collected and analyzed. We found that uplift in the southern Kyoto basin may have started in 2003–2004 and accelerated afterward. Subsidence along the ATTL is not clear in ERS or Envisat images except in its western part. Subsidence in a wedge-shaped zone between the ATTL and the Rokko fault is recognized till 2004 with ERS-1/2, but was not observed before the Kobe earthquake. Subsidence is concentrated into two spots and rates accelerated after 2003. These spots correspond to the areas of low terrace deposits during late Pleistocene.

Keywords

Active faults • ERS/Envisat • Ground deformation • InSAR • Kyoto basin • Osaka plain

1 Introduction

In 1960s, several cities located in alluvial plains in Japan experienced severe ground subsidence due to pumping out of groundwater (e.g. Ministry of Environment 2011). The Kyoto basin and the Osaka plain are typical examples. These plains are surrounded by faults, some of which are considered to be active (e.g. Okada and Togo 2000). Furthermore there may be buried faults beneath these plains. Ground subsidence correlates with the structure of soil layer that contains ground water. Since the motion of faults changes the ground structure, the spatial distribution of ground subsidence may change across faults (Bell et al. 2002).

Synthetic aperture radar interferometry (InSAR) has succeeded in imaging deformation due to earthquakes and other sources with a high spatial resolution (e.g. Massonnet et al. 1993; Fukushima et al. 2009; Takada and Fukushima 2013). Recent progress of InSAR time series analyses also enables us to reveal temporal variation in surface motion (Ferretti et al. 2001; Lanari et al. 2004; Hooper et al. 2004). There have been many studies of the interseismic deformation across plate boundary or active faults, ground subsidence in basins, volcanic deformations, and other deformation sources.

In our previous studies, we showed an area of uplift in the southern Kyoto basin and a zone of subsidence along an active fault, called the Arima-Takatsuki tectonic line (hereafter ATTL) (Hashimoto 2013, 2014). However, we could not reveal when these deformation started due to the limited time span of ALOS/PALSAR observations. In this study, we try to figure out the history of deformation during 19 years in these two regions using ERS-1/2 and Envisat data.

M. Hashimoto (✉)
Disaster Prevention Research Institute, Kyoto University, Uji, Kyoto 611-0011, Japan
e-mail: hashimoto.manabu.7e@kyoto-u.ac.jp

M. Hashimoto (ed.), *International Symposium on Geodesy for Earthquake and Natural Hazards (GENAH)*, International Association of Geodesy Symposia 145, DOI 10.1007/1345_2016_222

2 Deformation Detected by ALOS/PALSAR

Hashimoto (2014) analyzed SAR images acquired by ALOS/PALSAR to reveal ground deformation in the Osaka and Kyoto areas with the Gamma software (Fig. 1). These basins are bounded by active faults such as the ATTL and the Rokko Fault (e.g. Huzita 1962). Furthermore, a significant portion of a large active fault, the Uemachi fault, may be buried beneath the Osaka plain (e.g. Okada and Togo 2000). In this analysis, we performed a simple stacking of interferograms derived from 2-pass interferometry and conversion of LOS velocities from two different look directions into E-W and quasi-vertical velocity components.

In Hashimoto (2014), 20 PALSAR images from the path 414 (frame 680) acquired during the period from October 11, 2007 to October 19, 2010, and 15 images from the path 65 (frame 2920) during the period from January 9, 2007 to October 20, 2010 were processed to obtain line-of sight (hereafter LOS) velocities. Hashimoto (2014) used the Hole-filled SRTM3 DEM (Jarvis et al. 2008), which has a 3 arc-second resolution. 2-pass interferometry was applied to pairs of images with relatively short baselines. All interferograms were flattened before unwrapping. Then, unwrapped interferograms were stacked.

Combining two stacked interferograms deduced using images acquired on different orbits (Fig. 2a, b), we can estimate E–W and quasi-vertical components of average velocities (Fig. 2c, d). In this calculation, we take a reference point for the LOS velocities at (135.6 E, 34.6 N) considering the location in image as well as coherence of ERS-1/2 and Envisat interferograms. Figure 2d shows the distribution of quasi-vertical velocities in southern Kyoto basin and northern Osaka plain. We obtained uplift of about 10 mm/year in the southern part of Kyoto basin and 5 mm/year subsidence along the ATTL. Of course, these rates are relative to the motion of the reference point. It is noteworthy that the uplift region in south Kyoto is clearly cut by two active faults on its west side.

Two questions arise. When did these deformation sources start? What is their cause? The observations of ALOS/PALSAR in this region are limited to during 2007 to 2011. Therefore we look for other data over a longer time span to give answers to these questions.

3 Analysis of ERS 1/2 and Envisat Images

In order to reveal the history of ground deformation before 2007, we analyzed ERS 1/2 and Envisat images. Figure 1 shows the footprints of ERS-1/2 and Envisat (Track 117) as well as ALOS/PALSAR (Paths 65 and 414). ERS-1/2 and Envisat were equipped with C-band SAR. C-band SAR is not considered useful in regions with heavy vegetation such as Japan, but it is still useful in urban areas such as Kyoto and Osaka. The incidence angle of microwave of ERS-1/2 and Envisat is about 23°, while ALOS/PALSAR has a shallower incidence angle (38°). Therefore the radars of ERS-1/2 and Envisat sense the vertical component of surface displacement more than the horizontal ones.

The European Space Agency (ESA) provided ERS-1/2 and Envisat images. We collected and processed 14 ERS-1/2 acquired during Novemver, 1992 to September 2004, and 22 Envisat images during April 2003 to November 2007 (Fig. 3). In these analyses we simply applied 2-pass interferometry to 14 pairs of ERS-1/2 and 48 pairs of Envisat images and then stacked interferograms as has been done in Hashimoto (2014). We used the ASTER/GDEM, which has a 1 arc-second resolution (Tachikawa et al. 2011) to reduce the topographic phase and geocode images. We confirmed that there are no noticeable differences in the results of the analysis of ALOS/PALSAR using both ASTER/GDEM and SRTM3. Unfortunately, 4 ERS-2 images acquired during 2001 and 2002 cannot be properly analyzed due to their anomalous Doppler centroids. Furthermore, some interferograms of pairs including images, mainly acquired in summer, suffered from tropospheric disturbances. We discarded such interferograms in stacking. Finally, we mapped LOS velocities referring to the same reference point (135.6 E, 34.6 N) as in Fig. 2. In order to compare the results with Fig. 2, LOS decrease is positive in the following figures.

3.1 Results of Analysis of ERS-1/2 Images

We processed pairs of images with shorter baselines than 550 m and stacked interferograms (Fig. 3). The Kobe earthquake (Mw6.8) occurred in the region covered by images on January 17, 1995, and significant deformation was observed (Hashimoto et al. 1996; Ozawa et al. 1997). Therefore we separated pairs of ERS-1/2 images into three categories: preseismic, coseismic and postseismic interferograms (Fig. 4).

There are only two preseismic pairs. Figure 4a is a stacked interferogram of these two. We observed no clear deformation along the ATTL except in its western area. Around the western edge of the ATTL, a large LOS increase is found. LOS increases of ∼10 mm/year are also observed in the Kyoto basin. However, it is noteworthy that one pair has a baseline of 530 m, the longest in this study, and the coherence around the area between the Kyoto basin and the Osaka plain is relatively low.

There are six coseismic pairs, but the coherence of one pair was too low. Figure 4b is the stacked interferogram of the

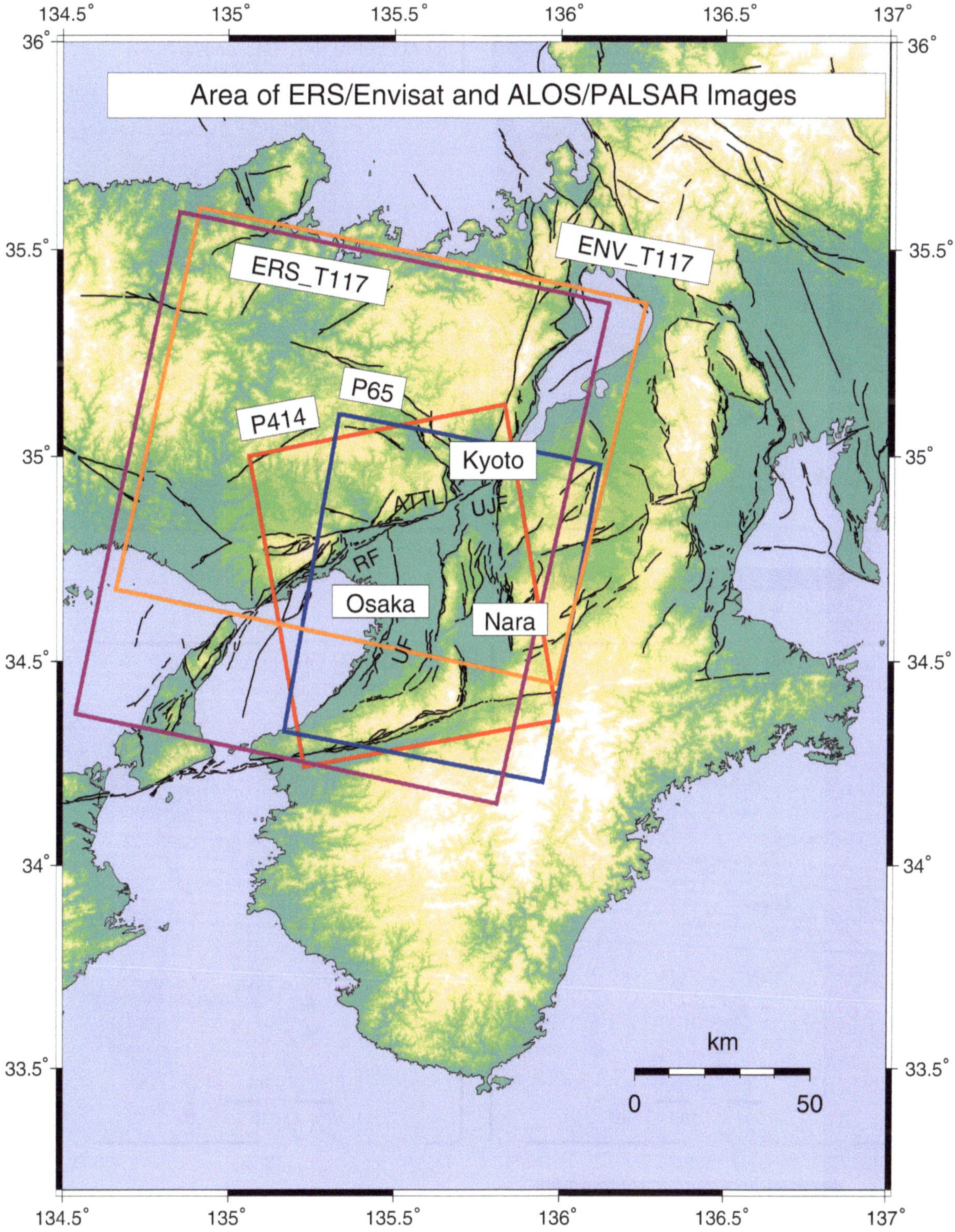

Fig. 1 Index map of the studied area. *Rectangles* are footprints of SAR sensors: *Red*, ALOS/PALSAR (P414, ascending); *Blue*, ALOS/PALSAR (P65, descending); *Pink*, ERS-1/2 (T117, descending); *Orange*, Envisat (T117, descending). Solid lines are surface trace of active faults (Geological Survey of Japan, 2005). ATTL, RF, UJF and UF indicate the Arima-Takatsuki Tectonic Line, Rokko, Ujigawa, and Uemachi faults, respectively. O and K show the location of the Osaka and Kyoto Prefectural Offices

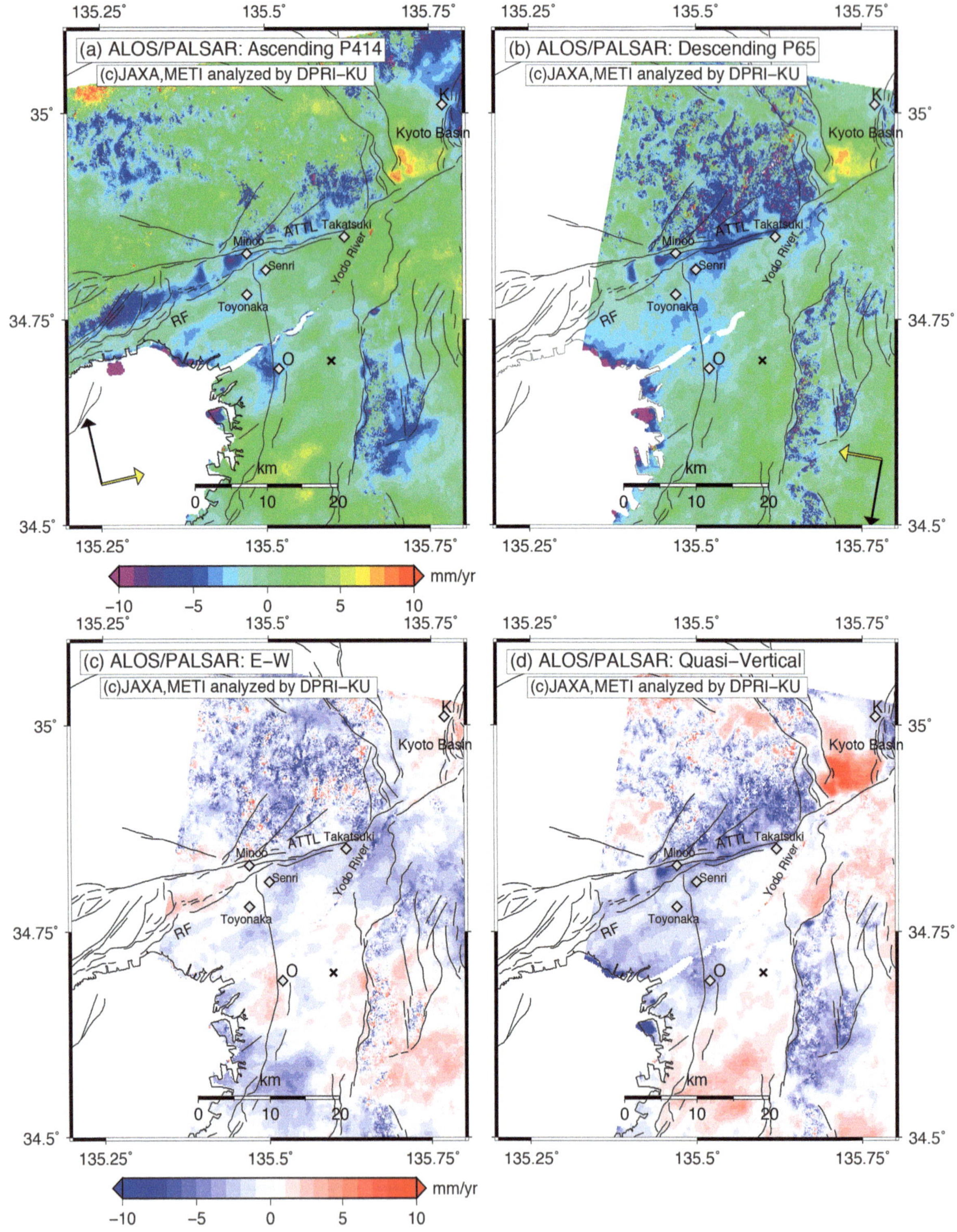

Fig. 2 Results of analysis of ALOS/PALSAR reproduced from the result of Hashimoto (2014). (**a**) Stacked interferogram from the ascending path P414. (**b**) Stacked interferogram from the descending path P65. *Yellow and black arrows* in these diagrams are the direction of emission of microwave and flight of satellite, respectively. (**c**) East–west component of velocity derived from stacked interferograms from the two directions. (**d**) Quasi-vertical component of velocity. The *cross* indicates the reference point (135.6 E, 34.6 N). See also the legend of Fig. 1

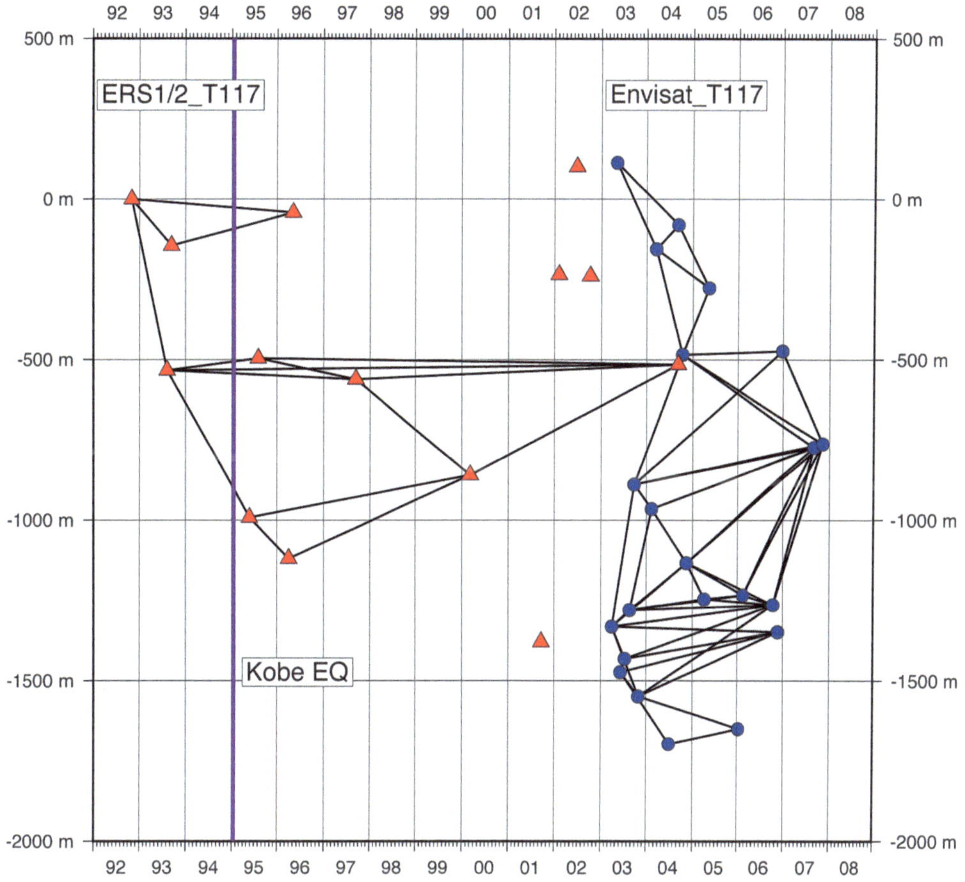

Fig. 3 Diagram of spatial and temporal baselines of ERS-1/2 (*red triangles*) and Envisat (*blue circles*) images. *Solid lines* are pairs of images processed in this study. *Purple vertical line* indicates the occurrence of the 1995 Kobe earthquake

five good coseismic pairs. Large coseismic deformation from the Kobe earthquake caused decorrelation in the western part of the image. We observed a decrease of LOS (>5 mm/year) in the northern part of the Kyoto basin. However, we consider this LOS decrease may be an artifact since all interferograms of pairs with the Aug. 8 1993 image show large LOS decrease in northern Kyoto basin. An LOS increase is found in a wedge-shaped zone between the ATTL and the Rokko fault zone, but no significant deformation is found in its eastern region.

Figure 4c shows the stacked interferogram of the post-seismic pairs. We recognized a large anomalous deformation possibly due to tropospheric disturbance in two interferograms of pairs with the image acquired on Sept. 10 1997. We excluded these interferograms and stacked 5 interferograms. The LOS increase in a wedge-shaped zone is much clearer than in the stacked coseismic interferograms. This LOS increase is consistent with the result obtained by Ozawa et al. (2004), who analyzed JERS-1 images after the Kobe earthquake. However, we cannot recognize a linear zone of LOS increase along the entire ATTL. We cannot recognize deformation in the Kyoto basin, either.

In summary, we cannot detect an LOS decrease that implies uplift in the southern Kyoto basin with ERS-1/2. An LOS increase that is consistent with subsidence is not found along ATTL except in its western area, either. Therefore these results suggest that the uplift in the Kyoto basin and subsidence along ATTL may have started after 2004.

3.2 Results of Analysis of Envisat Images

We processed 48 pairs of Envisat images acquired during 2003 and 2007. The longest perpendicular baseline is about 500 m (Oct. 18 2006 and Aug. 07 2007), but coherence of all processed interferograms is much better than that for ERS-1/2 interferograms. Therefore, there are not serious troubles in phase unwrapping in the area between the Kyoto basin and the Osaka plain. However, we have large artificial deformations in several interferograms possibly due to tropospheric disturbances. Especially interferograms of pairs with the image acquired on Aug. 29 2007 have large anomalous signal across the Osaka plain. According to the record of the Japan Meteorological Agency, there was a shower before

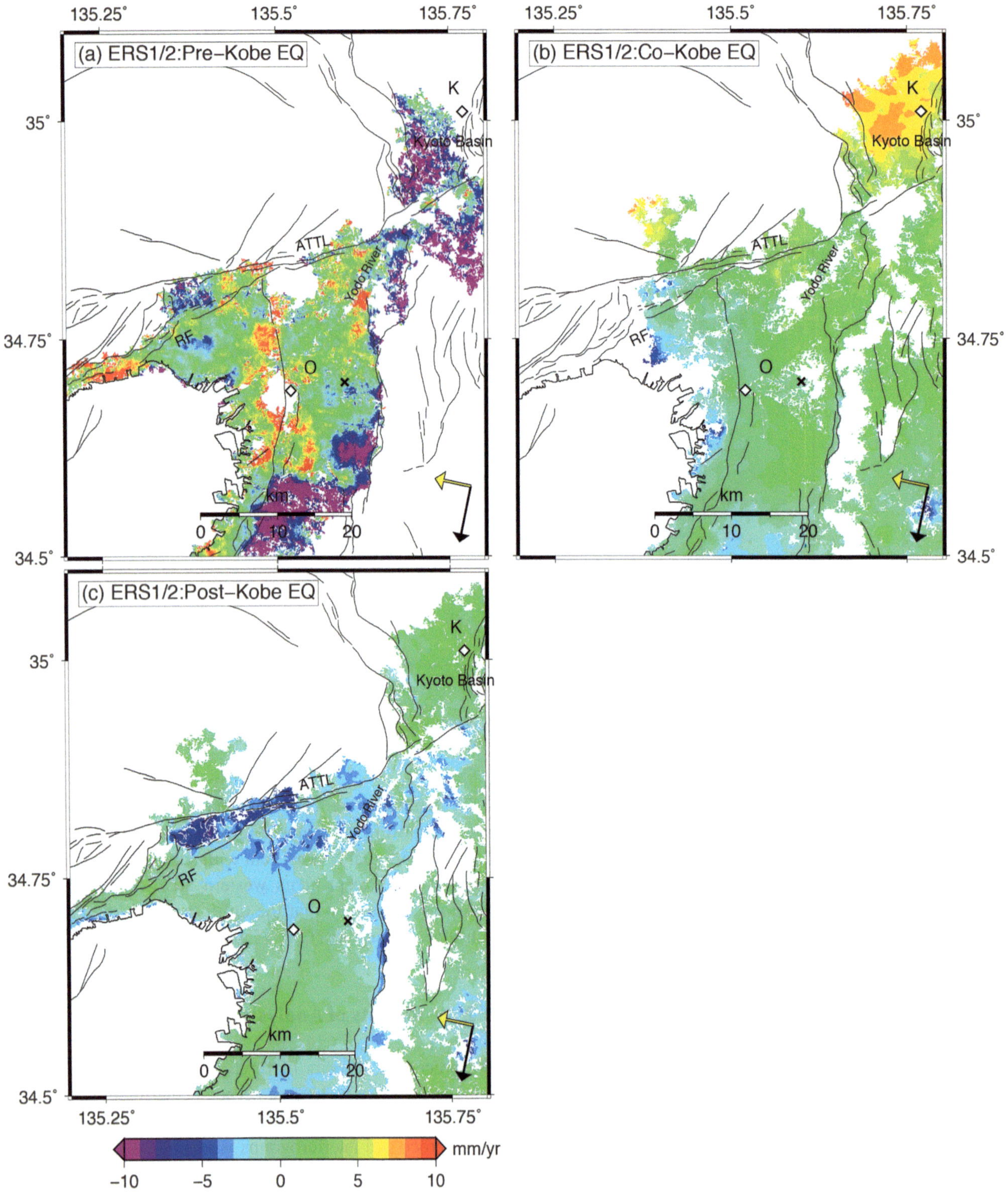

Fig. 4 Results of analysis of ERS-1/2 images. (**a**) Stacked interferogram of two pairs of images before the 1995 Kobe earthquake. (**b**) Stacked interferogram of six coseismic pairs. (**c**) Stacked interferogram of six pairs after the Kobe earthquake. See also the legend of Figs. 1 and 2

noon of this day (http://www.data.jma.go.jp/obd/stats/etrn/index.php), close to the acquisition of Envisat image. We discarded these images in stacking. In order to investigate the onset of uplift in the Kyoto basin, we separated the whole period into three periods and stacked interferograms for each time period: 2003–2004, 2004–2007 and 2005–2007 (Fig. 5).

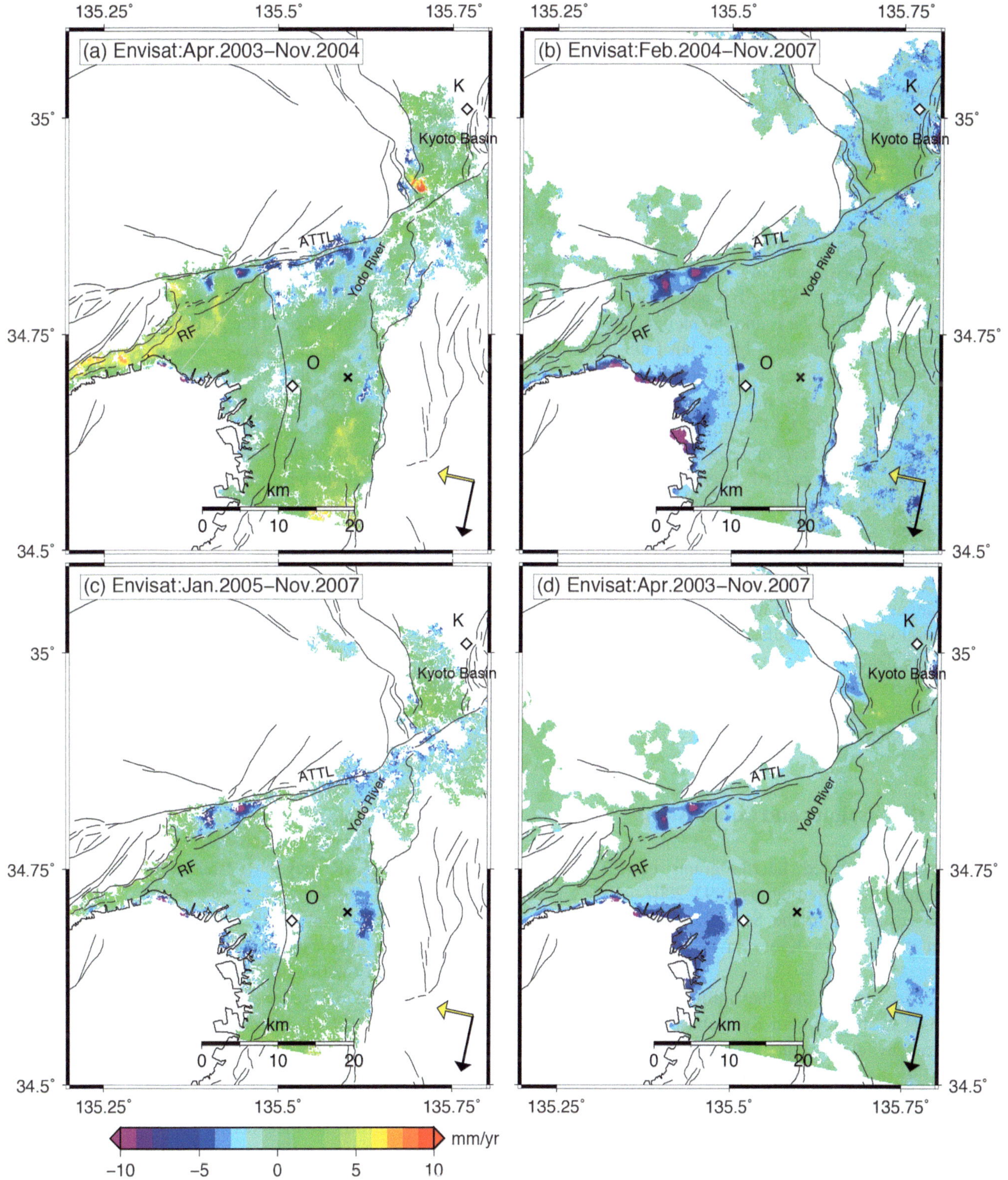

Fig. 5 Results of analysis of Envisat images. (**a**) Stacked interferogram of pairs during 2003–2004. (**b**) Stacked interferogram of pairs during 2004–2007. (**c**) Stacked interferogram of pairs during 2005–2007. (**d**) Stacked interferogram of pairs during 2003–2007. See also the legends of Figs. 1 and 2

Figure 5a is the stacked interferograms of 18 pairs of images during 2003–2004. An LOS increases of >5 mm/year are recognized in the wedge-shaped zone between two active faults recognized in Fig. 4c, but deformations are concentrated into two spots. An LOS increase is also found along the eastern portion of ATTL, but continuity is not clear due to low coherence. In the southern Kyoto basin a clear LOS decrease of ~10 mm/year is observed, which corresponds to the western portion of the uplift found by ALOS/PALSAR. We can note an LOS decrease of ~5 mm/year in the rest of uplift area detected by ALOS/PALSAR. However we must mention that there are several regions of low coherence in original interferograms that made us difficult unwrap them.

Figure 5b is a stacked interferogram of 22 pairs during 2004–2007. We observe an LOS increase of >10 mm/year in the wedge-shaped zone. No clear LOS increase is found along the eastern part of ATTL. It is noteworthy that the LOS decrease found in Fig. 5a cannot be recognized during this period in the southern Kyoto basin, but LOS decreases of ~5 mm/year are recognized in a wider area.

Figure 5c shows the result of stacking of eight interferograms during 2005–2007. Similar features are found to Fig. 5b, but an LOS increase appears in eastern part of ATTL.

Figure 5d is the stacked interferogram of 52 pairs during the whole period. Two spots of LOS increase of >10 mm/year are clearly seen in this figure, which suggests this is a persistent deformation during 2003 to 2007. Along the eastern part of the ATTL, we cannot observe significant LOS changes. LOS changes found in Fig. 5a–c may indicate its temporal variation or artifacts due to other effects such as troposphere disturbances. An LOS decrease of ~5 mm/year can be observed in the southern Kyoto basin. The shape of this LOS decrease is similar to that of uplift found by ALOS/PALSAR.

4 Discussions and Conclusions

We presented stacked interferograms of ERS-1/2 and Envisat images acquired during 1992–2007 to give answers to questions raised in the Sect. 2: timing and cause of uplift in the southern Kyoto basin and subsidence along ATTL.

We can conclude that the uplift may have started in 2003–2004 in the southern Kyoto basin, since we did not recognize any clear signal in interferograms before this period (Fig. 4c). It is reasonable to consider that the uplift was generated by the recovery of the groundwater table. The local government stopped pumping out groundwater in 2000 (Ministry of Environment 2011), which may have led to the recovery of groundwater table. However, there is a gap of three years between the cessation of pumping and the first detection of uplift. The fact that the rate of deformation during 2003–2007 (Fig. 5d) is smaller than that during 2007–2010 (Fig. 2d) suggests that pore spaces in the soil might have been refilled with groundwater till 2003 and then expanded.

Subsidence along the ATTL is not clear in stacked interferograms of ERS-1/2 and Envisat images (Figs. 4c and 5d). Therefore we conclude that this subsidence became significant after 2007, although we recognize similar features in some individual interferograms. We do not relate this subsidence to the movement of ATTL, since there is no detectable signal that implies a right-lateral motion of this fault in the map of converted E-W displacement (Fig. 2c) (Hashimoto 2014). A graben structure was found in the basement along ATTL by dense gravity surveys (Akamatsu et al. 2004). We speculate that the subsidence along ATTL is related to change in groundwater level and compaction of soil that was accumulated in the graben structure.

We found that LOS increases in a wedge-shaped zone between ATTL and the Rokko fault are recognized till 2004 with ERS-1/2, but was not observed before the Kobe earthquake. This fact suggests that this deformation might have been induced by stress changes due to the earthquake, although there might be a coincidence. LOS increases are concentrated into two spots and acceleration of their rates after 2003 with Envisat. These spots (rectangles in Fig. 6) correspond to low terrace deposits during late Pleistocene, which suggests the geology may control the surface movement in this area. We suspect that the difference in permeability of soil may affect the local deformation. Of course, we cannot reject the possibility that pumping of groundwater has been made locally. We need survey the information of utilization of groundwater, groundwater level changes, land use, etc.

In order to reveal the history of ground deformation in the Kyoto and Osaka basins during 19 years, we analyzed ERS-1/2 and Envisat images. Following conclusions are derived:

1. The uplift in the southern Kyoto basin observed by ALOS/PALSAR may have started in 2003–2004 and accelerated afterward.
2. The subsidence along the ATTL detected by ALOS/PALSAR is not clear in ERS or Envisat images, implying that this deformation became significant after 2007.
3. A subsidence in a wedge-shaped zone between the ATTL and the Rokko fault is recognized till 2004, but was not observed before the Kobe earthquake. This subsidence is concentrated into two spots and rates accelerated after 2003. These spots correspond to the areas of low terrace deposits during late Pleistocene.

We could not reveal temporal variation in these deformations with higher resolution, since we used stacking. We should apply time series analysis to the present dataset. There are lots of observations of GPS, leveling and ground water level for the purpose of monitoring of ground subsidence. We should compare all these data. In addition, modeling of groundwater flow and accumulation is essential for understanding of the process of deformation.

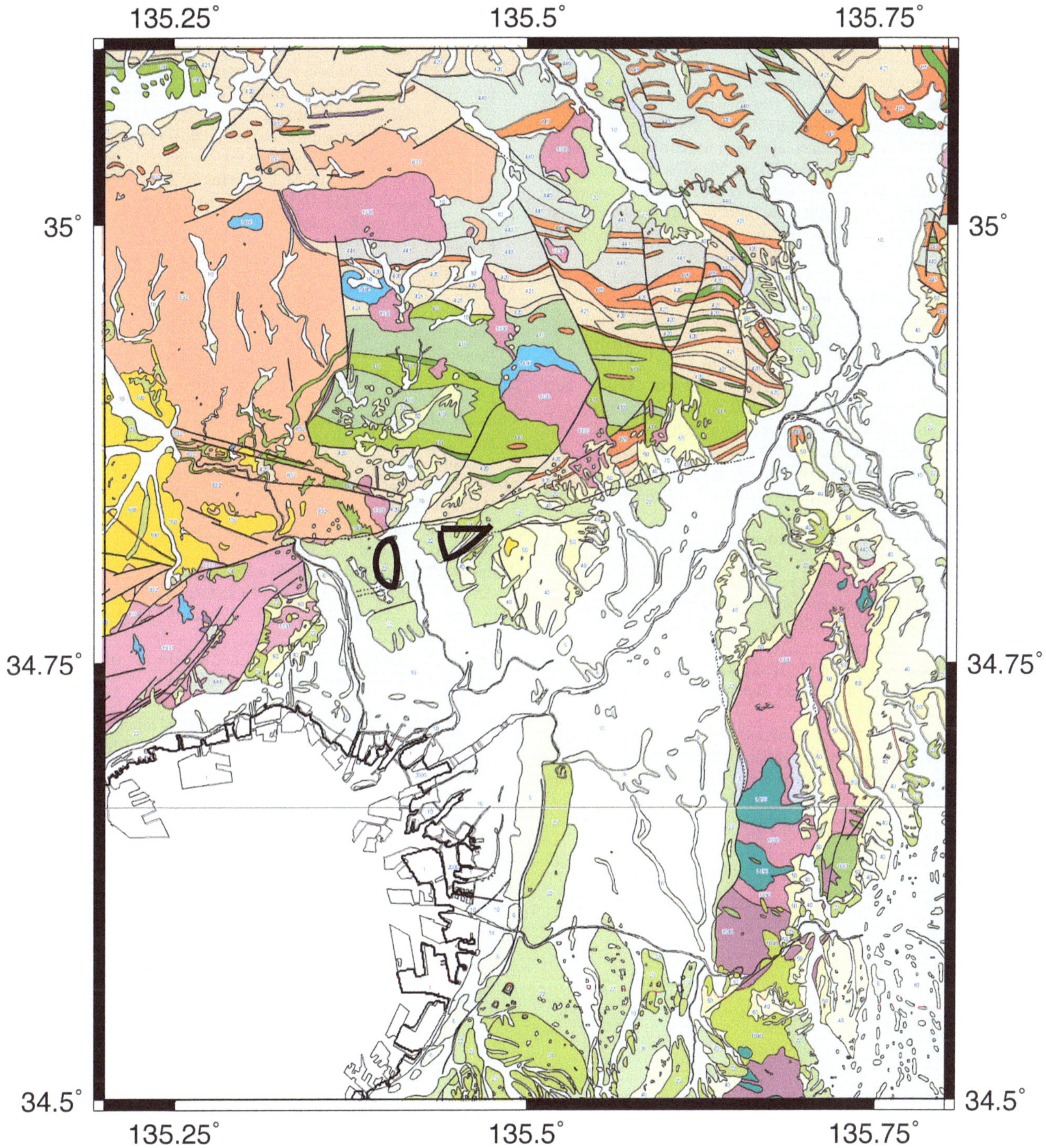

Fig. 6 Map of geology in and around the Kyoto basin and the Osaka plain reproduced from the Seamless digital geological map of Japan 1:200,000 (Geological Survey of Japan and AIST 2014). *Light green color* shows the river terrace deposit during Pleistocene. *Black rectangles* indicate zones of LOS increase (>5 mm/year) detected by Envisat (Fig. 5d)

Acknowledgements Analysis of PALSAR level 1.0 data were performed under "Intensive Observation of the Uemachi Fault" by the Ministry of Education, Culture, Sports, and Science and Technology. The ownership of ALOS/PALSAR data belongs to Ministry of Economy, Trade and Industry and Japan Aerospace Exploration Agency. ERS-1/2 and Envisat data were provided by European Space Agency under the proposal entitled "The 20 years history of ground deformation in the Kyoto basin and the Osaka plain" (#18630, PI: Manabu Hashimoto). We used Generic Mapping Tools ver.4.5.8 (Wells and Smith, 2012) to prepare illustrations.

References

Akamatsu J, Nakamura K, Nishimura K, Komazawa M (2004) (2–3) Crustal density structure using gravity data, in "FY2003 Report of I Prediction of Strong Ground Motion, the Study of Crustal Structure in Urban Area, the Special Project for Earthquake Disaster Mitigation in Urban Areas" by MEXT, pp 571–576 (in Japanese)

Bell JW, Amelung F, Ramelli AR, Blewitt G (2002) Land subsidence in Las Vegas, Nevada, 1935–2000: new geodetic data show evolution, revised spatial patterns, and reduced rates. Environ Eng Geosci VIII 3:155–174

Ferretti A, Prati C, Rocca F (2001) Permanet scatterers in SAR interferometry. IEEE Trans Geosci Remote Sens 39(1):8–20

Fukushima Y, Mori J, Hashimoto M, Kano Y (2009) Subsidence associated with the LUSI mud eruption, east Java, investigated by SAR interferometry. Mar Petrol Geol 26:1740–1750

Geological Survey of Japan, AIST (ed) (2005) Active fault database of Japan. http://riodb02.ibase.aist.go.jp/activefault/index_e.html?search_no=%20j008&version_no=1&search_mode=0

Geological Survey of Japan, AIST (ed) (2014) Seamless digital geological map of Japan 1: 200,000. Jan 14, 2014 version. Geological Survey of Japan, National Institute of Advanced Industrial Science and Technology

Hashimoto M (2013) Ground deformation related to active faults detected by persistent scatterer InSAR. In: Proceedings of the APSAR2013, Tsukuba, Japan Sept. 2013, TH2.R1.2

Hashimoto M (2014) Ground deformation in the Kyoto Basin and the Osaka Plain detected by ALOS/PALSAR, J. JSNDS 33–2:115–125 (in Japanese with English abstract)

Hashimoto M, Sagiya T, Tsuji H, Hatanaka Y, Tada T (1996) Coseismic displacements of the 1995 Hyogo-ken Nanbu Earthquake. J Phys Earth 44:255–279

Hooper A, Zebker H, Segall P, Kampes B (2004) A new method for measuring deformation on volcanoes and other natural terrains using InSAR persistent scatterers. Geophys Res Lett 31(23). doi:10.1029/2004GL021737

Huzita K (1962) Tectonic development of the median zone (Setouchi) of Southwest Japan since Miocene. J Geosci Osaka City Univ 6:103–144 (in Japanese)

Jarvis A, Reuter HI, Nelson A, Guevara E (2008) Hole-filled seamless SRTM data V4, International Centre for Tropical Agriculture (CIAT). http://srtm.csi.cgiar.org

Lanari R, Mora O, Manunta M, Mallorqui JJ, Bernardino P, Santosti E (2004) A small-baseline approach for investigating deformations on full-resolution differential SAR interferograms. IEEE Trans Geosci Remote Sens 42(7):1377–1386

Massonnet D, Rossi M, Carmona C, Adragna F, Peltzer G, Feigl K, Rabaute T (1993) The displacement field of the Landers earthquake mapped by radar interferometry. Nature 364:138–142

Ministry of Environment (2011) Information on ground subsidence in the Kyoto Basin, National Ground Environment Information Directory. http://www.env.go.jp/water/jiban/dir_h23/26kyouto/kyouto/index.html (in Japanese)

Okada A, Togo M (eds) (2000) Active faults in the Kinki district. University of Tokyo Press, Tokyo, 395 pp (in Japanese)

Ozawa S, Murakami M, Fujiwara S, Tobita M (1997) Synthetic aperture radar interferogram of the 1995 Kobe earthquake and its geodetic inversion. Geophs Res Lett 24:2327–2330

Ozawa T, Tobita M, Yarai H, Nishimura T, Murakami M, Ohkura H (2004) Postseismic deformation of the 1995 Hyogo-ken Nanbu Earthquake revealed by JERS-1/InSAR. In: Proceedings of the APSG Symposium (eds. C. Huang and Z. Qian), 105–111.

Tachikawa T, Hato M, Kaku M, Iwasaki A (2011) The characteristics of ASTER GDEM version 2, IGARSS, July 2011, http://www.jspacesystems.or.jp/ersdac/GDEM/ver2Validation/IGARSS2011_Proceedings_GDEM2.pdf

Takada Y, Fukushima Y (2013) Volcanic subsidence triggered by the 2011 Tohoku earthquake in Japan. Nat Geosci 6:637–641. doi:10.1038/ngeo1857

Wessel P, Smith WHF (2012) The Generic Mapping Tools (GMT) version 4.5.8 technical reference & Cookbook, SOEST/NOAA, http://gmt.soest.hawaii.edu/gmt4/

List of Reviewers

Hasanuddin Zainal Abidin
Raed Aldouri
Yosuke Aoki
Valerie Ballu
Rebecca Bendick
Yehuda Bock
Lukasz Bonenberg
Pierre Briole
David Chadwell
Raymond Durrheim
Przemysław Dykowski
Jeff Freymueller
Masayuki Fujita
Jianghui Geng
Shin-Chan Han
Emma Hill
Nicolas Houlie
Yuichi Imanishi
Francois Jouanne
Motoyuki Kido
Cecile Lasserre
Thorne Lay
Jack Loveless
Urs Marti
Takeshi Matsushima
Shinichi Miyazaki
Masayuki Murase
Kazunari Nawa
Takuya Nishimura
Yusaku Ohta
Mako Ohzono
Isabelle Panet
Matt Pritchard
Noelynna T. Ramos
Thomas Shoberg
Anne Socquet
Hiroaki Takahashi
Richard Walters
Lei Wang
Christopher Stephen Watson

M. Hashimoto (ed.), *International Symposium on Geodesy for Earthquake and Natural Hazards (GENAH)*,
International Association of Geodesy Symposia 145, DOI 10.1007/1345_2015

Author Index

M. Hashimoto (ed.), *International Symposium on Geodesy for Earthquake and Natural Hazards (GENAH)*,
International Association of Geodesy Symposia 145, DOI 10.1007/1345_2015

MIX
Papier aus verantwortungsvollen Quellen
Paper from responsible sources
FSC® C105338

If you have any concerns about our products,
you can contact us on
ProductSafety@springernature.com

In case Publisher is established outside the EU,
the EU authorized representative is:
Springer Nature Customer Service Center GmbH
Europaplatz 3, 69115 Heidelberg, Germany

Printed by Libri Plureos GmbH
in Hamburg, Germany